STRUCTURE AND DYNAMICS OF NON-RIGID MOLECULAR SYSTEMS

TOPICS IN
MOLECULAR ORGANIZATION AND ENGINEERING

Volume 12

The titles published in this series are listed at the end of this volume.

Structure and Dynamics of Non-Rigid Molecular Systems

Edited by

YVES G. SMEYERS

CSIC, Madrid, Spain

Springer-Science+Business Media, B.V.

Library of Congress Cataloging-in-Publication Data

Structure and dynamics of non-rigid molecular systems/ edited by Yves G. Smeyers.
p. cm. -- (Topics in molecular organization and engineering ; v. 12)
Includes index.
ISBN 978-94-010-4464-6 ISBN 978-94-011-1066-2 (eBook)
DOI 10.1007/978-94-011-1066-2
1. Molecular structure--Congresses. 2. Molecular dynamics--Congresses. I. Smeyers, Yves G. II. Title: Non-rigid molecular systems. III. Series.
QD461.S9244 1994
541.2'2--dc20 94-7726

Printed on acid-free paper

"The logo on the front cover represents the generative hyperstructure of alkanes", printed with permission from J.E. Dubois, Institut de Topologie et de Dynamique des Systèmes, Paris, France.

TABLE OF CONTENTS

PART III: DYNAMICS OF NON-RIGID MOLECULAR SYSTEMS

FOREWORD

This volume contains a selection of scientific papers related to the structure and dynamics of non-rigid molecules. This frontline topic was born a few decades ago, when Longuet-Higgins proposed his famous theory of Molecular Symmetry Groups (*Mol. Phys.* **6**, (1962) 457).

Unfortunately, since this early paper, very few publications have been devoted to the study of non-rigid molecules. Let us mention some books which dedicate some chapters to them: *Induced Representations in Crystals and Molecules*, by S. L. Altmann, Academic Publishers, 1977; *Molecular Symmetry and Spectroscopy*, by P. R. Bunker, Academic Publishers, 1979; and finally *Large Amplitude Motion in Molecules*, Vols. I and II, by several authors, Springer Verlag, 1979.

More recently an International Symposium on Non-Rigid Molecules was held in Paris, France, from 1–7 July 1982, the proceedings of which were published in the volume entitled *Symmetries and Properties of Non-Rigid Molecules. A Comprehensive Survey*, edited by J. Maruani *et al.*, Elsevier, 1983.

Finally, we should mention the very specialized work *The Permutational Approach to Dynamic Stereochemistry*, by J. Brocas *et al.*, McGraw-Hill, 1983.

The purpose of this book is to fill in this information on the structure and dynamics of non-rigid systems. To this aim, we have gathered a collection of recent papers written by the most qualified specialists in the world, covering a large field from van der Waals molecules to inorganic complexes and organic polyrotor molecules, as well as considering statistical and dynamic aspects.

We wish to express here our thanks to all the authors of the different chapters of this book, and especially to Professor R. S. Berry of the University of Chicago for agreeing to write the preface.

Y.G. SMEYERS

PREFACE

R. STEPHEN BERRY
Department of Chemistry and The James Franck Institute,
The University of Chicago,
Chicago, IL 60637,
U.S.A.

The successes of the structural concept of molecules, firmly established by such achievements as the demonstration in the mid–1870s of the structure of benzene and the simultaneous clarification of the molecular basis of optical activity, made into dogma the idea that 'molecules have structures'. Modern molecular spectroscopy grew up on this dogma; atoms comprising a molecule have precise positions, relative to one another, around which the atoms undergo vibrations of very small amplitude, relative to the distances between atomic centres. Hence molecules are like near-rigid, quivering Tinkertoys or Meccano models, capable of rotating almost as if they were truly rigid structures. This picture allows us to describe the vibrational and rotational motions as almost independent, separable motions. Whatever deviations there may be from separability can be considered small perturbations, interpretable in the traditional context of centrifugal distortions and Coriolis interactions. With the advent of quantum mechanics, these ideas all found expression in the language of discrete states, still identifiable as states of small-amplitude oscillation, or of near-rigid rotation, or of a combination of both. The quantum equations were tractable enough that it was not even necessary to call upon the Correspondence Principle to find the wave functions and the energy eigenvalues for the states of molecular rotor-vibrators.

All this made the concept of a molecule very different from the concepts, more or less simultaneously developed, of the atom and of the nucleus. The concepts of their 'structures' quickly becoming models of free-flowing collections of component particles moving relatively independently. The most popular model of the nucleus for a long time was that of a liquid drop. The corresponding model of atoms was the quantum analogue of a solar system. The excitations of a nucleus became the excitation of collective modes of a fluid; those of an atom became the single-particle excitations of the individual electrons, notably from one shell to another.

The lore of almost-rigid molecules developed despite chemists' awareness of tautomerism, of the capacity of some special systems to rearrange. One of the most elegant and historically important was the cyclopentane molecule, a $(CH_2)_5$ ring, with four carbons in a common plane and the fifth, out of that plane. The special tautomerism of this system allows the out-of-plane carbon to return to the common plane and another carbon to move out of the plane; because of the identity of equivalent atoms, the isomer produced by this tautomerism looks like the initial species, just rotated, clockwise or counter-clockwise, by $2\pi/5$ or $4\pi/5$. In fact, the process involves both a permutation and a rotation, so is called a 'pseudo-rotation'. More elaborate motions are involved when species such as the trigonal bipyramids of PF_5 or $Fe(CO)_5$ pseudo-rotate to exchange axial and equatorial pairs of atoms. Still more complex systems are still being analysed, as illustrated by the chapter by Brocas in this book, on heptacoordinated systems.

By the 1960s, and even more still later, as it became possible to excite molecules to specific, high-energy vibrational states and to probe the internal dynamics of tautomerizing or rearranging molecules, the idea grew that molecules may be not so rigid, and that many of their kinetic properties are consequences of rearrangements. For example, the pseudo-rotation of 5-coordinate phosphorus compounds, originally introduced to interpret the equivalence of the NMR resonance frequencies of all five fluorines of PF_5, turned out to be the explanation also for the mechanism of very important biochemical process of phosphate ester hydrolysis.

As the subject grew, its fundamental mathematical and quantum-mechanical aspects challenged theorists. This aspect is reviewed and developed in the first chapter, by Boldyrev. Some of the traditional concepts, such as those associated with normal modes of vibration of a nearly rigid molecule, lose their meaning and must be replaced in some cases with new concepts that can be used outside the confines of validity of the older ideas; Natanson explores this topic, both how the traditional ideas break down and how we might be able to extend and replace them. Experimental approaches to probe non-rigid species, by fluorescence of jet-cooled species and by photodissociation, are the bases of the chapters by Laane and collaborators, and by Delgado-Barrio and Beswick, respectively. The fundamental physical generalities of non-rigid systems are the subjects of the chapters by Smeyers, on the symmetries and group theory of the problem, and by Weeks and Levine, on the nature of the coupling of the anharmonic modes, a subject closely related to that examined by Natanson, but from quite a different viewpoint. The underlying potential surfaces of non-rigid molecules is the subject of the chapter by Cárdenas-Jirón and her collaborators.

Together, these chapters provide a picture of much of what is going on at the frontiers of the very lively field of non-rigid molecules. For truly it has become a topic in itself, with tangent lines to chemical kinetics, phase transitions, theory of potential surfaces, chaotic and regular dynamics and the fields in which non-rigid molecular processes play specific, crucial roles as in biochemistry. The book is probably immediately accessible to someone knowledgable in any of these subjects. It may require some background reading for others.

Only in relatively recent years have there been reviews of non-rigid molecular systems. Among them this writer and several others who have worked in the field have contributed. However, perhaps the most comprehensive, and one that would provide the most thorough background for this volume, is probably the 1986 monograph *Contemporary Theory of Chemical Isomerism* by Zdenek Slanina. That slim book reviews most of the concepts that Yves Smeyers and his co-authors have to assume, as they present the state of the art of non-rigid molecules.

R. STEPHEN BERRY
Spring 1994

THE STRUCTURE, SYMMETRY, AND PROPERTIES OF NON-RIGID MOLECULES

ALEXANDER I. BOLDYREV*
Institute of Chemical Physics,
Russian Academy of Sciences,
Kosygin str. 4,
Moscow V–334,
Russia

Abstract. Problems of molecular structure and a description of molecular properties of non-rigid (floppy) molecules are considered from the multiminima point of view. It is shown that polyatomic molecules can have many global minima on the potential energy surface and that the number of these global minima can be computed as the ratio of the order of the full permutation-inversion group, by the order of the symmetry point group of the molecule in its equilibrium configuration. The importance of non-rigid models for the description of structure, spectra and other molecular properties are demonstrated on the basis of the results of high quality *ab initio* calculations, the dynamic calculations of the ro-vibrational spectra, and experimental spectroscopic measurements at high resolution. Molecular species with large amplitude motions between formally chemically bonded atoms (e.g. LiCN, $LiBH_4$, $C_2H_3^+$, ArH_3^+ and others) have been considered as examples.

1. Introduction

In structural chemistry it is assumed that in studying the structure of molecules in the gas phase or in matrix isolation, the experimental structural parameters (bond lengths and valence and dihedral angles) are applicable to isolated molecules. However, in terms of a strict quantum description this is not correct. According to the main postulates of quantum mechanics, an isolated molecule can exist

* Present address: Department of Chemistry, The University of Utah, Henry Eyring Building, Salt Lake City, Utah 84112, U.S.A.

Y.G. Smeyers (ed.), Structure and Dynamics of Non-Rigid Molecular Systems, 1–45.

only in stationary states and, in such states, the molecule nuclear density function (assuming the Born–Oppenheimer approximation) is delocalized with respect to all absolute minima (the number of which for polytomatic molecules containing several identical nuclei is high; e.g., the ethane molecule has 240 absolute minima). In other words, the nuclear density function maxima in all global energy minima are equal for stationary states of an isolated molecule, and this molecule has a delocalized molecular structure (quantum molecular structure). However, this conclusion is not supported by the experience gained in studying the structure of molecules through various physical experimental techniques. According to the latter, the nuclear density function is localized in one minimum (classical molecular structure), and the molecule's structure and properties can be described in terms of quasi-rigid one-minimum models such as a "rigid rotator-harmonic oscillator" or a "semi-rigid rotator-anharmonic oscillator". To describe a molecule using these models, it is sufficient to know, out of the whole multidimensional multiminima potential energy surface (PES), only a small area around one absolute minimum, because it is precisely this area which determines the parameters of the quasi-rigid model (e.g., the bond lengths, valence and dihedral angles, quadratic, cubic and higher force constants).

This contradiction has generated a broad discussion in the literature [1–17]. It has been shown that the localization mechanism consists in the destruction phase coherence of the tunnel oscillator owing to quickly varying random changes of the tunnel splitting parameter γ occurring under the influence of collisions with other molecules [4, 16]. If the temperature of the molecular gas T is higher than the critical temperature $T > T_0$ (for two minimum PES $T_0 = \hbar\gamma/k$), then the nuclear density functions localizes in one minimum, and if $T < T_0$ then the nuclear density function is delocalized throughout all absolute minima. Normally, the value of T_0 is extremely small and even for the flexible NH_3 molecule, $T_0 =$ 0.5 K [4, 16]. It should be noted that a molecule whose nuclear density function is localized in one of the PES minima is not in a stationary state; and over some finite period of time t_0 (life time of the classical structure of a molecule) it will pass into another state with nuclear density function localization in another minimum. In other words, a molecule with nuclear density function localized in one minimum is not isolated. Indeed, in experimental investigations one works non-isolated molecules in non-stationary states. However, for so-called quasi-rigid molecules, the life time t_0 of the nuclear density function in one minimum is usually too large; and during the experimental study, the molecular nuclear density function will be localized in one minimum so one can ignore the existence of the other minima to explain the results of this experiment. For example, life times of the optical isomers of carbon compounds are very large and while these two kinds of isomers (+ and -) are indeed only two different minima on the PES of the same molecular systems, one can consider these isomers as individual molecules. But for general consideration, especially for description of the molecular properties in excited states and near the dissociation limit, as well as in case of floppy molecules, one should use multiminima models including not only all absolute minima, but also all local minima.

One can now define the differences between quasi-rigid and non-rigid molecules. If the barriers separating the minima are sufficiently high, then the tunnel splittings are low and the localized-state life time t_0 is large, and vice versa. The life time t_0 of the localized state of a quasi-rigid molecule is very long and therefore it can be described using the one-minimum quasi-rigid model at least at moderate temperature and for low-lying excited states. The localized-state life time in the non-rigid molecules is comparatively short, which is why these molecules are presented in the experiment as delocalized with respect to several if not all absolute minima. To describe non-rigid molecules, one should use new non-rigid models, which take into account many absolute minima and sometimes all absolute minima of the PES. From the above consideration it is clear that there are two types of molecular structure delocalization. The first is a quantum type, which occurs below critical temperature T_0. The second is due to classical jumping of the localized nuclear density function (classical structure) from one absolute minimum to another during the characteristic time of the experimental technique. The first type of delocalization is observable at very low temperature $T < T_0$, the second type may occur at high temperature or for excited rovibronic states of the floppy molecules.

2. The Symmetry of Non-Rigid Molecules

The groups of symmetry used to classify the state of a molecule are closely related to the models of molecular structure. The rigid ball-and-stick model offers a total description of symmetry by means of point symmetry groups. The switch to quasi-rigid (ball-and-spring) models due to small vibrational amplitude, allows one to keep the point groups to describe molecular symmetry.

However, the switch from quasi-rigid to a non-rigid delocalized model reveals the inapplicability of the point groups of equilibrium configuration. A new theory of molecular symmetry has been developed to describe the states of non-rigid molecules. The largest contribution to developing this theory was made by Longuet-Higgins [18], who has shown that the full group describing the internal symmetry of a molecule's nuclear subsystem is the full permutation-inversion (FPI) group $G^{(n)}$. This group is a direct product of the groups of identical nuclei permutations and the group of inversion (E):

$$G^{(n)} = S_l \otimes S_m \otimes S_k \otimes \cdots \otimes E$$

where S_l, S_m, S_k etc. are groups of permutation of l, m, k and etc. nuclei of the A, B, C, etc. types, respectively.

The order of the FPI group n is equal to $l! \times m! \times k! \times \cdots \times 2$ and it can be readily determined from the chemical bruto formula of the molecule. Thus, for

$$
\begin{array}{cccc}
\underset{H_{(2)}}{\overset{H_{(1)}}{\diagdown}}C_{(1)}=\underset{H_{(4)}}{\overset{H_{(3)}}{\diagup}}C_{(2)} &
\underset{H_{(2)}}{\overset{H_{(1)}}{\diagdown}}C_{(1)}=\underset{H_{(3)}}{\overset{H_{(4)}}{\diagup}}C_{(2)} &
\underset{H_{(3)}}{\overset{H_{(1)}}{\diagdown}}C_{(1)}=\underset{H_{(4)}}{\overset{H_{(2)}}{\diagup}}C_{(2)} &
\underset{H_{(3)}}{\overset{H_{(1)}}{\diagdown}}C_{(1)}=\underset{H_{(2)}}{\overset{H_{(4)}}{\diagup}}C_{(2)} \\
I & II & III & IV \\[4pt]
\underset{H_{(4)}}{\overset{H_{(1)}}{\diagdown}}C_{(1)}=\underset{H_{(3)}}{\overset{H_{(2)}}{\diagup}}C_{(2)} &
\underset{H_{(4)}}{\overset{H_{(1)}}{\diagdown}}C_{(1)}=\underset{H_{(2)}}{\overset{H_{(3)}}{\diagup}}C_{(2)} &
\underset{H_{(3)}}{\overset{H_{(2)}}{\diagdown}}C_{(1)}=\underset{H_{(4)}}{\overset{H_{(1)}}{\diagup}}C_{(2)} &
\underset{H_{(3)}}{\overset{H_{(2)}}{\diagdown}}C_{(1)}=\underset{H_{(1)}}{\overset{H_{(4)}}{\diagup}}C_{(2)} \\
V & VI & VII & VIII \\[4pt]
\underset{H_{(4)}}{\overset{H_{(2)}}{\diagdown}}C_{(1)}=\underset{H_{(3)}}{\overset{H_{(1)}}{\diagup}}C_{(2)} &
\underset{H_{(4)}}{\overset{H_{(2)}}{\diagdown}}C_{(1)}=\underset{H_{(1)}}{\overset{H_{(3)}}{\diagup}}C_{(2)} &
\underset{H_{(4)}}{\overset{H_{(3)}}{\diagdown}}C_{(1)}=\underset{H_{(2)}}{\overset{H_{(1)}}{\diagup}}C_{(2)} &
\underset{H_{(4)}}{\overset{H_{(3)}}{\diagdown}}C_{(1)}=\underset{H_{(1)}}{\overset{H_{(2)}}{\diagup}}C_{(2)} \\
IX & X & XI & XII
\end{array}
$$

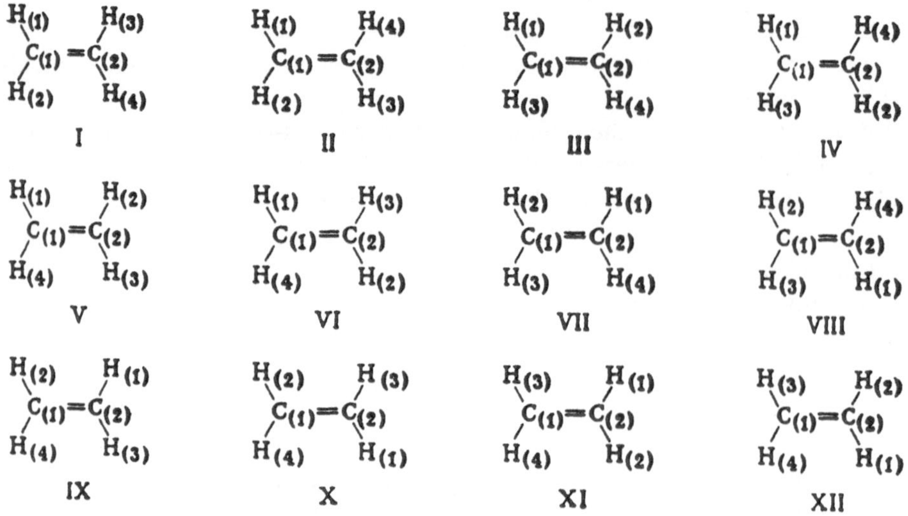

Figure 1. All symmetrically equivalent structures of the ethylene molecule.

NH_3, C_2H_6 and C_6H_6 we have the following order:

NH_3: $1! \times 3! \times 2 = 12$

C_2H_6: $2! \times 6! \times 2 = 2,880$

C_6H_6: $6! \times 6! \times 2 = 1,036,800.$

We will give no description here of the permutation and permutation-inversion groups, since this has been done in detail in monographs [19, 20]. One can also use other approaches to describing the symmetry of non-rigid molecules, such as those stated in [21–22], these are basically similar to the Longuet-Higgins' theory. Considering that the permutation-inversion symmetry has been widely accepted, we shall make use of precisely this approach.

An isolated molecule's Hamiltonian is invariant with respect to the operations of the $G^{(n)}$ group which is an accurate symmetry group of the molecule's nuclear subsystem. In principle, the vibrational–rotational levels can always be classified according to irreducible representations of the group $G^{(n)}$. However, it is seen that the order of the group $G^{(n)}$ can be very high, which makes its application rather difficult. Besides, the classification of the levels by the symmetry types of the group $G^{(n)}$ would make many levels "accidentally" degenerate. Actually this degeneracy is not accidental, but is called so because it is not caused by the group $G^{(n)}$. Such degeneracy is called the configurational degeneracy and it is due to the presence of more than one symmetrically equivalent configurations of corresponding absolute minima and large barriers between these minima.

Let us introduce the notion of symmetrically equivalent configuration using the example of the ethylene molecule. Figure 1 shows 12 symmetrically equivalent configurations of this molecule. There are configurations that can be achieved only through deformation of the molecule and no type of rotation of the rigid molecule in space can transfer one into another keeping the same numbers of the nuclei. Thus, configurations 1 and 2 can be transformed one into the other only through internal rotation around the double bond. However this rotation barrier is very high. The 12 symmetrically equivalent configurations correspond to 12 energy-equivalent minima. In the case of ethylene, the vibrational levels in each of the 12 minima will be 12 times degenerate (not exactly, but tunnelling splittings will be so small that we may not take them into account), because all the barriers are high and one can assume that all the absolute minima are isolated. In this manner the ethylene degeneracy is due to the PES shape in the area of all absolute minima.

However, as stressed above, the non-rigid molecules' absolute minima (either all or part of them) are separated by low barriers. In this case, the point group of the equilibrium configuration is not sufficient to classify all the levels that appear due to barrier tunnelling. In some cases, one needs a full permutation-inversion group, but sometimes one can do classification with its subgroups. Following Longuet-Higgins we shall call this minimal group, which allows us to classify entirely all the observed levels by symmetry, the group of molecular symmetry [18].

The above-mentioned symmetrically equivalent configurations permit the determination of the number of PES absolute minima, which is very important for developing multiminima models of the non-rigid molecules. One should recall that the full permutation-inversion group $G^{(n)}$ includes the following elements: inversion, permutations and permutations with inversion of identical nuclei. If we refer once again to the ethylene molecule (see Figure 1), then it can be established that it is all the elements of the point groups of the twelve absolute configurations (their number is $12 \times 8 = 96$, i.e. exactly equals the order of the FPI group) could be transferred to the elements of the full permutation-inversion group of the ethylene molecule. This is so because no permutation, no permutation with inversion of identical atoms, and no inversion will yield new configurations different from those obtained from these twelve configurations upon applying to them the elements of the point D_{2h}-group of the ethylene molecule's equilibrium configurations. Then, knowing the molecule's FPI group and the point group of its equilibrium configuration one can calculate the number of energy-equivalent minima, which is equal to the ratio of the order of the molecule's FPI group divided by the order of the point group of its equilibrium configuration (for linear molecules one should use C_{2v} group instead of D_{ooh} and C_s group instead of C_{oov}). For example, the order of the FPI group of the ethylene molecule is equal to $2! \times 4! \times 2 = 96$, while the order of the point group of the equilibrium configuration D_{2h} is equal to 8. That means that the number of absolute minima on its PES is 12, with 12 corresponding symmetrically equivalent configurations. Using this rule we can easily calculate the number of absolute minima for every molecule, if we know the point group symmetry of the geometrical configuration of the molecule at the absolute minimum (recall that FPI group is determined only by the bruto formula of a molecule which is the same for all isomers). For example, NH_3, C_2H_6, and C_6H_6 (benzene) have absolute minimum structures with C_{3v}, D_{3d} and D_{6h} point group symmetry. Then the number of respective absolute minima are:

$$NH_3: \quad 1! \times 3! \times 2/6 = 12/6 = 2$$

$$C_2H_6: \quad 2! \times 6! \times 2/12 = 2880/12 = 240$$

$$C_6H_6: \quad 6! \times 6! \times 2/24 = 1036800/24 = 43,200.$$

Henceforth, we shall use this rule in analysing PES shapes. Here it should be pointed out that this rule may help also to calculate the number of local minima and saddle points on the PES of a molecule. In these cases, one should use the order of the point group of local minima and of saddle points, respectively, while the order of the FPI group is the same as for calculation of the number of the absolute minima.

For non-rigid molecules, the choice of molecular symmetry group depends on the method of study.

A molecular symmetry group is obtained by excluding all non-realizable elements. A non-realizable element is such an element that transfers one symmetrically equivalent configuration into another through a barrier that cannot be tunnelled through in the time period characteristic of a physical investigation technique. Tunnelling splitting cannot be observed with a low resolution instrument: with high resolution, however, it can manifest itself. To analyse results, one should use two different groups of molecular symmetry for two cases, since the elements associated with tunnelling will be non-realizable in the first and realizable in the second case.

Let a molecule have n absolute PES minima. Then depending on the PES shape, three types of molecular symmetry groups are possible.

1. If all n absolute minima on PES are related by minima energy paths with low (surmountable) barriers, then all permutations, permutations with inversion of identical nuclei, and inversion are possible. The group of molecular symmetry is the full permutation-inversion group which is the molecular Hamiltonian's group. In this case, to find the eigenvalues of the vibrational–rotational Hamiltonian one has to take into account all n PES minima with n-fold degeneracy being lifted. The character of energy splittings is determined through reduction of reducible representations of the equilibrium configuration point group on the irreducible representation of FPI group.

2. If part of the energy-equivalent minima are separated by insurmountable barriers, then the symmetry elements corresponding to these rearrangements should be eliminated from the FPI group. In this case, the group of molecular symmetry is a subgroup of the FPI group. However, since transitions are possible between some

of the minima (part of the minima at least between two minima) the order of the group of molecular symmetry is higher than that of the equilibrium configuration point group. The way how to find non-rigid (molecular) group symmetry using the graph of the intramolecular non-rigid rearrangement was described in Ref. 15b. If a graph is used to represent the PES of a species in which high barrier do not allow of the minima to be reached from all other global minima (which is the most common case of floppy molecules), graph splits into a number of subgraphs (all of which are equivalent). The symmetry obtained from the subgraph [15b] than is what determines the symmetry of the vibrational/rotation wave function.

3. If all minima are separated by insurmountable barriers in such a way that no clear tunnelling is observed, then the group of molecular symmetry is the equilibrium configuration point group. In this case all experimentally observed vibrational levels can be classified by the irreducible representations of this group.

In this way, in order to determine the equilibrium configuration point group, it suffices to know the geometrical structure of the absolute minimum configuration; and to determine the full permutation-inversion group it suffices to know only the molecule's bruto-formula – to determine the group of molecular symmetry one should always know the shape of the PES over a wide range of internal coordinates. In order to use the point group of equilibrium configuration as the group of molecular symmetry one must certainly know that the absolute minimum is isolated from the others by high potential barriers. To use the FPI group as the group of molecular symmetry one must know that all the minima on the PES are connected by minimal energy pathways (MEP) with low barriers. And finally, one must know the PES shape in the case when the group of molecular symmetry is a subgroup of the FPI group of the molecule.

Let us now consider the problem of determining the molecular group using the example of the ethane molecule. The absolute minimum configuration (staggered) is known to have the point group of symmetry D_{3d}. The FPI group of the ethane molecule is a direct product of the group of permutation for six hydrogen atoms S_6, two carbon atoms S_2, and the group of inversion E, i.e.

$$G^{(\text{ethane})} = S_6 \otimes S_2 \otimes E,$$

where the order of this group (the number of elements) is equal to $6! \times 2! \times 2 = 2,880$ (this group is designated $G^{(2880)}$). If we want now to classify the vibrational levels in ethane with account taken of tunnel splitting due to internal rotation around C–C bond, the D_{3d} group is inadequate; while if we use the irreducible representations of the $G^{(2880)}$ group, the levels will turn out be highly degenerate. This is due to the fact that a large part of the $G^{(2880)}$ group's elements of symmetry are energetically forbidden. Thus, permutation of the different methyl groups' protons or inversion of the methyl group are impossible because of high energy barriers. If we eliminate all unrealizable elements from $G^{(2880)}$, then the remaining $G^{(36)}$ will be the group of ethane's molecular symmetry. Knowing the character table for $G^{(36)}$ [20] and the characters table for D_{3d}, one can establish the scheme of vibrational levels splitting for basic configuration of the ethane molecule, as a result of tunnelling under torsion barriers on the way of internal rotation. This scheme is as follows

$$\begin{array}{rcl} D_{3d} & & G^{(36)} \\ A_{1g} & \longrightarrow & A_1 + E_3 \\ A_{1u} & \longrightarrow & A_3 + E_3 \\ A_{2g} & \longrightarrow & A_2 + E_4 \\ A_{2u} & \longrightarrow & A_4 + E_4 \\ E_g & \longrightarrow & E_1 + G \\ E_u & \longrightarrow & E_2 + G \end{array}$$

It is irreducible representations of $G^{(36)}$ that enable us to classify wholly the ethane molecule's vibrational levels with account taken of the torsion tunnelling.

It follows from this analysis that in studying the structure of a molecule one must not confine the study to the area of one minimum on the PES, as is often done in the literature. One must study the PES shape over the area of all absolute minima and in the area of the molecule's pathways for transition from one minimum into another.

3. Potential Energy Surfaces for Non-Rigid Molecules

One can obtain the shape of potential energy surface (we continue to use the Born-Oppenheimer approximation) of non-rigid molecules through solving either a direct or reverse problem. In the first approach, one must solve the molecular Schrödinger equation in the adiabatic approximation for sufficiently wide range of coordinates of the nuclei. Here, it should be emphasized that to characterize the minima, maxima, and saddle points on a PES, one must know the first-, second-, and higher-order energy derivatives with respect to the coordinates. On the basis of these data, one can find the appropriate analytical form of the PES in the region of flexibility and then find a model non-rigid Hamiltonian. In the second approach, a PES is assumed to have a certain shape along non-rigid and quasi-rigid coordinates and, on this basis, one constructs a model Hamiltonian and calculates its theoretical spectrum. The model Hamiltonian's parameters are determined by obtaining a good fit between the theoretical and experimental spectra. If the assumed PES shape allows an adequate description of the experimental spectrum, then the shape of the PES and the model Hamiltonian are considered to be have been well chosen.

Both methods have their shortcomings. The direct approach, at the present-day level of *ab initio* techniques' accuracy in quantum chemistry, does not offer high, experiment-comparable accuracy in calculating the spectrum. In solving the reverse problem, one must guess the PES shape along the non-rigid coordinates. This can be done in cases such as the inversion of pyramidal molecules (e.g., ammonia and its derivatives) or rotation around single bonds, where chemical intuition helps the researcher. The nature of large-amplitude motions, however, can be complex, involving chemically bound atoms, and therefore in this case this approach is doomed to failure. Apparently, a combination of the two approaches is most promising; the PES shape (especially along non-rigid coordinates) is established using rather accurate *ab initio* calculations: then, on this basis, one constructs a model Hamiltonian. This Hamiltonian includes large-amplitude motions along the non-rigid coordinates, and harmonic or anharmonic, but small-amplitude vibrations along the other quasi-rigid coordinates. Then, the Hamiltonian's parameters can be determined from experimental data. In some cases, when experimental data on large-amplitude motion spectra are lacking, a rough frequency estimate can be made by using parameters from *ab initio* calculations. The *ab initio* approach has one more important advantage, relating, as it does, the PES shape of non-rigid molecules to the specificities of their electronic structure.

In what follows, a judgement on the PES shape will be made first of all on the basis of calculations, and theoretical results will then be compared to the experimental findings.

At present, a large number of structurally non-rigid molecules have been discovered with various forms of large-amplitude motions: internal rotations around single bonds [23, 24 and references therein], inversion of pyramidal AX_3 molecules [25, 26 and references therein], intramolecular pseudo-rotation in MX_k molecules [27 and references therein], cage rearrangements in cage boranes and carboranes [28 and references therein], intramolecular rotation of cations around anions in inorganic and complex salts [29–35], intramolecular motion of H^+ in carbocations [36–40], intramolecular anion degradations in $Li_2[BeH_4]$ and $Li_2[BeF_4]$ molecules [41], intramolecular motions of H^- in berryllo- and borohydrid anions [42], and others. This survey deals with non-rigid molecules for which not merely the fact of large-amplitude motion has been established but for which non-rigid symmetry questions have been considered and the large-amplitude motion spectrum investigated. The systems chosen with large-amplitude motions occurring between chemically bound, from the viewpoint of classical theory of valency, atoms, which is important for promoting our understanding of the structure of matter. To construct correct non-rigid models of a molecule, one must have an idea of the form of its PES in the region covering all minima separated by low barriers. This form will be conveniently presented in this survey with the help of graphs. The absolute minima corresponds to the graph's nodes and the minimal energy pathways of intramolecular non-rigid rearrangement to its edges.

4. Carbocations

The structure of carbocations has been for many years the subject of both experimental and theoretical studies, since carbocations are intermediate reactive species in numerous reactions, and knowing their structure is important for understanding reaction mechanisms.

The most thorough studies have been carried out on the PES of the simplest carbocations $C_2H_3^+$ and $C_2H_5^+$. As two examples, for these cations, one considers classical (c) and bridge (b) configurations:

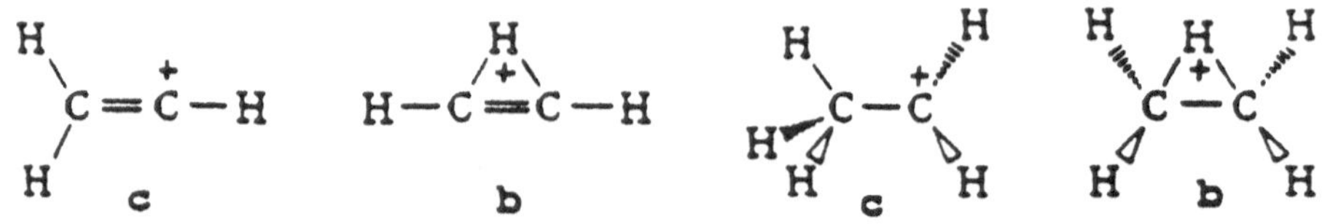

The vinyl cation $C_2H_3^+$, has been the subject of a number of theoretical studies using *ab initio* techniques [36]. At high *ab initio* level (MP4SDTQ [36e, i], CISD [36d, g, h, k] and CASSCF-CI [36j] with large basis sets) the bridged (b) $C_2H_3^+$ structure is lower than the classical (c) one by 2.5–4.6 kcal/mol. This theoretical result agrees with the experimental data. The observed vibrational–rotational pattern with the band origin at 3144.2 cm^{-1} has been identified as due to the antisymmetric CH stretching ν_7 band of the $C_2H_3^+$ ion with the *b*-structure [38]. Experimental estimation of the energy difference between *b*- and *c*-structures is > 6.0 kcal/mol, which is based on the upper limit of the observed level splitting < 0.005 cm^{-1} and an internal rotation model [37]. The experimental and theoretical *b*–*c* vinyl cation energy difference was discussed in [36k]. According to this analysis, the use of more elaborate basis sets (e.g., including a large (sp) basis, *g*-functions, and more *f*-functions) is estimated to improve the results by 0.7 kcal/mol. Therefore, the remaining discrepancy with experiment (assuming the later is right) might result from the incomplete treatment of electron correlation, or the possibility that the experimental analysis has overestimated the energy difference between *b* and *c*. From our point of view, several additional causes of this discrepancy should be taken into account. Firstly, zero point energy (ZPE) corrections to the *b*–*c* energy difference in *ab initio* calculations were made using a rigid rotator harmonic oscillator model, while the vinyl cation is a highly flexible molecule. Secondly, the geometries of both *b* and *c* were optimized at a lower correlated level than when the best *b*–*c* energy difference was obtained.

Is the classical (c) structure a minimum or saddle point on the intramolecular rearrangement? This is a question for dynamic consideration of the vinyl cation and has not yet been solved. Weber *et al.* [36] reported the barrier between *b*- and *c*-structures in $C_2H_3^+$ to be 1.6 kcal/mol. Pople [36i] found that both *b*-

and c-structures are minima and the transition structure between b and c lies only 0.4 kcal/mol above c at the MP2(full)/6–31G* level. However, at his more sophisticated computational level (but by using MP2(full)/6–31G* geometry), the transition structure lies below c. This suggests that the rearrangement barrier ultimately disappears and c itself becomes a saddle point on the potential surface. Similar results were found by Lindh *et al.* [36j]. They optimized geometries for b, c and transition structures at the CASSCF level using basis sets without polarization functions, and at this level they found a rearrangement barrier of 0.37 kcal/mol between b- and c-structures. However, at the CASSCF level with larger basis sets (with inclusion of polarization functions) the transition structure has again a lower energy than c. Further calculations with optimization of geometries at more sophisticated levels are required to solve this problem definitively.

Taking into account that the barrier of rearrangement $C_2H_3^+(b)$–$C_2H_3^+(c)$ is low, this carbocation should be regarded as non-rigid with respect to H-atom motion around the carbon dumbbell.

To describe the shape of the PES along the non-rigid coordinates of $C_2H_3^+$ let us make use of the graph in Figure 2. The $C_2H_3^+$ FPI group has order $2! \times 3! \times 2 = 24$, while the point group order of the absolute minima b-configuration and the saddle point c-configuration C_{2v} is 4. Then, the number of absolute minima and saddle points on the PES of $C_2H_3^+$ is 6. The non-rigid rearrangement of $C_2H_3^+$ is realized via an intraplane relay-transfer mechanism: first atom H_1 moves into the pivot configuration c_1

$$H_{(1)} \diagdown \atop H_{(2)} \diagup C_{(1)} = \overset{+}{C}_{(2)} - H_{(3)}$$

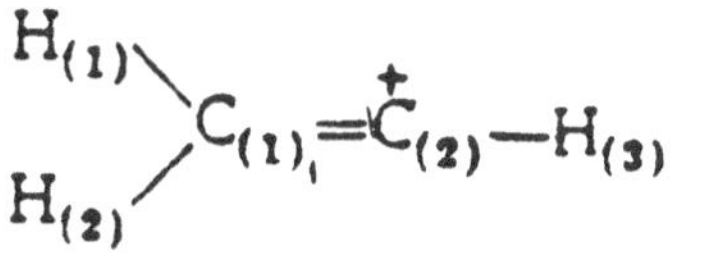

then when the cation converts configuration c_2

$$H_{(2)} - \overset{+}{C}_{(1)} = C_{(2)} \diagup^{H_{(1)}}_{\diagdown H_{(3)}}$$

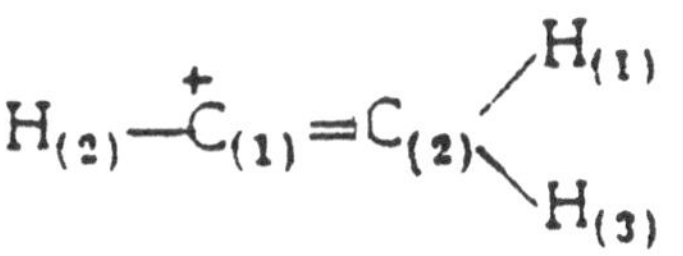

atom H_3 starts moving and so on. As is seen from the graph, this mechanism connects all the PES minima, since the molecular symmetry group of this ion is its FPI group. The character table for the permutation-inversion group $G^{(24)}$ is given in Refs. 37. Therefore the scheme of equilibrium configurational level splitting C_{2v} due to tunnelling under the barrier is to be (the statistical weights for two spin zero nuclei (e.g. ^{12}C) and three spin 1/2 (e.g. ^{1}H) are given in parentheses) [38]:

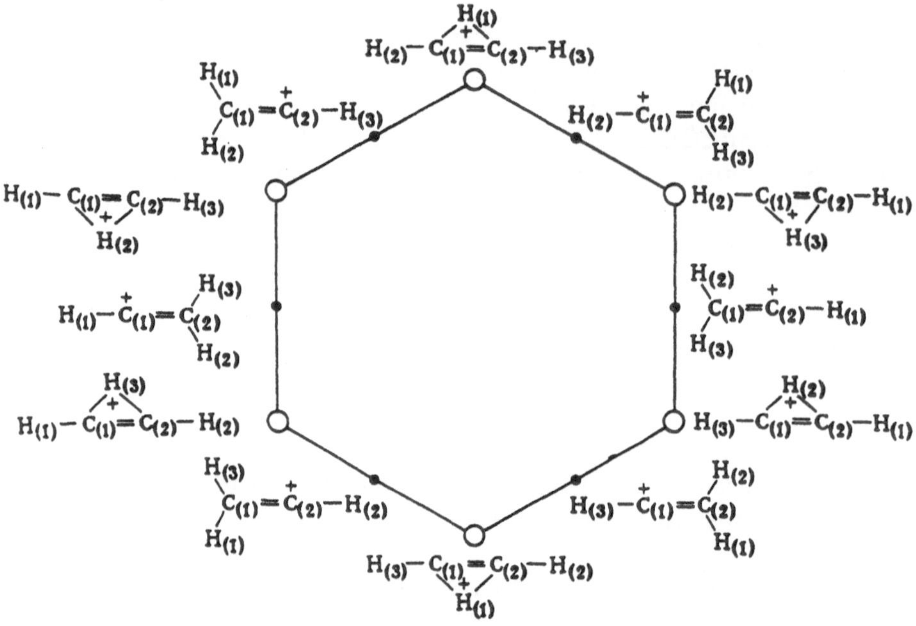

Figure 2. Graph of the intramolecular rearrangement of the vinyl cation.

C_{2v}		D_{6h}		$G^{(24)}$
a_1	$\rightarrow$	$a_{1g}(0)+b_{2g}(0)+e_{1g}(0)+e_{2g}(2)$	$\rightarrow$	$a^+_{1g}(0)+a^+_{2u}(0)+e^+_g(2)+e^+_u(0)$
a_2	$\rightarrow$	$a_{1u}(0)+b_{2u}(0)+e_{1u}(0)+e_{2u}(2)$	$\rightarrow$	$a^-_{1g}(0)+a^-_{2u}(0)+e^-_g(2)+e^-_u(0)$
b_1	$\rightarrow$	$a_{2u}(4)+b_{1u}(0)+e_{1u}(0)+e_{2u}(2)$	$\rightarrow$	$a^-_{2g}(4)+a^-_{1u}(0)+e^-_g(2)+e^-_u(0)$
b_2	$\rightarrow$	$a_{2g}(4)+b_{1g}(0)+e_{1g}(0)+e_{2g}(2)$	$\rightarrow$	$a^+_{2g}(4)+a^+_{1u}(0)+e^+_g(2)+e^+_u(0)$

Taking into account the statistical weights of split levels, one can see that only the b_1 and b_2 levels will be split into doublets. Their fine structure of vibrational levels, caused by tunnel effects, can be found in the high resolution vibrational spectra of vinyl cation.

Numerical calculations of the high-resolution vibration–rotation spectrum of the vinyl cation were made by Hougen [37a] and Bunker *et al.* [37b–d]. The simplest idealized model, in which the hydrogens are placed at the vertices of an equilateral triangle whose centre of gravity coincides with the centre of gravity of two carbon atoms located at the ends of a dumbbell. The hydrogens' triangle and the carbons' dumbbell are assumed rigid and this idealized model has only one large-amplitude coordinate ξ (see Figure 3). As was pointed by Hougen, this model preserved many features of the actual $C_2H_3^+$ problem: the idealized *b*- and *c*-structures have right C_{2v} symmetry; if the angle ξ (Figure 3) is allowed to change from 0 to 360°, (i.e., if the triangle of hydrogens is made to carry out

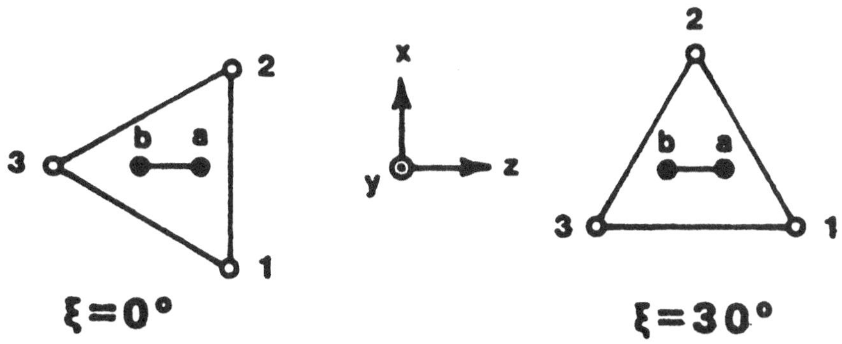

Figure 3. The idealized $C_2H_3^+$ structures used for internal rotor calculations, shown for two values of the large-amplitude coordinate ξ. The hydrogen (1, 2, 3) are located at the vertices of an equilateral triangle, and the carbons (a, b) are located at the ends of a dumbbell whose centre of mass coincides with that of the triangle.

one complete revolution about the C–C dumbbell), a set of all *c*- and *b*-structures is tracted out; the rotation constants of the idealized model are such that it is a nearly prolate symmetric top, with the near symmetric top axis essentially along the C–C bond and the angular momentum generated by the hydrogen motions perpendicular to the near symmetric top axis. The most important features of the actual $C_2H_3^+$ problem which are lost in this model arise from the facts that (i) the hydrogen atoms do not move around the circumference of a circle and (ii) the hydrogen atoms do not all move with the same angular velocity at each instant of time. The Principal-Axis-Method rotational-internal-rotation Hamiltonian has the following form for this model [37a]:

$$H = AJ_z^2 + BJ_x^2 + CJ_y^2 + Fp_\xi^2 - 2Cp_\xi J_y + V(\xi) \tag{1}$$

with the coordinate system in Figure 3. In the calculations by Hougen, it is assumed that the potential barrier to internal rotation has six-fold symmetry and that either the classical structure represented the energy minimum and the bridged structure represents the energy maximum or vice versa. The six-fold barrier to the internal rotation motion appropriate for this idealized model was introduced in the form

$$V(\xi) = 1/2[|V| + V\cos 6\xi], \tag{2}$$

where positive values of V correspond to a bridged equilibrium configuration and negative values of V to a classical equilibrium configuration. Bunker *et al.* expanded the internal-rotational potential $V(\rho)$ as

$$V(\rho) = \Sigma a_n \cos(6n\rho), \tag{3}$$

where the five a_n were determined by fitting to the five *ab initio* $V(\rho)$ values from the Ref. 36j at CEPA level (recall that at this level, the PES of the *b*–*c* rearrangement has a barrier). The maximum in this potential occurs at $\rho = 7.05°$ with $V_{max} = 1429\ \mathrm{cm}^{-1}$. Bunker *et al.* calculated internal-rotational tunnelling splittings using the extended quasi-linear molecular formalism (semi-rigid bender) [37b]. The Escribano-Bunker's semi-rigid bender Hamiltonian has the form [37d]:

$$H_{\text{srb}} = H_b + H^{(z)} + H^{(xy)} + H^{(y\rho)} \tag{4}$$

where

$$H_b = 1/2\mu_{\rho\rho}J_\rho^2 + 1/2[J\rho, \mu_{\rho\rho}]J_\rho + 1/2\mu^{1/4}[J_\rho, \mu_{\rho\rho}\mu^{-1/2}[J_\rho, \mu^{1/4}]] + V(\rho) \tag{5a}$$

$$H^{(z)} = 1/2\mu_{zz}J_z^2 \tag{5b}$$

$$H^{(xy)} = 1/2\mu_{xx}J_x^2 + 1/2\mu_{yy}J_y^2 + 1/2\mu_{xz}(J_xJ_z + J_zJ_x) \tag{5c}$$

$$H^{(y\rho)} = 1/2[J_\rho, \mu_{y\rho}]J_y + 1/2\mu_{y\rho}(J_yJ_\rho + J_\rho J_y) \tag{5d}$$

$$\mu_{xx} = I_{zz}/(I_{xx}I_{zz} - I_{xz}^2) \tag{6a}$$

$$\mu_{yy} = I_{\rho\rho}/(I_{yy}I_{\rho\rho} - I_{y\rho}^2) \tag{6b}$$

$$\mu_{zz} = I_{xx}/(I_{xx}I_{zz} - I_{xz}^2) \tag{6c}$$

$$\mu_{xz} = -I_{xz}/(I_{xx}I_{zz} - I_{xz}^2) \tag{6d}$$

$$\mu_{\rho\rho} = I_{yy}/(I_{yy}I_{\rho\rho} - I_{y\rho}^2) \tag{6e}$$

$$\mu_{y\rho} = -I_{y\rho}/(I_{yy}I_{\rho\rho} - I_{y\rho}^2) \tag{6f}$$

$$I_{\alpha\alpha} = \Sigma m_i(a_{i\beta}^2 + a_{i\gamma}^2) \tag{6g}$$

$$I_{xz} = -\Sigma m_i a_{ix}a_{iz} \tag{6h}$$

$$I_{y\rho} = -\Sigma m_i(a_{ix}a'_{iz} - a_{iz}a'_{ix}) \tag{6k}$$

$$I_{\rho\rho} = \Sigma m_i(a'_{i\alpha})^2 \tag{6l}$$

where α, β, $\gamma = x, y, z$, $a_{i\alpha}$ are the coordinates of nucleus i having mass m_i and $a'_{i\alpha} = da_{i\alpha}d\rho$. Calculated rotational energy levels (in cm^{-1}) for the vinyl cation with torsional splittings are given in Table I (the energy levels in column 3 for the vibrational ground state and in column 4 for $\nu_{\text{torsion}} = 1$). As seen from Table I, the calculated splittings in the rotational levels of the first excited state are about 5×10^{-3} cm^{-1} which are in accord with experimental observation [38]. In the experimental work by Crofton *et al.* [38], the spectrum has been observed in the 3.2 m region in air-cooled and water-cooled plasmas using C_2H_2:H_2:He mixtures and in liquid nitrogen-cooled plasmas using CH_4:H_2:He mixtures. The band origin at 3141 cm^{-1} has been the only one which can be assigned to $C_2H_3^+$. This band was attributed to the ν_7 fundamental frequency (antisymmetric C–H stretching model) of $C_2H_3^+$. This identification was based on the spectral pattern and the spectral intensity response to plasma chemistry. The spectrum is due to b vinyl cation structure, because of the spin statistical weight of the observed $\Delta K = 1$ K doublets, the values of the ground state constants, and absence of other equally

TABLE I. Calculated rotational energy levels (in cm^{-1}) for $^{12}C_2H_3^+$ with torsional splittings in (cm^{-1}) in parentheses [37b].

$J_{K_aK_c}$	Rigid rotor	Semirigid bender $v_{torsion} = 0$		Semirigid bender $v_{torsion} = 1$	
2_{20}	55.363	E_g^+	56.094	A_{2g}^+, E_g^+	53.895 (5.2×10^{-3})
2_{21}	55.363	A_{2g}^-, E_g^-	56.093 (3.3×10^{-5})	E_g^-	53.895
2_{11}	18.859	E_g^-	19.397	E_g^-, A_{2g}^-	19.204 (4.4×10^{-3})
2_{12}	18.592	E_g^+, A_{2g}^+	18.944 (3.0×10^{-5})	E_g^+	18.058
2_{02}	6.513	E_g^+	5.882	A_{2g}^+, E_g^+	5.123 (4.2×10^{-3})
1_{10}	14.428	E_g^+, A_{2g}^+	14.739 (2.8×10^{-5})	E_g^+	14.364
1_{11}	14.339	E_g^-	14.584	E_g^-, A_{2g}^-	13.970 (5.4×10^{-3})
1_{01}	2.171	A_{2g}^-, E_g^-	1.956 (2.8×10^{-5})	E_g^-	1.704
0_{00}	0.0	E_g^+	0.0	A_{2g}^+, E_g^+	0.0 (5.2×10^{-3})

The symmetry species indicate the energy order: e.g. for the 2_{21} pair (for $v_{torsion} = 0$), the A_{2g}^- level is above the E_g^- level. Band origin for the $v_{torsion} = 1 \leftarrow 0$ band is 404 cm^{-1}.

strong vibrational bands in the region between 3300 and 2700 cm^{-1}. Observed A–E state splitting in the vinyl cation spectrum gave the most definitive evidence for the tunnelling exchange of the apex proton and the end protons in the bridged structures. (While the A–E splittings in the ground state were found negligible compared with the spectral resolution in [38], these splittings were observed for the levels in excited states.) Some satellite lines are not definitely identified in

TABLE II. The splittings (in cm^{-1}) of the rotational levels in the v_7 state [37c].

J	K_a	K_c	$v(A_{2g} - E_g)$			
			obs.[(a)]	cal.[(b)]	calc.[(c)]	calc.[(d)]
2	1	1	0.207	-0.032	-0.065	-0.306
2	0	2	0.342	0.032	0.063	0.294
4	1	3	0.056	-0.014	-0.028	-0.145
4	0	4	0.063	0.015	0.031	0.146
6	1	5	-0.058	0.007	0.013	0.047
7	1	7	0.101	0.017	0.033	0.139
8	1	7	0.022	0.025	0.050	0.222
10	1	9	0.053	0.035	0.071	0.335
11	1	11	0.147	-0.004	-0.008	-0.036
12	1	11	0.079	0.035	0.070	0.343
14	1	13	0.043	0.022	0.045	0.219
15	1	15	-0.104	0.007	0.015	0.070

(a) Ref. [38].
(b) Using a pure V_6 potential with a barrier a height of 700 cm^{-1}.
(c) Using a pure V_6 potential with a barrier a height of 600 cm^{-1}.
(d) Using a pure V_6 potential with a barrier a height of 400 cm^{-1}.

the experimental spectrum because of overlapping transitions: but at least for the upper components of the $\Delta K = 1$ K doublets (relative to $J_{1,J-1}$ with J even in the excited state), the agreement between the R and P the transitions and intensity ratios of 2:1 were evidence to the authors of [38], that the observed splitting is due to the tunnelling. The tunnelling A–E splittings of the antisymmetric C–H stretch vibrations ν_7 of the HCCH part of the vinyl cation were calculated by Gomez and Bunker [37d] using a one-dimensional semi-rigid bender model (Table II). As seen from Table II, tunnelling splittings are strongly dependent on the values of potential barriers used in [37c]. Gomez and Bunker were completely unable to reproduce definitively the observed pattern of splitting. Neither the use of more parameters in the potential function nor the adjustment of the variation of the geometry with the one-dimension coordinate ρ, significantly improve the fit. They explain their lack of success by state perturbations such as $2\nu_4 + 2\nu_6$, $3\nu_4 + \nu_5$, $\nu_3 + \nu_5 + \nu_6$ and $\nu_6 + 2\nu_8$, which lie within 25 cm^{-1} of ν_7. Gomez and Bunker pointed out that it would be easier to interpret the splittings in the ground vibrational state or in the first excited torsional state ($\nu_9 = 404$ cm^{-1} from *ab initio* plus semi-rigid bender calculations) if some experimental means for measuring these splittings could be found.

According to the most accurate calculations [36e], the b configuration corresponds to the absolute minimum for $C_2H_5^+$ but the alternative *c*-configuration lies only $\approx$ 6 kcal/mol higher than *b*. This theoretical conclusion qualitatively agrees with the experimental data on the photoelectron spectrum of the C_2H_5 radical [39]. According to these data, the difference in energy between *b*- and *c*-configurations of $C_2H_5^+$ is roughly equal to the difference between the vertical and adiabatic IP of the C_2H_5 radical, which is $\approx$ 3 kcal/mol. This estimate is based on the facts that the main configuration of the ethyl radical is close to the *c*-configuration of $C_2H_5^+$ ion and the vertical electron abstraction from C_2H_5 yields an ion of $C_2H_5^+$ in the *c*-configuration. Adiabatic ionization of the C_2H_5 radical yields $C_2H_5^+$ ion in the main *b*-configuration.

The PES of the $C_2H_5^+$ rearrangement is considerably more complex. The order of the $C_2H_5^+$ FPI group is $2! \times 5! \times 2 = 480$ and the point group order of absolute minimum *b*-configuration C_{2v} is 4, with the number of absolute minima on the PES being 120. However, in case of $C_2H_5^+$ the *b*–*c* rearrangements alone are not able to connect all the PES minima; and therefore the ethanolike rotation is required in this ion's *c*-configuration. It is therefore necessary that the ethane type rotation be realized with a small barrier in the *c*-configuration of this ion. *Ab initio* calculations yield the value of 1 kcal/mol for the barrier of the ethane type rotation in the *c*-configuration of $C_2H_5^+$ [36e]. The total rearrangement involving both mechanism will connect all PES minima.

Let us consider the shape of the PES of $C_2H_3^+$ applying the localized MO (LMO) model [15a]. In the *c*-configuration, the $C_2H_3^+$ ion has three two-centre LMOs b(C–H) and a two-centre banana bond b(C–C), whereas in the *b*-configuration, the ion has two LMOs b(C–H), one LMO b(C–C), and a three-centre LMO t(C–H–C). As a result, the banana bond b(C–C) remains intact as *c* transforms into *b*, but one two-centre LMO b(C–H) transforms into a three-centre LMO t(C–H–C), which accounts for the close energies of the *c*- and *b*-configurations and for structural non-rigidity of $C_2H_3^+$. Similar analysis for $C_2H_5^+$ LMOs reveals that no destroying of the banana bond b(C–C) takes place in the course of its intramolecular rearrangement either. The only thing one observes is the transformation of one LMO b(C–H) to a three-centre LMO t(C–H–C). Thus, in terms of these notions, one can easily explain the structural non-rigidity of both cations. Apparently, with other protonated unsaturated hydrocarbons too, one should expect non-rigid motions of H atoms along conjugated carbon–carbon bonds.

Moreover, taking into account the close energies of the *c*- and *b*- configurations of the $C_2H_7^+$ ion (according to *ab initio* calculations the difference in energy between these configurations is $\approx$ 7 kcal/mol [36e]),

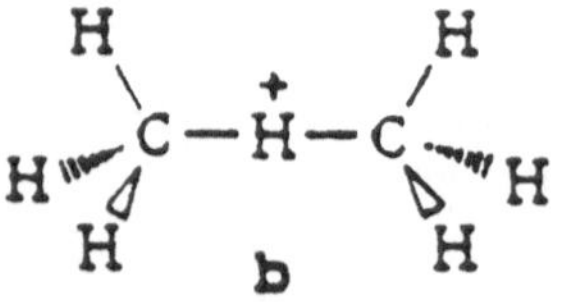
b

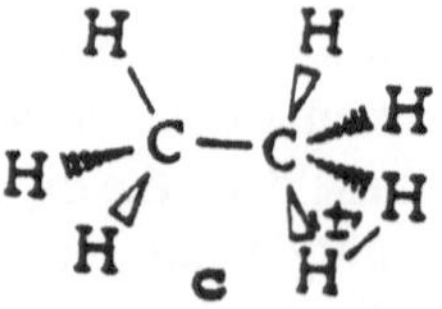
c

for saturated protonated hydrocarbons too one should expect low barriers for H^+ cation migration along carbon chains. In this manner, all the protonated hydrocarbons must be structurally non-rigid with respect to H^+ cation migration along carbon skeletons. This theoretical conclusion is supported by NMR spectra of protonated hydrocarbons [40], which suggest fast proton migration along a molecule's carbon backbone in such systems.

Similar non-rigid rearrangement were found with beryllo-($Be_2H_3^-$) and borohydrid-($B_2H_3^-$, $B_2H_5^-$) anions [42], which are isoelectronic to the carbocations discussed above. The H^- anion moves with low barriers around the beryllium and boron backbones. The PES shape of $Be_2H_3^-$ and $B_2H_3^-$ anions along the non-rigid coordinates is similar to that of $C_2H_3^+$ (the rearrangement mechanism is relay-transfer), and the PES shape of $B_2H_5^-$ is similar to that of $C_2H_5^+$ (the mechanism in relay-rotations).

Once again one can explain qualitatively the structural non-rigidity of the above-mentioned anions in terms of "LMO – chemical bond" analysis. As with carbocations, no destroying takes place with the anion bonds b(Be–Be) and b(B–B) as the *c*-configuration changes into the *b*-; and transformation of b(Be–H) or b(B–H) into t(Be–H–Be) or t(B–H–B) bonds, respectively, causes no substantial change in energy. Therefore, one can explain structurally the non-rigid rearrangement for all beryllo- and borohydride anions containing chains of Be or B atoms, with H^- anions moving along these chains.

5. Inorganic and Complex Acid Salts

According to notions which are widely accepted in chemistry, the molecules of inorganic $LMO_{(k+1)/2}$ ($LiBO_2$, $NaNO_3$, $KClO_4$ etc.) and complex $L[MX_{k+1}]$ ($LiBeF_3$, $NaBF_4$, KPF_6 etc.) acid salts consist of the oxygen acid residue $MO^-_{(k+1)/2}$ or MX^-_{k+1} (M being an atom from the second, third, or fourth period and X halogen or hydrogen, and k being the highest oxidation degree of atom M), which is considered to be a stable and rigid autonomous fragment: an outer cation L, connected to the anion mostly by ionic bond, can consequently be mobile in the outer sphere. This is indicated by the fact that the mean square vibration amplitudes of nuclear pairs l(L–M) and l(L–X) are 3 to 5 times larger than the vibration amplitude of pair l (M–X), l (L–M) being quite smaller than l(L–X). Besides, the structural non-rigidity of these salts is suggested by the "washing out" of molecular features in diffraction patterns (some peaks are absent on electron diffraction patterns). The absence of vibration frequencies of the cation against the anion (which, apparently, have very low values as well as intensities and cannot be detected by modern IR and Raman spectrometers) is one more fact in support of the possible cation mobility with respect to the anion [43]. The *ab initio* calculations performed over the last 15 years have given considerable support to the dynamic structure of these salts.

5.1. CYANIDES

The alkali metal cyanides were among the first non-rigid molecules for which *ab initio* calculations were used to establish a flattened PES shape [29]. The *ab initio* studies have demonstrated that various PES shapes are observed in the row of cyanides. Thus, the absolute minimum on the PES of lithium cyanide corresponds to a linear isocyanide configuration LiNC [29a, g, n, p]. The linear cyanide configuration LiCN corresponds to a local minimum which is 2.34 kcal/mol higher than the global minimum [29p]. The intermediate triangle configuration corresponds to a local minimum with relative energy 0.75 kcal/mol (with respect to the LiNC structure [29p]). The PES of sodium, potassium and rubidium cyanides have a different form. A triangular configuration corresponds to the absolute minimum for these molecules, while the linear isocyanide and cyanide configurations correspond to the tops of low barriers for sodium (3.8 kcal/mol and 3.1 kcal/mol [39e, f], respectively) and potassium (1.4 kcal/mol and 5.6 kcal/Mol [29d], respectively). Both linear configurations for rubidium isocyanide and rubidium cyanide are local minima with relative energies 1.2 kcal/mol and 5.0 kcal/mol [29l], respectively.

Found by means of *ab initio* calculations, these differences in the geometrical structure of configurations corresponding to the absolute minima of the PES of lithium, sodium, and potassium cyanides agree with experimental data obtained with microwave and rotational spectra of the molecules in the gas phase [30]. However, despite some differences, one common feature typical of all alkaline metal cyanides is that the barriers for L^+ cation orbital rotation around CN^- are relatively low, and all of these molecules can be regarded as non-rigid.

If we substitute an alkali metal atom by hydrogen, the PES shape undergoes substantial changes. The absolute minimum is the linear cyanide configuration HCN. The linear isocyanide configuration HCN is a local minimum lying 11.2 kcal/mol higher than the absolute one [44]. Both minima are separated by a high barrier of 23.7 kcal/mol whose apex corresponds to a triangular configuration so that at low temperature both isomers can be considered as independent quasi-rigid molecules. The interesting question is, for which cyanides should we expect low barriers to orbital motion of the cation with respect to the anion (i.e. which cyanides must be non-rigid)? A simple qualitative model has been suggested [29c] to explain the high potential barrier on the PES of HCN–HNC rearrangement and its sharp decrease on that of LiCN–LiNC. The model is based

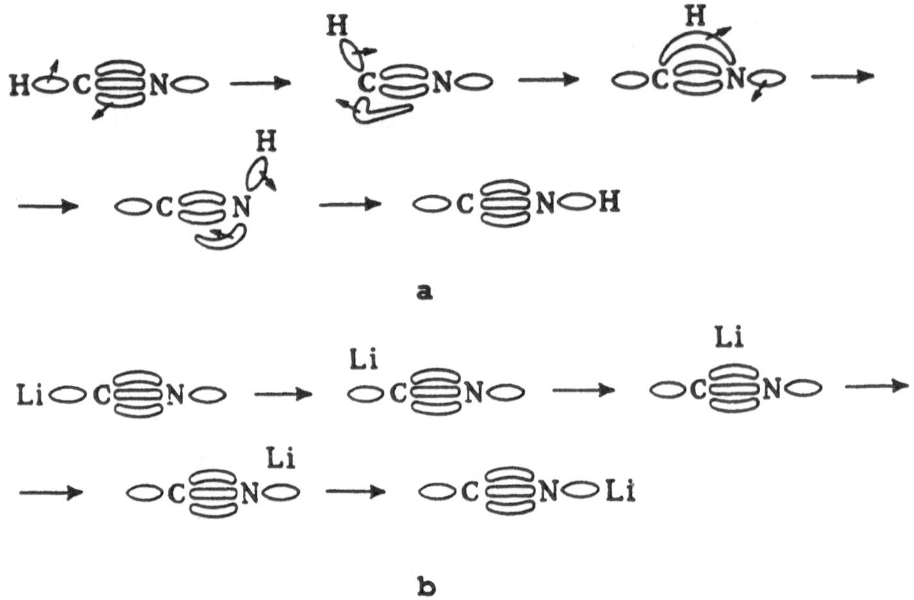

Figure 4. Scheme of the changes of localized MO during HCN–HNC and LiCN–LiNC intramolecular rearrangements.

on analysing changes in the LMOs (which can be roughly compared to traditional chemical bonds) in the course of L motion around CN (Figure 4) [29c].

The linear HCN (or HNC) configuration has three types of valence LMOs: one LMO of C–H bond type, b(C–H) (or N–H bond b(N–H)); three LMO of banana C–N bond types, b(C–N); one LMO for the lone pair of N atom 1(N) (or the lone pair of C atom 1(C)). In addition to the traditional description of the triple C≡N bond as one σ-bond formed by sp-hybrid AOs of the C and N atoms and two π-bonds by pure p_x- and p_y-AOs of C and N, one can present the triple bond as three identical banana C–N bonds formed by sp^3-hybrid AO of C and N atoms. If no additional restrictions are imposed when constructing LMOs from canonical MO (for instance, if σ- and π-MOs are localized separately), then one obtains LMOs which correspond to the picture of banana bonds. In what follows, we shall regard the triple bond as consisting of three banana bonds; although transition to the LMO obtained with σ-MO and π-MOs taken separately leads us, in the long run, to the same conclusions. The triangular configuration of the HCN barrier peak has the following set of LMOs: one LMO of the three-centred bond C–H–N, t(C–H–N); two LMO of pairs of atoms C, 1(C), and N, 1(N) (Figure 4a illustrates all three LMOs) and two LMO of banana C-N bond types. The barrier is caused by one stable banana C–N bond being destroyed in the angular configuration. The energies of three bonds C–H, N–H, and C–H–N are close. With Li(CN) there is no breakage of the banana C–N bond in area of triangular configurations. The alkali metals' cations slide around the CN^- anion exerting comparatively weak perturbation of its electron structure (Figure 4b), which explains the flat PES pattern of LiCN–LiNC, NaCN–NaNC and other similar rearrangements.

On the basis of this qualitative picture one can assume that the stronger the perturbation effect exerted by the atom L on the banana b(C–N) bond in the cyclic L(CN) configuration, the higher the potential barrier for L migration from LCN to LNC. In other words the barrier must increase along the row Li < Al < H < F type as the electron affinity L^+ or the first ionization potential (IP) of L increases. This qualitative conclusion is supported by the result of quantum chemical calculations of the PES of Cu [45] motion around the CN group. In this case the Cu atom IP = 7.7 eV, which is higher than IP(Li) = 5.4 eV, but considerably lower than IP(H) = 13.6 eV. The barrier for Cu atom motion around the CN group is 13 kcal/mol [45], which is higher than 2.3 with Li(CN), but considerably lower than 23.7 with H(CN). Thus, simple notions about the destabilizing effect of the atom in cyclic configuration L(CN) and numerical assessment of barriers obtained through calculations, allow us, using the values of the first IP of univalent atoms and groups, to define the range of cyanides whose transition barriers of the rotation L^+ around CN^- must be comparatively low. Among these are first of all the cyanides of alkaline and alkaline earth metals, transition metals at the beginning of the periods, lanthanides and actinides, and also some L groups with the first IP < 5–6 eV.

The analysis done makes it possible to characterize the height of the potential barrier which separates the LAB and LBA isomers, but is fails to provide an assessment of these isomers' relative stability. The molecules can be rigid with respect to L migration around AB even in the absence of the barrier, provided the difference of the isomers' total energies is sufficiently large. The question of the relative stability of the LAB and LBA isomers has been considered in [46]. For all the above-mentioned cyanides with an L atom IP < 5–6 eV, the two linear configurations LAB and LBA are close in energy; and it is precisely with these isomers that one should expect the manifestation of migrational non-rigidity of the L atom or L group around the cyanide backbone CN.

Similar consideration can be used to describe the differences between migrationally non-rigid $Li(N_2)^+$ and $Li(CO)^+$ systems and migrational rigidity in HN_2^+ and $H(CO)^+$. In ion-molecular complexes $Li(N_2)^+$ and $Li(CO)^+$, a cation of Li^+ moves around N_2 and CO backbones, causing weak perturbation in their electronic structures; and this makes them non-rigid. Whereas with $H(N_2)^+$ and $H(CO)^+$, the proton destroys one banana bond, b(N–N) or b(C–O), in cyclic configurations, which gives rise to high barriers. In this case too, by analogy with cyanides, one can easily encompass the range of non-rigid ion-molecular complexes.

Several types of potentials and model Hamiltonian were used to calculate the large amplitude motion spectrum of alkali metal cyanides. The Hamiltonians' eigenvalues were found using variational [29b, h–j, o] techniques. Let us consider the most accurate calculations presented in [29i].

The model Hamiltonian for lithium and potassium cyanides was as follows:

$$\mathcal{H}(\theta,\mathcal{R}) = -\frac{\hbar^2}{2\mu_1 R^2}\frac{\partial}{\partial R}\left(R^2\frac{\partial}{\partial R}\right) - \frac{\hbar^2}{2}\left(\frac{1}{\mu_1 R^2}+\frac{1}{\mu_2 r^2}\right)\frac{1}{\sin\theta}\frac{\partial}{\partial\theta}\left(\sin\theta\frac{\partial}{\partial\theta}\right)+V(R,\theta), \quad (7)$$

where R is the distance between Li and the mass centre of CN; θ – the angle between R and CN; r – the length of the C–N bond, which was assumed to be fixed; and $\mu_1^{-1} = m_L^{-1}+(m_C+m_N)^{-1}, \mu_2 = m_C^{-1}+m_N^{-1}$. The potential $V(R,\theta)$ was constructed as an expansion in Legendre polynomials:

$$V(R,\theta) = \sum_{\lambda=0}^{9} P_\lambda(\cos\theta)V_\lambda(R). \quad (8)$$

The expansion coefficients (8) were obtained by the least squares methods using numerical PES points from *ab initio* calculations.

Brocks and Tennyson [29i] assumed, on the basis of *ab initio* data, that the PES of Li(CN) has two minima corresponding to the linear LiNC and LiNC configurations and a triangular saddle point on the pathway of the intramolecular rotation [29g]. The isocyanide LiNC structure is a global minimum. The cyanide LiCN structure is a local minimum lying 2386 cm^{-1} above LiNC; and a saddle point lies 3377 cm^{-1} higher than the main structure. While this is not the most exact *ab initio* PES, let us consider their results. The energies of the Li(CN) deformational levels for $J = 0$ are given in Table III. The calculated values of the bottom vibrational frequencies in the area of LiNC minimum $\nu_2 = 754$ cm^{-1} and $\nu_3 = 127$ cm^{-1} are in satisfactory agreement with the experimental $\nu_2 = 680$ cm^{-1} and $\nu_3 = 119$ cm^{-1} obtained from the LiNC IR-spectra in inert matrices [30a]. Similar frequencies in the area of the LiCN minimum have the values of $\nu_2 = 688$ cm^{-1} and $\nu_3 = 166$ cm^{-1}. In [29i] the 70 lowest vibrational states of LiNC were obtained for $J = 0$. The first 30 vibrational levels belong to the area of the main LiNC configuration. The 31st level belongs to the area of the LiCN isomer. Levels between the 31st and the top of the barrier (37 in all) belong mainly to the area of the main LiNC configuration, but some of the levels are localized in the LiCN isomer area. Some of the highest subbarrier states are already substantially delocalized with respect to the two minima; and the Li nuclear density functions has noticeable amplitude in the neighbouring minimum. Thus, for instance, the 58th vibrational level possesses an energy 3133 cm^{-1} over the absolute minimum. It is localized mainly in the local minimum area, but the probability density of the Li cation presence in the area of LiNC configuration amounts to 5%. It is interesting to note that although some states below the barrier are greatly delocalized, other higher states are found to be localized and it is difficult to predict for which states one is to expect free rotation or "polytope' states. It is clear that at 1500 K, the temperature of e.g. electron diffraction studies, the delocalization of the nuclear density function of molecules in the "over-barrier" states will be considerable. Undoubtedly, a molecule in such states cannot be classified as cyanide or isocyanide. The classical valency theory that assumed all the interactions in molecule are divided into valency and non-valency interactions is simply inapplicable for describing molecules in such states. It is impossible to describe such states within the framework of a one-minimum "rigid rotator harmonic oscillator" picture. It should be noted that harmonic approximations lead to substantial errors in the case of subbarrier states, lying much lower than the barrier. Thus, with $n_{\nu 3} = 10$ ($\Delta E = 975$ cm^{-1} at 3377 cm^{-1} barrier) the deformation frequency error in harmonic approximation is circa 30%. However, the lowest several levels may be described within the harmonic oscillator approximation.

With non-rigid molecules, one can expect a sharp dependence of the average geometrical parameters on both the excited state number and temperature. This is what actually takes place. Thus, with LiNC where for the main vibrational state already $\langle \varphi_{\text{Li–N–C}} \rangle = 11°$, at $n_{\nu 3} = 9$ the average angle $\langle \varphi_{\text{Li–N–C}} \rangle = 28°$ [29i]. This means that in electronic diffraction studies in the gas phase the Li(CN) molecule will have a corner effective configuration.

5.2. LMX_2 MOLECULES

One can include in this type of molecule the dimers of alkali-halides $(LX)_2$. However, according to the available data, the barriers for these molecules' intramolecular rearrangements are high and they should be regarded as quasi-rigid [47].

Among LMO_2 molecules, where M = B, Al, Ga and L = Li ÷ Cs, the PES shape has been studied for $LiBO_2$ [48a–c, f], $NaBO_2$ [48d–f], $LiAlO_2$, $NaAlO_2$, $LiBS_2$, LiOBS, LiOAlS [48g]. According to *ab initio* calculations at the MP2(full)/6–31+G* level for $LiBO_2$ [48c, f], the PES absolute minimum corresponds to a linear configuration m, while a rhombic configuration b

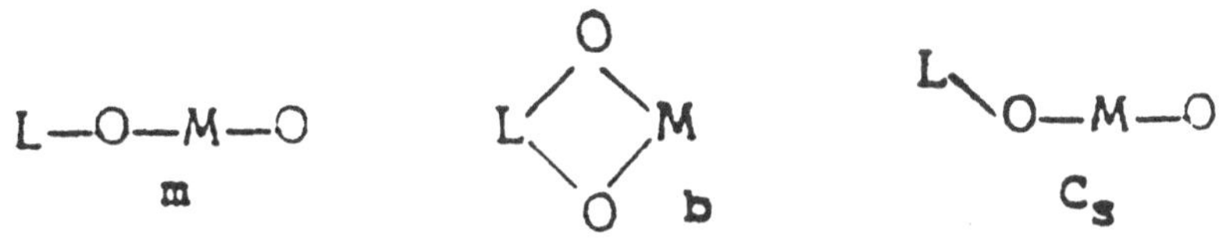

corresponds to the saddle point for Li^+ motion around BO_2^-. In the case of $NaBO_2$, two local minima were found the correlated MP2(full)/6–31+G* *ab initio* level [48e, f]. The linear structure m is a global minimum and the rhombic b is a local one. These *ab initio* conclusions agree with the experimental data [49h], where

TABLE III. Li(CN) $J = 0$ and $J = 1$ levels in the region of the barrier [29a].

$J = 0$					$J = 1, k = 1$				
Level Number	a	Label v_s, v_b	Energy	Relative Amplitude (%)	Level Number	a	Label v_s, v_b	Energy	Relative Amplitude(%)
56	NC		3073.2	0.1	53	NC		3086.9	0.3
57	NC	3,12	3101.8	0.5	54	NC		3133.8	4.7
58	CN	0,6	3132.7	4.9	55	CN	1,1	3145.2	0.3
59	NC		3167.2	5.0	56	NC		3160.5	0.7
60	NC	4,2	3185.2	0.2	57	NC		3168.7	0.7
61	NC		3202.2	10.2	58	CN		3231.8	20.8
62	NC		3237.6	4.1	59	NC		3251.2	41.2
63	NC		3245.7	2.1	60	CN	0,7	3272.9	59.4
64	CN	0,8	3297.3	78.5	61	NC		3288.8	0.7
65	CN	1,2	3301.6	17.0	62	CN		3317.6	19.1
66	NC		3333.1	31.5	63	CN		3332.5	23.0
67	CN		3349.0	67.0	64	CN		3368.0	68.6
68	NC	4,4	3376.0	1.0					

Under heading a, NC/CN denotes that the largest amplitude (and labelling) is around $\theta = 180°/0°$. Energies (in cm^{-1}), are relative to the LiNC ground state. Labels have been assigned when possible. The barrier to isomerization is 3377 cm^{-1}. As a measure of the degree of tunnelling, the ratio of the maxima of absolute amplitude on each side of the barrier are given. For a predominantly isocyanide state, this ratio is $|\Psi_{max}(< 55°)/\Psi_{max}(> 55°)|$ and for a cyanide state the inverse is given.

a doublet of closely spaced doublets have been observed in vibrational spectra of $LiBO_2$ and $NaBO_2$ in matrix isolation for ^{18}O enriched salts, as expected for the m main structure. Since there are two absolute minima and one saddle point for $LiBO_2$, and two absolute, one local minima and two saddle points for $NaBO_2$, the graph describing the motion of the cation around the anion is very simple and there is no need to present it here. The most accurate barrier estimate for L^+ migration around BO_2^-, obtained in [48c], is 6.9 kcal/mol at the MP4/6–311+G*//MP2(full)/6–31+G* level. The energy difference between the m- and b-structures for $NaBO_2$ is only 1.7 kcal/mol at the MP4/6–31+G*//MP2(full)/6–31+G* level [48e]. While the barrier between m- and b-structures was not calculated, we expect this barrier should be small enough to consider both these molecules as flexible. It will be interesting to study heavy alkali metal metaborates (with K, Rb and Cs), which may be more stable in the b configuration (see for example cyanides) and should also be highly flexible. Electron diffraction studies of $NaBO_2$, KBO_2, $RbBO_2$, and $CsBO_2$ made it possible to conclude that the main configuration was an asymmetrical C_s [49a–c, e–g] intermediate between m and b, similar to the main configuration of methaboron acid HBO_2. The IR spectra of $LiBO_2$ in inert matrices also bear witness to the fact that the main configuration is C_s-symmetry [49d, h]. This conclusion contradicts the results of *ab initio* calculations, where the C_s configuration does not correspond to a stationary point on the PES. The reason for these contradictions lies exactly in the structural non-rigidity of methaborates to vibrations of the cation L^+ with respect to the anion BO_2^-. Thus, on the basis of the main m-configuration of $LiBO_2$ and using *ab initio* force fields in the harmonic approximation (with allowance made of the perpendicular shifts of Li^+), and at the temperature of electron diffraction studies ($T = 1550 \pm 50°$ K), an average corner structure with $\varphi_{Li–O–B} = 135°$ was obtained in [48a].

Both $LiAlO_2$ and $NaAlO_2$ molecules have linear m global minima. Structures b in both cases are local minima and lie above the m-structure by 12.6 kcal/mol ($LiAlO_2$) and by 5.7 kcal/mol ($NaAlO_2$) at the MP4/6–31G*//HF/6–31G*+ZPE level. As we know from high level calculations of $LiBO_2$, larger basis sets and more sophisticated correlated methods may lead to softer PES, so we expect lower barriers on the intramolecular PES of these molecules.

Potential energy surfaces are dramatically changed when oxygen is substituted by sulphur. For $LiBS_2$ and $LiAlS_2$, structures b are found to be a global minimum [48g]. Frequencies were not calculated for linear m-structures and therefore we do not know whether these structures are local minima or not. Structures m lie energetically higher than b by 20.8 kcal/mol ($LiBS_2$) and 6.2 kcal/mol ($LiAlS_2$) at the MP4/6–31G*//HF/6–31G* level. Mixed linear structures LiSBO, LiOBS, LiSAlO and LiOAlS have also been studied [48g]. The linear structures LiOBS and LiOAlS have been found more stable than LiSBO (17.9 kcal/mol) and LiSAlO (28.9 kcal/mol) at the MP4/6–31G*//HF/6–31G* level. Frequencies and the cyclic

b-structures were not studied. Calculated relative energies between m- and b-structures may not be reliable, because, in many cases, HF/6–31G* geometries were used and the basis sets are not large enough to provide accurate estimates. Further calculations at more sophisticated levels are needed to obtain reasonable data for ro-vibrational calculations within flexible model Hamiltonians.

Studies of the PES shape of LBO_2 molecules have revealed two PES minima and low barriers for heavy alkaline metal salts. This is a non-rigid rearrangement in which the cation moves around the anion to connect the two PES minima. Therefore the group $D_{\infty h}$ – the symmetry group of the isolated anion – is the molecular symmetry group for the polytopic states of these molecules.

5.3. LMX_3 MOLECULES

According to experimental data [50] and theoretical calculations [31], the absolute minima on the PES of all $LM'O_3$ and LMX_3 molecules, where L = Li ÷ Cs, M′ = N, P, As; M = Be, Mg, Ca; X = H and F, corresponded to main b-configurations.

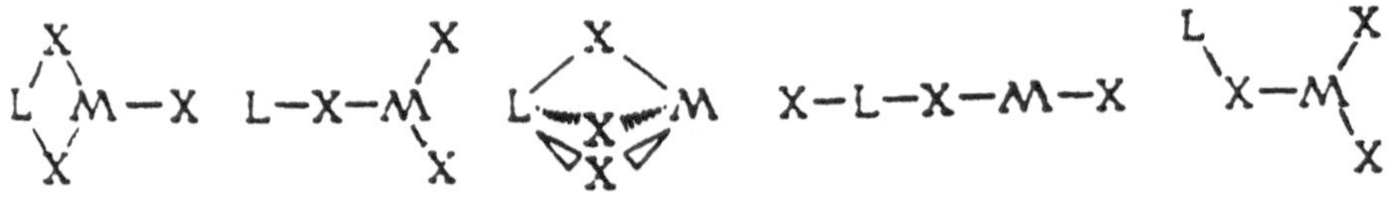

Let us determine the number of the absolute minima on the basis of the rule given in the Introduction. The order of the FPI group of these molecules is $1! \times 1! \times 3! \times 2 = 12$ and the order of the point group (C_{2v}) of the main configuration b is 4. Therefore the PES of these molecules exhibit 3 absolute minima. On the basis of the main b configuration, one can assume the following types of intramolecular rearrangements:

(a) migration of cation L^+ around MX_3^- anion

$$b- \;-m- \;-b' \tag{9}$$

$$b- \;-t- \;-b' \tag{10}$$

(b) rearrangement with intramolecular anion destruction

$$b- \;-m^* - \;-b. \tag{11}$$

Figure 5 shows the graphs of intramolecular rearrangements via mechanisms (9)–(11). It may be seen that in the third mechanism (11) each b-configuration forms two loops, but there are no ways to connect different b-configurations. This mechanism, describing the intramolecular destruction of MX_3^- anion, leads only to local non-rigidity and large vibration amplitudes of L, M, and X atoms in one minimum. On the other hand, the first and second mechanisms, corresponding

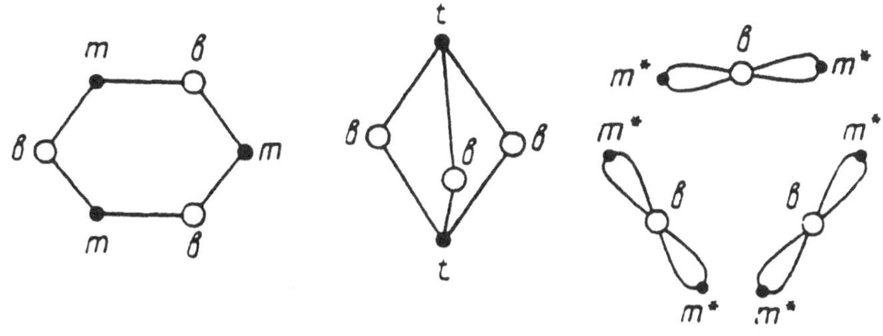

(9) (10) (11)

Figure 5. Graphs of the (9)–(11) intramolecular rearrangement mechanism of the LMX_3 molecules.

to L^+ migration around MX_3^-, allow conversion of LMX_3 molecules from *b*-configuration into the other two. In this manner both mechanisms ensure global non-rigidity.

Frequency calculations [31h, k, m] show that *b*- and *t*-structures are minima and the m-structure is a saddle point second order for $LiBeH_3$ and $NaBeH_3$ both at HF/6–31+G* and MP2(full)/6–31G* levels. At the highest current *ab initio* level QCISD (T)/6–311++G(2df,2pd)//MP2(full)/6–31++G**+ZPE, structures m and t lie above b by 19.4 kcal/mol and 12.4 kcal/mol ($LiBeH_3$), and by 14.0 kcal/mol and 8.3 kcal/mol ($NaBeH_3$), respectively [31m]. For both these molecules, as well as for other $LBeH_3$ (L = K, Rb, Cs) species, (10) is the mechanism of the intramolecular rearrangement. We localized a barrier on this mechanism for $LiBeH_3$, which is only 0.1 kcal/mol at MP2(full)/6–31++G**+ZPE level: therefore energy differences between *b*- and *t*-structures are good estimations for the barriers of the intramolecular non-rigidity for $LiBeH_3$ and $NaBeH_3$. No barriers were found for $m \rightarrow b$ or $m \rightarrow t$ rearrangements.

$LiMgH_3$ and $NaMgH_3$ also have global minima at *b*-structures and local minima at *t*-structures. However, for these species, m-structures are saddle points (one imaginary frequency) at the MP2(full)/6–31++G* level [31h, k, m]. The relative energies of t- and m-structures with respect to b are 21.1 and 19.6 kcal/mol ($LiMgH_3$), and 15.1 and 13.8 kcal/mol ($NaMgH_3$) at QCISD(T)/6–311++G(2df,2pd)//MP2(full)/6–31++G**+ZPE level [31m]. Therefore, (9) is the mechanism of the intramolecular rearrangements for $LiMgH_3$, $NaMgH_3$ and other $LMgH_3$ (L = K, Rb, Cs) molecules. Analysis of the vector of the imaginary frequencies of the m-structures of $LiMgH_3$ and of $NaMgH_3$ shows that this vector leads to *b*-structures: therefore saddle points on the intramolecular PES of these species have the m-structure. Because the *ab initio* level in these calculations of complex LMH_3 hydrides is very high, we expect that the calculated barriers of the intramolecular rearrangements of these species are quite accurately estimated. These barriers are too high to observe tunnelling splittings due to jumping LMH_3

hydrides from one global minimum to another.

Four complex fluorides $LiBeF_3$ [31c, d–g, k], $NaBeF_3$ [31k], $LiMgF_3$ [31k] and $NaMgF_3$ [31k] have also been studied. The *b*-structures were found to be the global minimum for all these species. This agrees with experimental interpretations of the IR spectrum of $LiBeF_3$ [50g] and gas phase electron diffraction data for $KBeF_3$ [50a]. High symmetry *m*- and *t*-structures of $LiBeF_3$ are both second-order saddle points. The relative energies of these two structures are 25.1 and 17.8 kcal/mol (at MP2/6–31G*//HF/6–31G*+ZPE), respectively [31k]. For $NaBeF_3$, the *m*-structure is a second-order saddle point and the *t*-structure is a local minimum. The corresponding relative energies are 17.4 and 6.6 kcal/mol (at MP2/6–31G*//HF//6–31G**+ZPE), respectively [31k]. Barriers for intramolecular rearrangements for both of these molecules are not known. However, the decreasing of the relative energy of *t*-structure from 17.8 to 6.6 kcal/mol along $LiBeF_3$–$NaBeF_3$ shows that $KBeF_3$, $RbBeF_3$ and $CsBeF_3$ molecules may be flexible.

For $LiMgF_3$ and $NaMgF_3$, only the *b*- and *t*-structures have been studied [31k]. Bidentate structures *b* are global minima and tridentate structures *t* are local minima with relative energies 22.2 kcal/mol for LiMgF3 and 13.0 kcal/mol for $NaMgF_3$ (at the MP2/6–31G*//HF/6–31G*+ZPE level). Saddle points for cation motions around anions are not known for these molecules.

Six molecules of LMO_3 type: $LiNO_3$ [31a, c, k, n, o], $NaNO_3$ [31n, o], $LiPO_3$ [31d, j, k, n, p], $NaPO_3$ [31n, o], $LiAsO_3$ [31l] and $LiVO_3$ [31l] have been studied by *ab initio* tools.

Ab initio frequencies analysis indicates that bidentate structures *b* of $LiNO_3$, $NaNO_3$, $LiPO_3$, $NaPO_3$ and $LiAsO_3$ have positive vibrational frequencies and therefore these forms correspond to the minima of the respective energy surfaces. The main *b*-structure for LNO_3 and LPO_3 came also from the ^{18}O isotopic enrichment experiments of Ogden and co-workers [50g, h, i]. These authors demonstrated that the highest-frequency fundamental observed for argon-matrix isolated KNO_3 and $NaPO_3$ appeared as a doublet of triples in accord with the prediction for the bidentate structure *b*. Comparison between *ab initio* calculated [31i, j, n, o] and experimental frequencies and frequency shifts [50g, h] for *b*-structures of LNO_3, LPO_3 and $LAsO_3$ shows a good agreement and provides one more indication on the main *b*-structures for these species. Monodentate structures *m* of $LiNO_3$, $NaNO_3$ and $LiAsO_3$ molecules have two negative eigenvalues along the b_2 in-plane and b_1 out-of-plane bending modes, thus revealing the monodentate structures to be second-order saddle points on the full energy surfaces. The monodentate structures of $LiPO_3$ and $NaPO_3$ have one negative eigenvalue related to the b_2 bending mode of the metal; and consequently these structures are first-order saddle points with respect to the migration of the alkali metal. The pyramidal structures *t* of $LiNO_3$, $NaNO_3$, $LiPO_3$, $NaPO_3$ and $LiAsO_3$ have three imaginary frequencies corresponding to the a_1 and *e* symmetry modes of the metal, which is consistent with a third-order saddle point.

Calculated relative energies of the m- and t-structures with respect to b: 20 kcal/mol and 30 kcal/mol for $LiNO_3$; 17 kcal/mol and 18 kcal/mol for $NaNO_3$; 14 kcal/mol and 45 kcal/mol for $LiPO_3$; 13 kcal/mol and 29 kcal/mol for $NaPO_3$ and 10 kcal/mol and 49 kcal/mol for $LiAsO_3$ (all data at HF/6–31G*) [31l, o], respectively, provide an upper boundary for barriers of the mechanisms (9) and (10). However, more extended calculations including the search for first-order saddle points and using a more sophisticated *ab initio* level are needed for more reliable estimates of the barriers in these species.

The only $LiVO_3$ molecule with a central transition metal atom have been studied *ab initio* [31m]. The bidentate binding planar structure b was found to be the most stable. Monodentate m and tridentate t-structures are second-order and third-order saddle points. The calculated relative energies of m- and t-structures with respect to the most stable b are 3 and 26 kcal/mol (MP2/6–31G*//HF/6–31G*+ZPE), respectively.

Since all the experimental data for LMX_3 molecules were interpreted in terms of one-minimum quasi-rigid model, it will be interesting to review the interpretation using a non-rigid three-minima model where internal MX_3^- and L–M vibrations are quasi-rigid, while the interplane motion of a cation around an anion is a large amplitude non-rigid motion. The non-rigid group symmetry for all LMX_3 molecules is the FPI group of these species; and this group is isomorphic with the point group D_{3h}, the symmetry group of the isolated MX_3^- anion.

A similar potential energy surface was found for the ArH_3^+ cation [51b]. Three structures m, b and t were optimized at

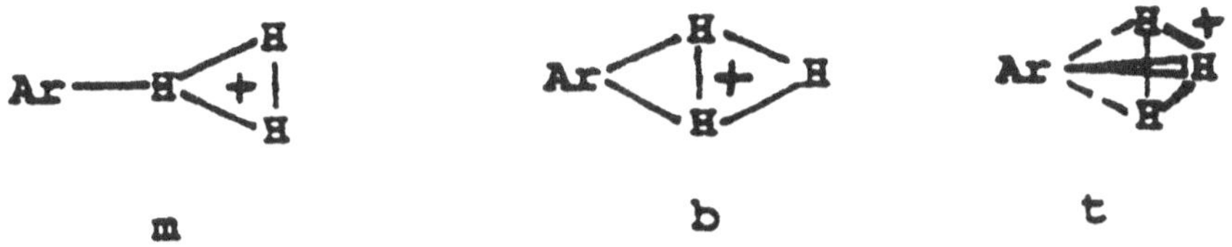

MP2/[(13s10p3d/7s5p3d)$_{Ar}$+(5s2p/3s2p)$_H$]. Structure m is a minimum, while structures b and t are transition states with imaginary modes 414i and 319i (e) respectively. Motion along each of these modes leads to structure m. The barrier heights between two equivalent forms of structure m through transition states b and t are 3.9 kcal/mol and 5.5 kcal/mol, respectively. The H_3^+ unit of structures m, b and t shows a non-substantial deviation from that obtained for H_3^+ itself. The largest distortion is in m-structure, where the angle $< H_tH_bH_t$ decreases by 7.9° and bond distance H_b–H_t increases by 0.006 Å.

Figure 5 shows that non-rigid rearrangements with an outersphere cation motion connect all three absolute minima on the PES of LMX_3 and ArH_3^+ molecules. Consequently, the group of molecular symmetry is the FPI $G^{(12)}$ group, which is isomorphic to the D_{3h} point group of an isolated MX_3^- anion or the H_3^+ cation.

Thus, the scheme of main configuration level splitting due to the tunnel effect will be:

C_{2v}		D_{3h}
a_1	– – –	$a_1' + e'$
a_2	– – –	$a_1'' + e''$
b_1	– – –	$a_2' + e'$
b_2	– – –	$a_2'' + e''$.

Atoms in the inorganic and complex salts LMX_3 are too heavy to observe tunnelling splittings; but ArH_3^+ and ArD_3^+ are good opportunities to see these tunnelling splittings. Experimentally, tunnelling splittings in ArH_3^+ and ArD_3^+ were observed by high resolution submillimetre-wave spectroscopy [51a].

The flexible model calculations were carried out only for ArH_3^+ and ArD_3^+ species [51a, c]. In the most sophisticated calculations [51c], the rotational frequencies and internal-rotational splittings in the spectra of these ions were calculated using the semirigid bender Hamiltonian:

$$H = H_t + H^{(a)} + H^{(bc)} + H^{(c\rho)} \tag{12}$$

where

$$\begin{aligned} Ht &= 1/2\mu_{\rho\rho}J_\rho^2 + 1/2[J_\rho, \mu_{\rho\rho}]J\rho \\ &\quad + 1/2\mu^{1/4}[J_\rho, \mu_{\rho\rho}, \mu^{-1/2}[J_\rho, \mu^{1/4}]] + V(\rho) \end{aligned} \tag{13a}$$

$$H^{(a)} = 1/2\mu_{aa}J_a^2 \tag{13b}$$

$$H^{(bc)} = 1/2\mu_{aa}J_b^2 + 1/2\mu_{cc}J_c^2 \tag{13c}$$

$$H^{(c\rho)} = 1/2\mu_{c\rho}(J_cJ_\rho + J_\rho J_c) \tag{13d}$$

The 4 × 4 matrix is the inverse of the I matrix and the non-vanishing elements of the I matrix are:

$$I_{aa} = 1/2m_Hr^2 \tag{14a}$$

$$I_{bb} = 1/2m_Hr^2 + MR^2 \tag{14b}$$

$$I_{cc} = m_Hr^2 + MR^2 \tag{14c}$$

$$I_{c\rho} = m_Hr^2 \tag{14d}$$

$$I_{\rho\rho} = m_Hr^2 + M(dR/d\rho)^2. \tag{14e}$$

Here m_H and m_{Ar} are the masses of H (or D) and Ar, respectively, and $M = 3m_Hm_{Ar}/(m_{Ar} + 3m_H)$. In Equation (13) J_a, J_b and J_c are the components of the total angular momentum, and J_ρ is the momentum conjugate to the ρ given by

$$J_\rho = -i\hbar d/d\rho. \tag{15}$$

The internal rotation of H_3^+ is described by the angle ρ. A complete internal rotation is described when ρ goes from 0 to 360°. An equilateral triangular shape for H_3^+ was assumed. The internal-rotation potential $V(\rho)$ was taken of the form:

$$V(\rho) = 1/2V_3(1 - \cos 3\rho). \qquad (16)$$

This Hamiltonian only treats rotation and internal rotation explicitly. The five ordinary vibrational degrees of freedom (the three vibrations of the H_3^+ triangle, the Ar–H_3^+ stretch, and an out-of-plane vibration of the Ar atom) are considered as having been averaged over ρ (in other words, excluded from further consideration). As pointed out by Escribano and Bunker [51c], in fitting the differences between the eigenvalues of this Hamiltonian to the observed transition frequencies it was found impossible to determine $dR/d\rho$ with any significance and therefore R was taken as a constant (independent of ρ), as a result of which the second and the third terms in the right-hand side of Equation (13a) vanish. It was also found impossible to determine r(H–H) or r(D–D), so these were taken as the zero-point bond lengths of the bare H_3^+ and D_3^+ ions. It was also found necessary to add centrifugal distortion terms to the Hamiltonian. The final set of parameters adjusted in the fitting were R, V_3 and the centrifugal distortion constants D_{JK}, d_1 and d_2. Separate fittings were performed for ArH_3^+ and ArD_3^+ data since the parameters are effective parameters as explained above. The resulting effective torsional potential functions obtained for ArH_3^+ and ArD_3^+, together with some calculated torsional energy levels are presented in Figure 6. A very good agreement between experimental [51a] and calculated [51c] frequencies have been obtained.

5.4. LMX_4 MOLECULES

PES studies of LMX_4 and $LM'O_4$ molecules L = Li÷Cs, M = B, Al, etc., M′ = Cl, Br, etc., X = H, Hal, with the help of quantum chemical calculations [32, 33] and experimental techniques [34, 35, 52] have demonstrated that the absolute minima correspond to either t or b configurations

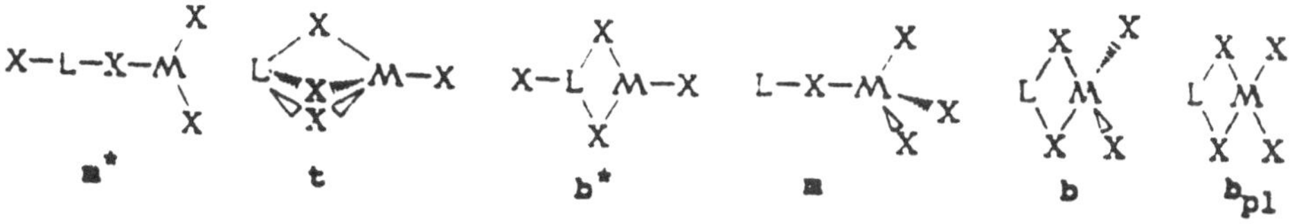

Let us determine the number of absolute PES minima for a LMX_4 molecule assuming the main t- and b-configurations. The order of the molecule's FPI group is $1!\times 1!\times 4!\times 2 = 48$, and the point groups of the t- and b- main configurations have orders 6(C_{3v}) and 4(C_{2v}), respectively. Then the PES exhibits 8 absolute minima with the main t-configuration and 12 minima with the main b-configuration. A priori, one can offer several mechanism of intramolecular rearrangements of

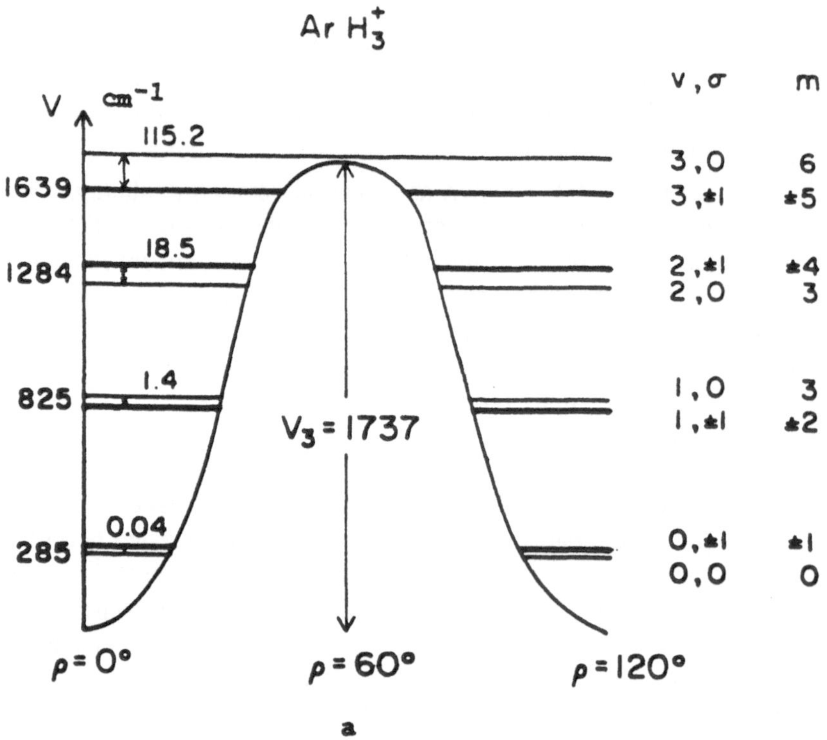

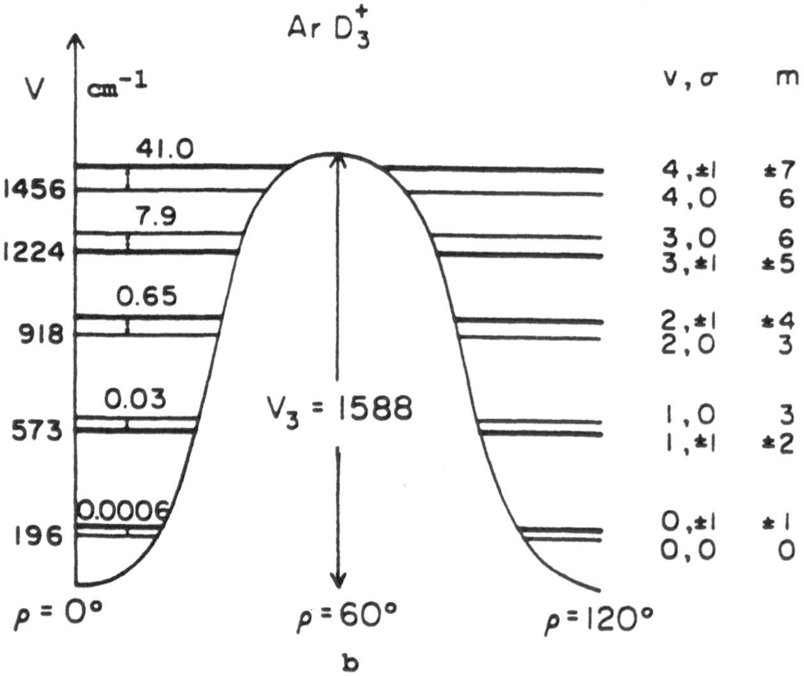

Figure 6. The effective internal-rotational potential and the internal-rotational energy levels for ArH_3^+ (a) and ArD_3^+ (b). The internal-rotational splittings are indicated (in cm^{-1}) above each pair of split levels [51c].

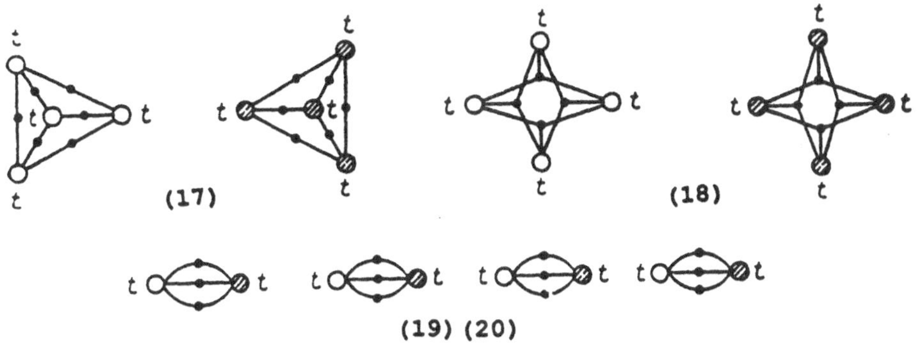

Figure 7. Graphs of the (17)–(20) intramolecular rearrangement mechanisms of the LMX_4 molecules with the main t-structure.

LMX_4 molecules. Assuming the main t-configuration:

(a) cation migration in the molecule's outer sphere

$$t - - - b - - - t' \quad (17)$$

$$t - - - m - - - t' \quad (18)$$

(b) inversion of tetrahedral anion through the planar configuration

$$t - - - b_{pl} - - - t' \quad (19)$$

(c) intramolecular destruction of the MX_4^- anion

$$t - - - b^* - - - t'. \quad (20)$$

Similarly, assuming the main b-configuration, one can suggest the following mechanism for the LMX_4 molecule:

(a) a cation moves around a rigid anion

$$b - - - t - - - b' \quad (21)$$

$$b - - - m - - - b' \quad (22)$$

(b) inversion of MX_4^- through the planar configuration

$$b - - - b_{pl} - - - b \quad (23)$$

(c) intramolecular destruction of the MX_4^- anion

$$b - - - m^* - - - b'. \quad (24)$$

Figures 7 and 8 show the graphs describing the above-mentioned mechanism. Let us analyse these mechanisms and the chemical and physical consequences they may lead to.

Mechanisms (20) and (24) describe rearrangements with intramolecular destruction; it may be seen from Figures 7 and 8 that they, unlike the LMX_2 and LMX_3 molecules, connect inversely different main configurations. Low barriers on these rearrangements (but a large barrier assumed for all others) should lead to two consequences: 1) the NMR spectrum of LMX_4 will have two X signals;

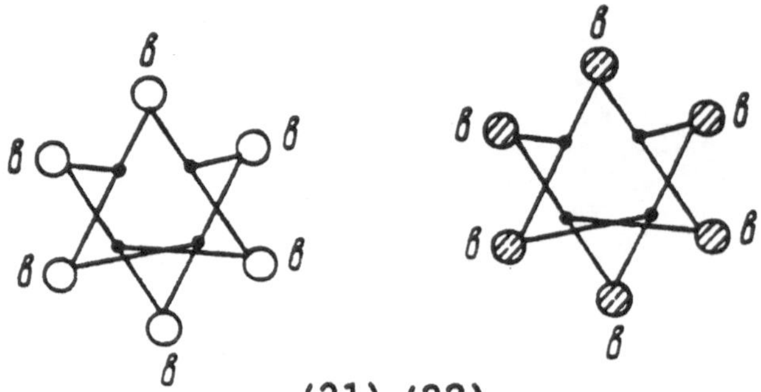

(21) (22)

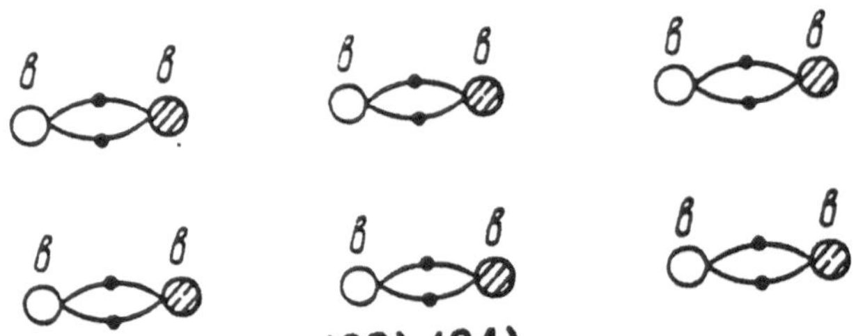

(23) (24)

Figure 8. Graphs of the (21)–(24) intramolecular rearrangement mechanisms of the LMX_4 molecules with the main *t*-structure.

2) racemization will be observed with optically active $LMXX'X''X'''$ molecules. Inversion mechanisms (19) and (23) also connect only inversely different main configurations. However, in the course of inversion rearrangements the MX_4^- anions in LMX_4 are presented as autonomous fragments. It is evident that in the case of low barriers, one should expect the same consequences as for the above mechanism with anion destruction. On the contrary, mechanisms (17), (18) and (21), (22) connect all the main configurations belonging to one of the two types of inversely different tetrahedral anions; and there are no paths that would connect inversely different main configurations. If the barriers are low, one should expect the following: (1) NMR spectra will exhibit only one peak from the X ligands; (2) it will be possible to isolate individually the optical isomers of $LMXX'X''X'''$.

The barrier for mechanisms (20) and (24) may be estimated only from *ab initio* calculations for $LiBH_4$ [31b], where configurations m^* and b^* were found to lie higher than the main one by more than 45 kcal/mol, which actually means that such rearrangements are forbidden. It can be assumed that the intramolecular destruction of anions will be energy-forbidden for other $LMHal_4$ and LMO_4 molecules also. For this reason the single-charged anion MX_4^- in LMX_4 molecules can be considered as a stable autonomous structural fragment. The inversion barriers of isolated BH_4^- and AlH_4^- anions are very high: 140 kcal/mol and

61 kcal/mol at the HF/DZ+P level [53], respectively. Despite the fact that in $LiBH_4$ and $LiAlH_4$ salts the barriers decrease to 112 kcal/mol and 42 kcal/mol at HF/DZ+P, respectively [53], they are, nevertheless, high enough for considering MH_4^- anions as quasi-rigid fragments in the composition of salts. This conclusion is also valid for LMF_4 and LMO_4 species. More extensive studies have been carried out for cation migration around MX_4^- anions via mechanisms (17), (18) and (21), (22). For all LMF_4 and LMH_4 molecules, the energy of the monodentate configuration m is much higher than b and t, and therefore mechanisms (18) and (22) for these molecules may be excluded from consideration.

According to the MP2(full)/6–31G* *ab initio* calculations [54], the t-structure is a global minimum (this agrees with the recent experimental conclusions from analysing the observed rotational constants of $LiBH_4$ in the gas phase [34c]); and the b-structure is a saddle point on the PES of $LiBH_4$. The calculated barrier at the highest *ab initio* level for the mechanism (17) in $LiBH_4$ is 5.8 kcal/mol at the QCSID(T)/6–31+G(d,p)//MP2(full)//6–31G*+ZPE level. $NaBH_4$ also has its global minimum at a t-structure in accordance with experimental data in the gas phase [34b]. Substitution of Li by Na reduces the t–b energy separation to 4.0 kcal/mol at the QCSID (T)/6–31+G(d,p)//MP2(full)/6–31G*+ZPE level and the b-structure is again a saddle point [54]. We expect that KBH_4, $RbBH_4$ and $CsBH_4$ molecules are more flexible with respect to internal rotation of the cation around the anion and provide a better perspective for observing tunnelling due to jumping from one global minimum to another.

The bidentate and tridentate structures of the molecules $LiAlH_4$ and $NaAlH_4$ are both minima; and the energy separation between the b and t minima is very low. In fact, both structures for $LiAlH_4$ are practically degenerate (at the MP4/6–31G*//MP2/6–31G*+ZPE level, the b-structure is 0.1 kcal/mol more stable that t [32n, o]). However for $NaAlH_4$, the structure b is more stable than t by 0.8 kcal/mol (at the same level [32n, o]). From these data $LAlH_4$ molecules are the most appropriate species for the experimental observation of tunnelling splittings due to molecules jumping from one global minimum to another. However, for developing the flexible Hamiltonian and the flexible model for these molecules the whole potential energy surface including all saddle points should be studied. There is no experimental data for these species in the gas phase.

Normal coordinate calculations for LBF_4 molecules (M = Li, Na, K) show that both b- and t-structures are minima [32m–p] having all real vibrational frequencies, whereas m is a second-order saddle point structure having a degenerate vibrational mode with an imaginary frequency [32m]. However, when M = Rb the nature of b-structure changes from being a minimum to a saddle point [32m]. The b-structure was found to be the main structure for $LiBF_4$ and as was the t-structure for $NaBF_4$, KBF_4 and $RbBF_4$. At the MP2/6–31+G//HF/6–31+G level, the energies of t relative to b are 4.0 kcal/mol, -0.3 kcal/mol, -3.0 kcal/mol and -2.9 kcal/mol for $LiBF_4$, $NaBF_4$, KBF_4 and $RbBF_4$, respectively [32m]. However, after increasing the basis set to 6–311+G*, the relative stability of the t-structure increases by 2.2 kcal/mol for $LiBF_4$ and by 1.4 kcal/mol for $NaBF_4$. We expect that this trend will be more pronounced at higher *ab initio* levels. Francisco and Williams [32m] have found barriers of 4.2 kcal/mol and 1.4 kcal/mol for conversion from bidentate to tridentate structures at the MP2/6–31+G//HF/6–31+G level for $LiBF_4$ and $NaBF_4$, respectively. No barriers were found at this level for KBF_4 and $RbBF_4$. However, these are only preliminary results and further *ab initio* calculations at a more sophisticated level are needed to obtain reliable barriers of the intramolecular rotations in these species. The IR spectra of matrix isolated KBF_4 and $CsBF_4$ have been attributed to the monodentate structure m (C_{3v}) [52e]. However this assignment does not agree with the conclusions of *ab initio* calculations, because the monodentate C_{3v} symmetry structures of LBF_4 are not low-energy structures but second-order saddle points. Reassignment of the experimental IR spectra according to the results of *ab initio* calculations may be made and satisfactory agreement achieved [32o].

Ab initio frequency analyses have shown that the b- and t-structures of $LiAlF_4$ and $NaAlF_4$ are local minima and the m-structures for both of these molecules are second-order saddle points. The bidentate structure b is more stable than t for $LiAlF_4$ (by 4.3 kcal/mol at the MP4/6–31G*//HF/6–31G* level) and the tridentate structure t is more stable then b for $NaAlF_4$ (by 1.5 kcal/mol at the same level [31k, 32o, p]). The experimental IR spectrum of $NaAlF_4$ was attributed to the bidentate b main structure: however there is not good agreement between calculated and experimental data for this structure. Moreover as we pointed out, according to *ab initio* results, the t-structure is more stable for this molecule. More sophisticated calculations are needed to solve this problem.

Among the oxygen-containing molecules LMO_4, the PES was studied for $LiClO_4$ only [32k, n]. The bidentate (b) and the tridentate (t) structures were found to be global and local minima (no imaginary frequencies at HF/3–21G*). The monodentate (m) structure is a second-order saddle point (two imaginary frequencies). At the MP2/6–31G*//HF/6–31G*+ZPE/HF/3–21G* level, the t- and m-structures lie 8.5 kcal/mol and 18.8 kcal/mol above the main b-structure, respectively [32n]. At higher *ab initio* levels, one can expect that the relative energy of the t-structure with respect to b should be smaller, because correlation energy is larger in more compact structures. We expect that $LClO_4$ molecules with heavy alkali metal atoms L may have even lower b–t energy separations and may be flexible with respect to internal outersphere rotation cation L^+ around the anion ClO_4^-.

Extensive analysis of the PES of LMX_4 molecules allows one to formulate the following non-rigid model of these molecules: the anion is a rigid autonomous fragment with the cation moving around it along a sphere. The anion vibrations and the L–M vibration can be regarded as quasi-rigid, whereas the cation two-dimensional outersphere motion is non-rigid.

Let us determine the molecular symmetry group of the non-rigid molecules LMX_4. The order of the FPI group of LMX_4 type of molecule is $1! \times 1! \times 4! \times 2 = 48$, i.e. the group is $G^{(48)}$. This group is isomorphic to the point group O_h. However, as follows from the above states, the FPI group is not the group of molecular symmetry for LMX_4 molecules, since all elements of these groups, containing inversion or permutation with inversion, are forbidden by high barriers. The same is indicated by the graphs in Figures 6 and 7 (they are not completely connected). Having excluded all the unrealizable elements, we find that the molecular symmetry group of non-rigid LMX_4 molecules is $G^{(24)}$, which is isomorphic to the T_d-symmetry group of an isolated rigid anion. Then the scheme of vibrational level splitting induced by the tunnel effect for the molecules with the main configuration t could be as follows:

$$\begin{array}{rcl} C_{3v} & & T_d \\ a_1 & \longrightarrow & a_1 + f_2 \\ a_2 & \longrightarrow & a_2 + f_1 \\ e & \longrightarrow & e + f_1 + f_2 \end{array}$$

and for the main configuration b:

$$\begin{array}{rcl} C_{2v} & & T_d \\ a_1 & \longrightarrow & a_1 + e + f_2 \\ a_2 & \longrightarrow & a_2 + e + f_1 \\ b_1 & \longrightarrow & f_1 + f_2 \\ b_2 & \longrightarrow & f_1 + f_2 \end{array}$$

From the analysis of PES and symmetry of LMX_{k+1} and $LMO_{(k+1)/2}$, one can formulate the following general conclusions: (1) in the salts under study, the anion is a quasi-rigid fragment while the cation can move around it with low barriers, and so, to describe the structure, one must make use of multiminima non-rigid models; (2) the group of molecular symmetry, the irreducible representations of which should be used to classify the states of these molecules, is the point group of the isolated MX_{k+1}^- or $MO_{(k+1)/2}^-$.

The large amplitude motion spectra of LMX_4 ($LiBH_4$, $NaBH_4$, $LiCH_4^+$ and $LiBF_4$) molecules were investigated in Refs. 33 using the simplest non-rigid model. Within this model the MX_4^- anion is tetrahedral, with the Li^+ cation moving in the outersphere of the anion. The problem in this case is two-dimensional and the Hamiltonian for $J = 0$ (effects related to rotation of the molecule as the whole are absent) has the form [34]:

$$H(\theta\varphi) = -B_0\Delta_{\theta,\varphi} + V(\theta, \varphi) \tag{25}$$

where

$$B_0 = \frac{\hbar}{2}\left(\frac{1}{mr_0^2} + \frac{1}{Mr_0^2} + \frac{1}{I_0}\right),$$

$$\Delta_{\theta,\varphi} = \sin^{-1\theta}\frac{\partial}{\partial\theta}\sin\theta\frac{\partial}{\partial\theta} + \sin^{-2\theta}\frac{\partial}{\partial\varphi}.$$

For hydrides, the potential $V(\theta,\varphi)$ has the form $t^{(3)}Y_{A1}^{(3)}$, where the spherical harmonic $Y_{A1}^{(3)} = (Y_2^{(3)} - Y_{-2}^{(3)})$ is:

$$Y_{A1}^{(3)} = -1/4(105/\pi)^{1/2}\cos\theta\sin^2\theta\sin 2\varphi \qquad (26)$$

The parameter $t^{(3)}$ was selected to reproduce the potential barrier from earlier *ab initio* calculations [32a]. The spherical harmonic $Y_{A1}^{(3)}$ provides a qualitatively correct description of the PES shape for LBH_4 and LCH_4^+, with the minima above the faces, saddle points above the edges and the maxima above the vertices of the tetrahedron MX_4. In the case of LMF_4 and LMO_4, the potential $V(\theta,\varphi)$ has the form:

$$V(\theta,\varphi) = t^{(0)}Y_{A1}^{(0)} + t^{(3)}Y_{A1}^{(3)} + t^{(4)}Y_{A1}^{(4)}. \qquad (27)$$

A constant term, $t^{(0)}Y_{A1}^{(0)} = t^{(0)}(4\pi)^{1/2}$ is added to the potential function in order that the minimum energy be zero. The shape of $Y_{A1}^{(3)}$ is given by expression (26), while the harmonic $Y_{A1}^{(4)}$ has the form:

$$Y_{A1}^{(4)} = -1/32(21/\pi)^{1/2}[35\cos^4\theta - 30\cos^2\theta + 3 + 5\sin^4\theta\cos 4\varphi]. \qquad (28)$$

The potential of (27) reproduces qualitatively correctly the PES shape for LMF_4, with minima above the edges, saddle points above the faces, and maxima above the vertices of the tetrahedral MF_4^- anion.

The energy levels for the Hamiltonians with potentials (26) and (27) were found by the variation method. An orthonormalized set of spherical harmonics adapted to the T_d symmetry was used as the basis. The parameters d(L–M), d(M–H), B and E of our Hamiltonian were taken to be equal to 1.997 Å, 1.250 Å, 4.986 cm^{-1}, 1053 cm^{-1} for LiBH4; 2.350 Å, 1.250 Å, 4.351 cm^{-1}, 900 cm^{-1} for $NaBH_4$ and 2.291 Å, 1.088 Å, 5.957 cm^{-1} and 614 cm^{-1} for $LiCH_4^+$ [33a]. Table IV lists the calculated energies of the levels of $LiBH_4$, $NaBH_4$, and $LiCH_4^+$ that lie below the respective barrier. First, we would like to draw attention to the qualitative level of these data, because low level *ab initio* data were used as parameters. A rough description of the levels lying under the barrier can be provided by assuming that they are localized in separate minima, with no account being taken of the tunnel effect, and by assuming the potential to be harmonic. In the case of LBH_4, the energy states can be assigned indices n and l – the main and angular quantum numbers of a two-dimensional oscillator (at a given n, $l = n, n-2, \ldots, 1(0)$) and one can also add to the strict selection rules by symmetry T_d an approximate rule for dipole transitions: $\Delta n = 1$, $\Delta l = 1$. If one takes

TABLE IV. Calculated energies of the underbarrier non-rigid deformational level of $LiBH_4$, $NaBH_4$ and $LiCH_4^+$ (cm^{-1}) [33a].

Level	Harmonic Oscillator	C_{3v}	T_d	$LiBH_4$ E_k	$LiBH_4$ ΔE_{tun}	$NaBH_4$ E_k	$NaBH_4$ ΔE_{tun}	$LiCH_4^+$ E_k	$LiCH_4^+$ ΔE_{tun}
			A_1	240		209		198	
1	0	A_1			6×10^{-5}		8×10^{-3}		1×10^{-2}
			F_2	240		209		198	
2			E	474		413		390	
3	1	E	F_2	474	3×10^{-3}	413	4×10^{-3}	390	4×10^{-1}
4			F_1	474		413		390	
5			A_1	687		597		548	
	2	A_1			8×10^{-2}		9×10^{-2}		4
6			F_2	687		597		552	
7			E	704		613		572	
8		E	F_2	704	6×10^{-2}	613	7×10^{-2}	574	4
9			F_1	704		613		576	
10	3		E	891		774			
11		E	F_2	891	1.6	775	1.8		
12			F_1	891		776			
13			A_1	926		805			
		A_1			1.1		1.2		
14			F_2	927		806			
15			F_1	930		810			
		A_2			0.1		0.2		
16			A_2	930		810			

into account the anharmonicity of the potential in each individual C_{3v} symmetry minimum, the $n+1$-fold degenerate level in the harmonic approximation will split into two types of symmetry C_{3v}: $l = 0 \rightarrow A_1$, $l > 0$, (mod 3) = 0 $\rightarrow A_1 + A_2$, $l(\text{mod } 3) = 1 \rightarrow E$, $l(\text{mod } 3) = 2 \rightarrow E$. Taking account of the tunnel effect will make each degenerate level, corresponding to the group of states localized in different minima but possessing the same energy and symmetry type C_{3v}, split into the levels of certain types of non-rigid symmetry T_d: $A_1 \rightarrow A_1 + F_2$, $A_2 \rightarrow A_2 + F_1$ and $E \rightarrow E + F_1 + F_2$.

From Table IV it may be seen that with hydrides the harmonic picture is distorted already for the bottom levels. The underbarrier levels correspond to anharmonic vibrations. The tunnel splittings ΔE_{tun} of these levels grow quickly as one passes from the bottom to near-barrier levels. Close to the barrier and higher, the tunnel splittings become comparable with the transition frequencies while the levels' pattern is completely dissimilar to that of the harmonic oscillator. Highly excited states are described by almost free rotation of the BH_4 fragment and are distributed as a free rotator's levels. For LMH_4 molecules, the value of B_0 is rather high (which corresponds to a small effective tunnelling mass), the main contribution being made by the low momentum of inertia I_0 of the BH_4 fragment (in $LiBH_4$ it is more than 80%). Since the potential barriers of the BH_4 groups in other ionic borohydrides similar CH_3BeBH_4, $Be(BH_4)_2$, $Al(BH_4)_3$ and $Zr(BH_4)_4$ are as low as in $LiBH_4$ and the effective tunnelling mass is largely determined by the BH_4-group: the tunnelling splitting for these borohydrides may be similar to that of $LiBH_4$.

The first attempt to find tunnelling splittings in the gas phase $NaBH_4$ molecule was made by Kawashima *et al.* [34b]. However, the spectrum in the ground vibrational state did not show any splitting due to the tunnelling motion of BH_4^-. The discrepancy between theoretical prediction of the tunnelling splitting of 8×10^{-5} cm^{-1} in Ref. 33a and the absence of any fine structure in the experimental study of $NaBH_4$ may be due to an overestimate of the tunnelling splittings in our work. We used a barrier of 900 cm^{-1} (on the basis of old *ab initio* calculations at HF/DZ level [31c]), while the most accurate recent value for the barrier is 1386 cm^{-1} at QCISD(T)/6–31+G(d,p) [53]. With the higher value of the barrier, the tunnelling splitting in $NaBH_4$ should be even smaller $\approx 10^{-8}$ cm^{-1} in accordance with our calculations using a Hamiltonian (25). The *ab initio* study of the potential energy surface of $LAlH_4$ has shown that the energy difference between b- and t-structures is even smaller than for LBH_4 molecules, and therefore $LiAlH_4$, $NaAlH_4$ and other $LAlH_4$ species are the best species for experimental observation of the tunnelling splitting. Another appropriate system for experimental observation of this kind of tunnelling splitting, even in the ground state, should be the $LiCH_4^+$, $NaCH_4^+$, KCH_4^+, $RbCH_4^+$ and $CsCH_4^+$ cations and similar ion-molecular systems. Recently, tunnelling splittings were observed for the $C_2H_3^+$ [38] and ArH_3^+ [51a, d] cations. Another molecule with an extremely low barrier for intramolecular BH_4 motion may be $Be(BH)_2$. According to the best *ab initio* calculations, the energy difference between two lowest configurations (b, b) (bidentate, bidentate coordination BH_4 unit to Be) and (t, t) (tridentate, tridentate coordination BH_4 unit to Be) is only 1 kcal/mol [54]. However, more detailed *ab initio* study, including a search of the saddle points, should be made before dynamic calculations will be carried out.

For LMF_4 fluorides (and oxygen-containing salts LMO_4) there are practically no tunnel splittings at the lowest levels ($< 10^{-8}$ cm^{-1}); only when approaching the saddle point do the splittings reach $0.001 \div 0.1$ cm^{-1}. This is due to the fact that the tunnelling mass of the fluorides is large. For these species, the "rigid rotator harmonic oscillator" model can be used to provide an accurate enough description of the deformational vibrations of the lower vibrational levels. The main consequence of the non-rigidity for excited levels below the barrier is the

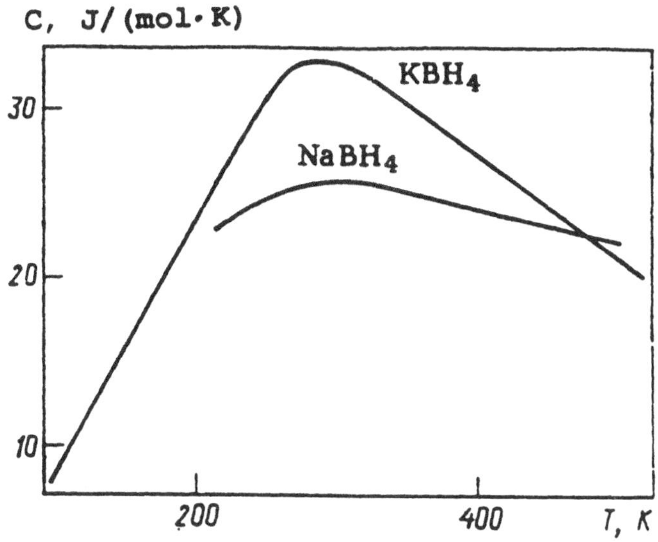

Figure 9. Temperature dependence of the rotational heat capacity of the BH_4^- anions of the crystal $NaBH_4$ and KBH_4 salts [55].

low frequency of vibrations and their considerable anharmonicity. However, to describe the levels near and above the barrier one should make use of a non-rigid model with inhibited two-dimensional rotation. Fluorides are different from hydrides in that they have more levels lying below the barrier and, as compared to hydrides, their level pattern undergoes abrupt changes in moving from the underbarrier to overbarrier levels.

The multiminima structure of potential energy surface is important for calculations of thermodynamic functions. It should be noted that the importance of taking into account large amplitude motions in statistical calculations of entropy and heat capacity was understood as early as the 1930s. Thus, it was shown [55] that the experimental data on the heat capacity and entropy of ethane could be reconciled with the theoretical calculations made on the basis of statistical thermodynamics only if one took into account the inhibited internal rotation. The structural non-rigidity often manifests itself in the unusual temperature dependencies of thermodynamic functions. For example, for crystalline borohydrides of sodium and potassium [56], transition from a state where the anion vibrates about the equilibrium position to a state with internal anion rotation gives rise to a maximum in the rotational heat capacity curve (Figure 9).

Intensive theoretical and experimental study of non-rigid molecules during the last 10 years has shown that one-minimum descriptions and corresponding rigid rotator-harmonic oscillator structural models are not applicable to these species. New multiminima descriptions with flexible motion models should be developed for every class of non-rigid molecule, and molecular symmetry and bond theory should be modified for these species.

Acknowledgments

We are grateful to Jack Simons for discussions and for his comments on this manuscript.

References

[1] R.G. Woolley: *Adv. Phys.* **25** (1967) 27.
[2] H. Primas: *Int. J. Quant. Chem.* **1** (1967) 493.
[3] H. Primas: *Helv. Chem. Acta* **51** (1968) 1037.
[4] I.B. Bersuker: *Theor. Exper. Khim.* **5** (1969) 293.
[5] I.B. Bersuker, I.Ya. Ogurtsov and Ya.I. Sharapov: *Theor. Exper. Khim.* **9** (1973) 451.
[6] H. Primas: *Theor. Chim. Acta* **39** (1975) 127.
[7] R.G. Woolley: *Chem. Phys. Lett.* **73** (1976) 73.
[8] R.G. Woolley and B.T. Sutcliffe: *Chem. Phys. Lett.* **45** (1977) 393.
[9] R.G. Woolley: *Chem. Phys. Lett.* **55** (1978) 443.
[10] H. Primas: *Adv. Chem. Phys.* **38** (1978) 1.
[11] P. Claverie and S. Dinner: *Isr. J. Chem.* **19** (1980) 54.
[12] R.G. Woolley: *Structure and Bonding*, Springer, Berlin (1982), Vol. 28.
[13] P. Claverie and G. Jona-Lasino: *Phys. Rev.* **33A** (1986) 2245.
[14] Y.G. Smeyers: *Current Aspect of Quantum Chemistry*, R. Carbó (Ed.), Elsevier, Amsterdam (1981), Vol. 21, pp. 293–312.
[15] (a) A.I. Boldyrev: *Modern Problems of Physical Chemistry*, Ya. M. Kolotyrkin (Ed.), Khimia, Moscow (1988), Chapter I, pp. 6–47. (b) V.V. Nefedova, A.I. Boldyrev, and J. Simons: *J. Chem. Phys.* **98** (1993) 8801.
[16] A.I. Boldyrev and V.A. Onyschuck: *Zh. Fiz Khim.* **62** (1988) 1474.
[17] R.G. Woolley: *Molecules in Physics, Chemistry and Biology*, J. Maruani (Ed.), Kluwer Academic Publishers, London (1988), Vol. 1, Chapter 2, pp. 45–89.
[18] H.C. Longuet-Higgins: *Mol. Phys.* **3** (1963) 445.
[19] I.G. Kaplan: *Symmetry of Manyelectron Systems*, Nauka, Moscow (1969).
[20] P.R. Bunker: *Molecular Symmetry and Spectroscopy*, Academic Press, New York (1979).
[21] B.J.J. Dalton: *Chem. Phys.* **54** (1971) 4745.
[22] (a) H. Frei, A. Bauder and H.H. Günthard: *Top. Curr. Chem.* **81** (1979) 1. (b) G.S. Ezra: *Symmetry Properties of Molecules*, Lecture Notes in Chemistry, Vol. 28, Springer, Berlin (1982).
[23] (a) Y.G. Smeyers: *Fol. Chim. Theor. Lat.* **5** (1977) 27. (b) Y.G. Smeyers: *Fol. Chim. Theor. Lat.* **6** (1978) 139. (c) Y.G. Smeyers and M.N. Bellido: *Int. J. Quant. Chem.* **19** (1981) 553. (d) Y.G. Smeyers: *J. Mol. Struct.* **107** (1984) 3. (e) Y.G. Smeyers and A. Hernández-Laguna: *Structure and Dynamics of Molecular Systems*, R. Daudel *et al.* (Eds.), D. Reidel, Dordrecht (1985), pp. 23–40. (f) Y.G. Smeyers: *Fol. Chim. Theor. Lat.* **13** (1985) 159. (g) Y.G. Smeyers and A. Hernández-Laguna: *Int. J. Quant. Chem.* **29** (1986) 553. (h) Y.G. Smeyers and A. Niño: *J. Comput. Chem.* **8** (1987) 380. (i) Y.G. Smeyers, A. Niño and D.C. Moule: *J. Chem. Phys.* **93** (1990) 5786.
[24] (a) B.P. van Eijk and F.B. van Duijneveldt: *J. Mol. Spectrsc.* **102** (1983) 273. (b) B.P. van Eijk and E. van Zoeren: *J. Mol. Spectrosc.* **111** (1985) 138. (c) R. Fantoni, K. van Helvoort, W. Knippers and J. Reuss: *Chem. Phys. Lett.* **110** (1986) 1. (d) K.J. Abed and S. Clough: *Chem Phys. Lett.* **142** (1987) 209.
[25] (a) J. Berkovitz and J.P. Greene: *J. Chem. Phys.* **81** (1984). (b) G.T. Fraser, D.D. Nelson Jr., A. Charo and W. Klemperer: *J. Chem. Phys.* **82** (1985) 2535. (c) M.H. Begemann and R.J. Saykally: *J. Chem. Phys.* **82** (1985) 3570. (d) C. Yamada and E. Hirota: *Phys. Rev. Lett.* **56** (1986) 923. (e) D. Christen, R. Minkwitz and R. Nass: *J. Am. Chem. Soc.* **109** (1987)

7020. (f) T. Nakanaga and T. Amano: *Chem. Phys. Lett.* **134** (1987) 195. (g) H. Petek, D.J. Nesbitt, J.C. Ourutsky, C.S. Gudeman, X. Yang, D.O. Harris, C.B. Moore and R.J. Saykally: *J. Chem. Phys.* **92** (1990) 3257.

[26] (a) V. Spirko and D. Papousek: *Mol. Phys.* **36** (1978) 791. (b) C. Pouchan: *Chem. Phys. Lett.* **197** (1985) 326. (c) S. Yubushita and M.S. Gordon: *Chem Phys. Lett.* **117** (1985) 321. (d) T.J. Sears, P.R. Bunker, P.R. Davis, S.A. Johnson and V. Spirko: *J. Chem Phys.* **83** (1985) 2676. (e) M. Welti, T.H. Ha and E. Pretsch: *J. Chem Phys.* **83** (1985) 2676. (f) D.A. Dixon: *J. Chem. Phys.* **83** (1985) 6055. (g) M.T. Nguyen and T.K. Ha: *Chem. Phys. Lett.* **123** (1986) 123. (i) J. Moc, J.M. Rudzinski, Z. Latajka and H. Ratajczak: *Chem. Phys. Lett.* **168** (1990).

[27] (a) R.B. King: *Inorg. Chim. Acta* **49** (1981) 237. (b) R.B. King: *Inorg. Chem.* **20** (1981) 363. (c) R.B. King: *Theor. Chim. Acta* **59** (1981) 25. (d) R.B. King: *Inorg. Chim. Acta* **57** (1982) 79. (e) R.B. King: *Theor. Chim. Acta* **63** (1983) 103. (f) R.B. King: *Inorg. Chem.* **24** (1985) 1716. (g) R.B. King, D.H. Rouvray: *Theor. Chim. Acta* **69** (1986) 1. (h) R.B. King: *Inorg. Chem.* **25** (1986) 506. (i) R.B. King: *Molecular Structure and Energetics*, VCH Publishers Inc., New York (1986), Vol. 1, p. 123. (j) R.B. King: *Int. J. Quant. Chem.* **20** (1986) 227. (k) R.B. King: *J. Math. Chem.* **1** (1987) 249. (l) R.B. King: *J. Comput. Chem.* **8** (1987) 341.

[28] (a) D.F. Gaines, D.E. Coons and J.A. Heppert: *Advances in Boron and Boranes*, F.F. Liebmann, A. Greenberg and R.E. Williams (Eds.), VCH Publishers, New York (1988), Chapter 5, p. 91. (b) M.L. McKee: *J. Phys. Chem.* **93** (1989) 3426. (c) B.M. Gimarc, B. Daj, D.S. Warren and J.J. Ott: *J. Am. Çhem. Soc.* **112** (1990) 2597. (d) A.M. Mebel and O.P. Charkin: *Chemistry of Inorganic Hydrides*, N.T. Kuznetsov (Ed.), Nauka, Moscow (1990), p. 43. (e) N.T. Kuznetsov and K.A. Solntsev: *Chemistry of Inorganic Hydrides*, N.T. Kuznetsov (Ed.), Nauka, Moscow (1990), p. 5. (f) D.M.P. Mingos and D.J. Wales: *Electron Deficient Boron and Carbon Clusters*, G.A. Olah, K. Wade and R.E. Williams (Eds.), Wiley, New York (1991), Chapter 5, p. 143.

[29] (a) E. Clementi, H. Kistenmacher and H. Popkie: *J. Chem. Phys.* **58** (1973) 2460. (b) V.A. Istomin, N.F. Stepanov and B.I. Zhilinskii: *J. Mol. Spectrsc.* **67** (1977) 265. (c) A.I. Boldyrev, O.P. Charkin, K.V. Bozhenko, N.M. Klimenko and N.G. Rambidi: *Zh. Neorg. Khim.* **24** (1979) 612. (d) P.E.S. Wormen and J. Tennyson: *Chem. Phys.* **75** (1981) 1245. (e) M.L. Klein, J.D. Goggard and D.G. Bounds: *Chem. Phys.* **75** (1981) 3901. (f) C.J. Marsden: *Chem. Phys.* **76** (1982) 6451. (g) R. Esser, J. Tennyson and P.E.S. Wormer: *Chem. Phys. Lett.* **89** (1982) 223. (h) J. Tennyson and B.T. Sutcliffe: *J. Chem. Phys.* **77** (1982) 4061. (i) G. Brocks, J. Tennyson and A. van der Avoid: *J. Chem. Phys.* **80** (1984) 3223. (k) S.C. Farantos and J. Tennyson: *J. Chem. Phys.* **82** (1985) 800. (l) E. van Leuken, G. Brocks and P.E.S. Woemer: *Chem. Phys.* **110** (1986) 365. (m) J. Tennyson, G. Brocks and S.C. Frantos: *Chem Phys.* **104** (1986) 399. (n) P.v.R. Schleyer, A. Severyn, A.E. Reed and P. Hobza: *J. Comput. Chem.* **7** (1986) 666. (l) G. Brocks: *Chem. Phys.* **116** (1987) 33.

[30] (a) Z.K. Ismail, R.H. Hauge and J.L. Margrave: *J. Chem. Phys.* **57** (1972) 5173. (b) Z.K. Ismail, R.H. Hauge R. H. and J.L. Margrave: *J. Mol. Spectrsc.* **45** (1973) 304. (c) J.J. van Vaals, W.L. Meerts and A. Dymanus: *Chem. Phys.* **82** (1983) 385. (d) T. Törring, J.P. Bekoy, W.L. Meerts, J. Hoett, E. Tiemann and A. Dymanus: *J. Chem. Phys.* **73** (1980) 4875. (e) J.J. van Vaals, W.L. Meerts and A. Dymanus: *J. Chem. Phys.* **77** (1982) 5245. (f) J.J. van Vaals, W.L. Meerts and A. Dymanus: *J. Mol. Spectrosc.* **106** (1984) 280. (g) J.J. van Vaals, W.L. Meerts and A. Dymanus: *Chem. Phys.* **86** (1984) 147.

[31] (a) J. Almlof and A.A. Ischenko: *Chem. Phys. Lett.* **61** (1979) 79. (b) A.I. Boldyrev, O.P. Charkin, N.G. Rambidi and A.I. Avdeev: *Chem. Phys. Lett.* **50** (1977) 239. (c) V.G. Zakzhevskii, A.I. Boldyrev and O.P. Charkin: *Chem. Phys. Lett.* **70** (1980) 147. (d) V.G. Zakzhevskii, A.I. Boldyrev and O.P. Charkin: *Zh. Neorg. Khim.* **25** (1980) 2414. (e) L.P. Sukhanov, A.I. Boldyrev and O.P. Charkin: *Koord. Khim.* **6** (1980) 1631. (f) E.-U. Wurthwein, M.-B. Krogh-Jespersen and P.v.R. Schleyer: *Inorg. Chem.* **20** (1981) 3663. (g) A.I. Boldyrev, V.G. Solomonik and O.P. Charkin: *Chem. Phys. Lett.* **68** (1982) 51. (h) L.P. Sukhanov and A.I. Boldyrev: *Zh. Fiz. Khim.* **58** (1984) 654. (i) S.P. Konovalov and V.G. Solomonik: *Koord. Khim.* **26** (1986) 463. (j) V.V. Sliznev and V.G. Solomonik: *Zh. Strukt. Khim.* **26** (1986) N5, 19. (k) F. Ramondo, L. Bencivenni and V. Di Martino: *Chem. Phys.* **158** (1991) 41. (l) F. Ramondo, L. Bencivenni, N. Sanna and S.N. Cesaro: *J. Mol. Struct.*,

THEOCHEM **253** (1992) 121. (m) A.I. Boldyrev and J. Simons, to be published. (n) J.S. Francisco and I.H. Williams: *Chem. Phys.* **120** (1988) 389. (o) F. Ramondo, L. Bencivenni, R. Caminiti and F. Grandinetti: *Chem. Phys.* **145** (1990) 27.

[32] (a) A.I. Boldyrev, O.P. Charkin, N.G. Rambidi and V.I. Avdeev: *Chem. Phys. Lett.* **44** (1976) 20. (b) J.D. Dill, P.v.R. Schleyer, J.S. Binkley and J.A. Pople: *J. Am. Chem. Soc.* **99** (1977) 6159. (c) A.I. Boldyrev, O.P. Charkin, N.G. Rambidi and V.I. Avdeev: *Zh. Strukt. Khim.* **19** (1978) 203. (d) A.I. Boldyrev and O.P. Charkin: *Zh. Strukt. Khim.* **20** (1979) 969. (f) L.A. Curtiss: *Chem. Phys. Lett.* **68** (1979) 225. (g) R. Bonaccorsi, E. Scrocco and J. Tomasi: *Theor. Chim. Acta* **52** (1979) 113. (h) V.G. Zakzhevskii, A.I. Boldyrev and O.P. Charkin: *Chem. Phys. Lett.* **73** (1980) 54. (h) V. Barone, G. Dolcetti, F. Lelj and N. Russo: *Inorg. Chem.* **20** (1981) 1687. (i) V. Kello, M. Urban and A.I. Boldyrev: *Chem. Phys. Lett.* **106** (1984) 455. (j) O.P. Charkin, D.G. Musaev and N.M. Klimenko: *Koord. Khim.* **11** (1985) 409. (k) S.P. Konovalov and V.G. Solomonik: *Zh. Strukt. Khim.* **26** (1985) 12. (l) D.J. DeFrees, K. Raghavachari, H.B. Schlegel, J.A. Pople and P.v.R. Schleyer: *J. Phys. Chem.* **91** (1987) 1857. (m) J.S. Francisco and I.H. Williams: *J. Phys. Chem.* **94** (1990) 8522. (n) F. Ramondo, L. Bencievenni, R. Caminiti and C. Sadun: *Chem. Phys.* **151** (1991) 179. (o) M. Spoliti, N. Sanna and V. Di Martino: *J. Mol. Struct., THEOCHEM* **258** (1992) 83. (p) G. Scholz and L. Curtuss: *J. Mol. Struct., THEOCHEM* **258** (1992) 251.

[33] (a) L.Ya. Baranov and A.I. Boldyrev: *Chem. Phys. Lett.* **96** (1983) 218. (b) L.Ya. Baranov and A.I. Boldyrev: *Mol. Phys.* **54** (1985) 989. (c) L.Y. Baranov and A.I. Boldyrev, unpublished results. (d) E. Hirota: *J. Mol. Spectrosc.* **153** (1992) 447. (e) N. Ohashi and J.T. Hougen: *J. Mol. Spectrosc.* **153** (1992) 429.

[34] (a) F.S. Pianlto, A.M.R.P. Bopegedera, W.T.M.L. Fernando, R. Hailey, L.C. O'Brien, C.R. Brazier, P.C. Keller and P.F. Bernath: *J. Am. Chem. Soc.* **112** (1990) 7900. (b) Y. Kawashima, C. Yamada and E. Hirota: *J. Chem. Phys.* **94** (1991) 7707. (c) Y. Kawashima, and E. Hirota: *J. Chem. Phys.* **96** (1992) 2460. (d) Y. Kawashima and E. Hirota: *J. Mol. Spectrosc.* **153** (1992) 466.

[35] (a) T.J. Marks and J.R. Kolb: *Chem. Rev.* **77** (1977) 263. (b) G. Gundersen, L. Hedberg and K. Hedberg: *J. Chem. Phys.* **59** (1973) 3777. (c) D.F. Gaines, J.L. Walsh, J.H. Moris and D.F. Hillenbrand: *Inorg. Chem.* **17** (1978) 1516.

[36] (a) W.A. Lathan, W.J. Hehre and J.A. Pople: *J. Am. Chem. Soc.* **93** (1971) 808. (b) A.C. Hopkinson, K. Yates and I.G. Csizmadia: *J. Chem. Phys.* **55** (1971) 3835. (c) P.C. Hariharan, W.A. Lathan and J.A. Pople: *Chem. Phys. Lett.* **21** (1973) 309. (d) J. Weber, M. Yoshimine and A.D. McLean: *J. Chem. Phys.* **64** (1976) 4159. (e) R.A. Krishnan, R.A. Whiteside, J.A. Pople and P.v.R. Schleyer: *J. Am. Chem. Soc.* **103** (1981) 5649. (f) K. Hirao and S. Yamabe: *Chem. Phys.* **89** (1984) 237. (g) G.P. Raine and H.F. Schaefer III: *J. Chem. Phys.* **81** (1984) 4043. (h) T.J. Lee and H.F. Schaefer III: *J. Chem. Phys.* **85** (1986) 3347. (i) J.A. Pople: *Chem. Phys. Lett.* **137** (1987) 10. (j) R. Lindh, B.O. Roos and W.P. Kremer: *Chem. Phys. Lett.* **139** (1987) 407. (k) C. Liang, T.P. Hamilton and H.F. Schaefer III: *J. Chem. Phys.* **92** (1990) 3653.

[37] (a) J.T. Hougen: *J. Mol. Spectrosc.* **123** (1987) 197. (b) R. Escribano, P.R. Bunker and P.C. Gomez: *Chem. Phys. Lett.* **150** (1988) 60. (c) P.C. Gomez and P.R. Bunker: *Chem. Phys. Lett.* **165** (1990) 351. (d) R. Escribano and P.R. Bunker: *J. Mol. Spectrosc.* **122** (1987) 325.

[38] M.W. Crofton, M.F. Jagod, B.D. Rehfuss and T.J. Oka: *Chem. Phys.* **91** (1989) 5139.

[39] F.A. Houle and J.L. Beauchamp: *J. Am. Chem. Soc.* **101** (1979) 4967.

[40] G.A. Olah and P. v. R. Schleyer (Eds.): *Carbonium Ions*, Wiley, New York (1968–1976), Vol. I–V.

[41] (a) A.I. Boldyrev, L.P. Sukhanov, V.G. Zakzhevskii and O.P. Carkin: *Chem. Phys. Lett.* **79** (1981) 421. (b) L.P. Sukhanov, A.I. Boldyrev and O.P. Charkin: *Chem. Phys. Lett.* **97** (1983) 373. (c) Yu.B. Kirillov, V.L. Bugaenko, V.L. Grishkin and A.I. Boldyrev: *Koord. Khim.* **62** (1988) 2983.

[42] (a) Yu.B. Kirillov, A.I. Boldyrev, N.M. Klimenko and O.P. Charkin: *Koord. Khim.* **8** (1982) 1203. (b) Yu.B. Kirillov, A.I. Boldyrev, N.M. Klimenko and O.P. Charkin: *Khoord Khim.* **9** (1983) 326. (c) Yu.B. Kirillov and A.I. Boldyrev: *Koord. Khim.* **11** (1985) 749. (d) Yu.B. Kirillov, A.I. Boldyrev, *Dokl. AN SSSR* **272** (1983) 635.

[43] (a) N.G. Rambidi: *J. Mol. Struct.* **28** (1975) 77, 89. (b) M. Hargittai and I. Hargittai: *Geometries of Coordination Molecules in Gas Phase*, Mir, Moscow (1976). (c) N.G. Rambidi: *Zh. Strukt. Khim.* **23** (1982) N6, 113.

[44] J.N. Murrell, S. Carter and L.O. Halonen: *J. Mol. Spectrosc.* **93** (1982) 307.

[45] D.G. Musaev, A.I. Boldyrev, O.P. Charkin and N.M. Klimenko: *Koord. Khim.* **10** (1984) 938.

[46] O.P. Charkin and T.S. Zyubina: *Koord. Khim.* **12** (1986) 1011.

[47] (a) P.A. Kollman, J.F. Liebman and L.C. Allen: *J. Am. Chem. Soc.* **92** (1970) 1142. (b) C.P. Baskin, C.F. Bender and P.A. Kollman: *J. Am. Chem. Soc.* **95** (1973) 5868. (c) M. Rupp and R. Ahlrichs: *Theor. Chim. Acta* **46** (1977) 117. (d) H. Umeyama and K. Morokuma: *J. Am. Chem. Soc.* **99** (1977) 1316. (e) J.D. Dill, P.v.R. Scheleyer, J.S. Binkley and J.A. Pople: *J. Am. Chem. Soc.* **99** (1977). (f) A.I. Boldyrev, V.G. Solomonik, V.G. Zakzhevskii and O.P. Charkin: *Chem. Phys. Lett.* **73** (1980) 58. (g) P.N. Swepton, H.L. Seller and L. Shafer: *J. Chem. Phys.* **74** (1981) 2372. (h) V.G. Solomonik, V.M. Ozerova and A.I. Boldyrev: *Zh. Neorg. Khim.* **27** (1982) 1891.

[48] (a) S.P. Konovalov and V.G. Solomonik: *Zh. Neorg. Khim.* **29** (1984) 1655. (b) M.T. Ngugen: *J. Mol. Struct., THEOCHEM* **136** (1986) 371. (c) A.I. Boldyrev and P.v.R. Schleyer: *Chem. Phys. Lett.* **172** (1990) 193. (d) F. Ramondo, L. Bencivenni and C. Sadun: *J. Mol. Struct., THEOCHEM* **209** (1990) 101. (e) A.I. Boldyrev and P.v.R. Schleyer P. v. R., unpublished results. (f) J.S. Francisco and I.H. Williams: *Chem. Phys.* **160** (1992) 255. (g) L. Bencivenni, M. Pelino and F. Ramondo: *J. Mol. Struct., THEOCHEM* **253** (1992) 109.

[49] (a) P.A. Akishin and V.P. Spiridonov: *Zh. Strukt. Khim.* **2** (1961) 63. (b) P.A. Akishin and V.P. Spiridonov: *Zh. Strukt. Khim.* **3** (1963) 267. (c) Yu.S. Ezhov, S.M. Tolmachev, V.P. Spiridonov and N.G. Rambidi: *Teplofitz. Vys. Temp.* **6** (1968) 68. (d) K.S. Seshadri, L.A. Niman and D. White: *J. Mol. Spectrsc.* **30** (1969) 128. (e) Yu.S. Ezhov, S.M. Tolmachev and N.G. Rambidi: *Zh. Strukt. Khim.* **13** (1972) 972. (f) Yu.S. Ezhov and S.A. Komarov: *Zh. Strukt. Khim.* **16** (1975) 662. (g) S.A. Komarov and Yu.S. Ezhov: *Zh. Strukt. Khim.* **16** (1975) 899. (h) R. Teghil, B. Janis and L. Bencivenni: *Inorg. Chim. Acta* **88** (1984) 115.

[50] (a) V.P. Spiridonov, E.V. Erokhin and Yu.A. Brezgin: *Zh. Strukt. Khim.* **13** (1972) 321. (b) A.N. Hodchenkov, V.P. Spiridonov and P.A. Akishin: *Zh. Strukt. Khim.* **6** (1964) 765. (c) K.A. Gingerich and F. Miller: *J. Chem. Phys.* **63** (1975) 1211. (d) L. Bencivenni and K.A. Gingerich: *J. Mol. Struct.* **98** (1983) 195. (e) V.A. Kulikov, V.V. Ugarov and N.G. Rambidi: *Zh. Struct. Khim.* **22** (1981) N3, 168. (f) K.P. Petrov, A.I. Kolesnikov, V.V. Ugarov and N.G. Rambidi: *Zh. Strukt. Khim.* **21** (1980) N4, 198. (g) I.R. Beattie, J.S. Ogden and D.D. Price: *J. Chem. Soc. Dalton* (1979) 1460. (h) S.N. Jenny and J.S. Ogden: *J. Chem. Soc. Dalton* (1979) 1465.

[51] (a) M. Bogey, H. Bolvin, C. Demuynck and J.L. Destombes: *Phys. Rev. Lett.* **58** (1987) 988. (b) E.D. Simandiras, J.F. Gaw and N.C. Handy: *Chem. Phys. Lett.* **141** (1987) 166. (c) R. Escribano and P.R. Bunker: *Chem. Phys Lett.* **143** (1988) 439. (d) M. Bogey, H. Bolvin, C. Demuynck, J.L. Destombes and B.P. van Eijck: *J. Chem. Phys.* **88** (1988) 4120.

[52] (a) S.J. Cyvin, B.N. Cyvin and A. Snelson: *J. Phys. Chem.* **75** (1971) 2609. (b) R. Huglen, S.J. Cyvin and H.A.Z. Oye: *Naturforsch.* **34a** (1979) 1118. (c) K. Petrov, V.A. Kulikov, V.V. Ugarov and N.G. Rambidi: *Zh. Strukt. Khim.* **21** (1980) 71. (d) E. Vajda, I. Hargittai and J. Tremmel: *Inorg. Chim. Acta* **25** (1977) 143. (e) R.L. Hunt and B.S. Ault: *Spectrochim. Acta* **37A** (1981) 63.

[53] A.I. Boldyrev and O.P. Charkin: *Zh. Strukt. Khim.* **26** (1985) N3, 158.

[54] M. Buhl and P.v.R. Schleyer, unpublished results.

[55] J.D. Kemp and K.S. Pitzer: *J. Chem. Phys.* **4** (1936) 749.

[56] D. Smith: *J. Chem. Phys.* **60** (1974) 958.

DO WE REALLY KNOW HOW TO DEFINE NORMAL VIBRATIONS IN NON-RIGID MOLECULAR SYSTEMS?

G. A. NATANSON*
Chemical Dynamics Corporation
Upper Marlboro, MD 20772,
U.S.A.

Abstract. The paper analyses ambiguities which exist both in a choice of slowly deforming geometries serving as reference points for vibrations in floppy molecules and in a definition of the large-amplitude internal variables near the selected reference configurations. Since the cited ambiguities change both the frequencies and the shape of normal vibrations, the vibrational analysis becomes an ill-defined problem even in the harmonic approximation. The author formulates some general criteria for selecting a "preferable" coordinate system, which would make it possible to give an intrinsic meaning to the term "harmonic vibrations of a floppy molecule".

A schematic view of a molecular system as a set of nuclei vibrating near some slowly deforming geometries [1] is now considered as a routine way of treating floppy molecules with several atoms. The reference geometries are most often chosen in a more or less intuitive fashion, whereas vibrations themselves are defined via the Eckart–Sayvetz transformation [1, 2], to a large extent, under influence of the well-known work of Hougen, Bunker, and Johns [3]. Such a choice of molecular variables makes both frequencies and shape of the normal vibrations unambiguously determined for all allowed geometries ρ of the so-called "static" molecular model $\underset{\sim}{\bar{a}}(\rho) \equiv \{\underset{\sim}{a}_1(\rho), \underset{\sim}{a}_2(\rho), \ldots, \underset{\sim}{a}_N(\rho)\}$ serving as reference points for vibrations. This approach leaves unanswered two questions:

1. *Is there any intrinsic way of defining the static molecular model* $\underset{\sim}{\bar{a}}(\rho)$*?*

* Present address: Computer Sciences Co., 10110 Aerospace Rd., Seabrook, MD 20706, U.S.A.

Y.G. Smeyers (ed.), Structure and Dynamics of Non-Rigid Molecular Systems, 47–66.

2. *Should the Eckart–Sayvetz conditions be considered as preferable means to describe nuclear vibrations near the given static molecular model?*

The first question was initially addressed by Natanson and Adamov [4, 5]. We proved that combining the Eckart–Sayvetz conditions with the requirement for the potential V to have a local minimum with respect to vibrational coordinates at any geometry, allowed for the model, unambiguously determines the geometry of a model with a single internal degree of freedom. In particular, if the orientation of the static molecular model $\underset{\sim}{\bar{a}}\,(\rho_1)$ is chosen to satisfy the relation [3]:

$$\sum_d m_d\, \underset{\sim}{a}_d(\rho_1) \times \frac{d\, \underset{\sim}{a}_d}{d\rho_1} \equiv \underset{\sim}{0} \tag{1}$$

(to eliminate the zeroth-order momentum coupling between overall rotations and large-amplitude internal motion), then one comes to the equation [4, 5]:

$$m_d \frac{d\, \underset{\sim}{a}_d}{d\rho_1} = \beta(\rho_1) \left. \frac{\partial V}{\partial\, \underset{\sim}{r}_d} \right|_{\underset{\sim}{\bar{r}} \,=\, \underset{\sim}{\bar{a}}\,(\rho_1)} \tag{2}$$

which determines the gradient-following path in the space of mass-weighted Cartesian coordinates $\sqrt{m_d}\,\underset{\sim}{r}_d$. (The coefficient $\beta(\rho_1)$ in (2) can be eliminated by means of the appropriate choice of the variable ρ_1.) Later, under influence of works of Hofacker [6], Marcus [7], and Fukui [8] in the theory of chemical reactions, this analysis was extended by the author [9] to a much broader family of coordinate transformations eliminating kinematic coupling between large- and small-amplitude motions along the selected reference path in the space of nuclear geometries. It was shown that the "intrinsic" geometry $\sigma_0(\rho_1)$ of the static molecular model ("the intrinsic internal path" in our terms [9]) is determined by the covariant equation:

$$\frac{d\sigma_0^{\mu}}{d\rho_1} = \mathbb{G}^{\mu\mu'}\left[\sigma_0(\rho_1)\right] \left. \frac{\partial V}{\partial \sigma^{\mu'}} \right|_{\sigma=\sigma_0(\rho_1)}, \tag{3}$$

where $\sigma \equiv \{\sigma^1, \sigma^2, \ldots, \sigma^{3N-6}\}$ is an arbitrary set of internal variables and

$$\mathbb{G}^{\mu\mu'}(\sigma) = \sum_d \frac{1}{m_d} \frac{\partial \sigma^{\mu}}{\partial\, \underset{\sim}{r}_d} \bullet \frac{\partial \sigma^{\mu'}}{\partial\, \underset{\sim}{r}_d}. \tag{4}$$

The equation was earlier suggested by Marcus [7] for the reaction path, apparently from a similar prospective. In this connection the author also refers the reader to a very interesting discussion independently presented by Rowe and Ryman [10]. In terms of the latter work Equation (3) describes "a fall line" under the metric

$$d\sigma^2 = \mathbf{T}_{\mu\mu'}(\sigma) d\sigma^{\mu} d\sigma^{mu'}, \tag{5}$$

where the functions $\mathbf{T}_{\mu\mu'}(\sigma)$ form the $(3N-6) \times (3N-6)$ matrix inverse of $\underset{\approx}{\mathbb{G}}\,(\sigma)$. It is worth mentioning in this connection that some authors [11, 12] incorrectly indicate the metric matrix in (5) as the block of the full $3N \times 3N$ metric matrix restricted to the space of internal variables which is not the same [9].

Note that the question of the orientation of the model $\underset{\sim}{\bar{a}}\ (\rho_1)$ in space for different values of the variable ρ_1 becomes completely irrelevant. In particular, as it was emphasized by Pickett [13], the use of the condition (1) for molecules with internal rotation leads to multivalued solutions which make it difficult to impose the necessary boundary conditions on rotational wave functions.

Within the covariant approach developed by the author [9], the set of internal variables in question was divided into two groups represented by p large-amplitude internal variables $\rho \equiv \{\rho_1, \rho_2, \ldots, \rho_p\}$ and $\Gamma = 3N - 6 - p$ vibrational coordinates $Q \equiv \{Q_1, Q_2, \ldots, Q_\Gamma\}$. One of the most important conclusions made by the author [9] is that harmonic vibrational frequencies evaluated at geometries allowed for the given "intrinsic" static model depend on the choice of large-amplitude variables ρ. This led the author to the second of the aforementioned questions. Although the analysis performed in [9] indeed revealed some remarkable features of the Eckart–Sayvetz coordinate system, none of them could be used a decisive argument making such a choice preferable for separating large- and small-amplitude internal motions. More recently [14] the author discovered a very interesting attribute of the cited coordinate system, which did make a difference; and the current publication intends to summarize the author's latest conclusions concerning this rather non-trivial problem: how to define nuclear vibrations near a slowly deforming geometry.

Before going to the main subject of this paper – the choice of the large-amplitude internal variable ρ_1 in a valley surrounding a reference path $\sigma_0(\rho_1)$ in the space of molecular geometries σ, it is worth making some additional comments concerning the first of the aforementioned questions – how to choose the optimum geometries $\sigma_0(\rho_1)$ used as the reference points for nuclear vibrations. A certain revision of the author's earlier analysis [9] seems especially useful, taking into account a recent interest [15, 16] in the so-called "gradient extremals" [17] ("the optimum ascent path" in terms of Basilevky and Shamov [18]). In particular, the author feels necessary to slightly correct his earlier assertion [19] that 'use of the "gradient extremal" as the reaction path necessarily results in a force that pushes the system from the selected origin toward the internal intrinsic path'.

By definition any variable Q_β is referred to as a vibrational coordinate, if it is equal to zero in any geometry $\sigma_0(\rho_1)$. In addition, the coefficients of kinematic coupling between large- and small-amplitude internal variables are required to vanish in the reference geometries $\sigma_0(\rho_1)$:

$$g^{\beta\rho}(\rho_1, 0) = \sum_d \frac{1}{m_d}\left[\frac{\partial Q_\beta}{\partial\, \underset{\sim}{r}_d} \bullet \frac{\partial\ \rho_1}{\partial\, \underset{\sim}{r}_d}\right]_{\sigma=\sigma_0(\rho_1)} \equiv 0. \tag{6}$$

In [9] the author refers to the latter requirement as the *local* Hofacker–Marcus conditions to distinguish it from much more restrictive constraints requiring for small- and large- amplitude internal variables to be globally separable [6, 7]. It is probably more appropriate, as done below, to refer to (6) as the Hofacker–Marcus–Quade conditions, since it was Quade [20] who first represented the necessary condition for an approximate separation of large- and small-amplitude internal degrees of freedom in such a form. As proven by the author [9], the Hofacker–Marcus–Quade conditions are fulfilled for an arbitrary set of vibrational coordinates if they hold at least for one set of coordinates Q, provided that one uses exactly the same large-amplitude internal variable ρ_1.

Another important result proven by the author [9] is that combining the Hofacker–Marcus–Quade conditions with the extremum condition

$$\left.\frac{\partial V}{\partial Q_\beta}\right|_{\sigma=\sigma_0(\rho_1)} = 0 \tag{7}$$

leads to Equation (3) for the internal intrinsic path. Note that "intrinsic" semirigid models are necessarily mass-dependent and hence the potential obtained by freezing vibrations is not invariant under isotopic substitution. For example, Villarreal, Bauman, and Laane [21] tried to fit the observed far-infrared spectra of cyclopentene, cyclopentene–1-d_1, cyclopentene–1,2,3,3-d_4, and cyclopentene-d_8 by using a one-dimensional potential, and found out that the parameters in the potential function vary by as much as 8%. Later [22] these discrepancies were eliminated by using a two-dimensional semirigid model with the so-called "ring-puckering" and "ring-twisting" internal degrees of freedom. Our guess is that one can achieve a comparable accuracy by drawing its own internal intrinsic path for each isotopic modification and then solving the one-dimensional Schrödinger equation for the appropriate one-dimensional section of the two-dimensional potential surface.

It is worth mentioning in this connection that the simplifications recently attributed by Shida, Almöf, and Barbara [16] to the steepest-descent path:

1. *the bath is always orthogonal to the reaction path;*
2. *there are no rotational or translational components induced*

have no direct relation to the subject. First, the vibrational axes can be always directed "perpendicular" to the reference path, regardless of its particular choice. As for the second simplification, it sounds rather obscure and possibly implies relation (1) coupled with the requirement for the centre of mass of the semirigid model $\underset{\sim}{\bar{a}}\,(\rho_1)$ to be at the origin of the laboratory frame for any allowed value of the variable ρ_1. However, as discussed in detail in the author's earlier paper [9], which was apparently overlooked by Shida, Almöf, and Barbara [16], one can satisfy both requirements for any reference path – this is just the question of how to orientate a semirigid model of the given geometry in space for different values of the variable ρ_1. In particular, it can be done for geometries forming the gradient extremal. The only essential advantage of the internal intrinsic path is that its use makes it possible to eliminate the linear term in the expansion of the potential function as a Taylor series in vibrational coordinates Q [9] – cf. Equations (3.2) and (3.9) in [16] and especially the comments accompanying the latter equation.

Vibrational coordinates can be always chosen in such a way that

$$g^{\beta\beta'}(\rho_1, 0) = \sum_d \frac{1}{m_d} \left[\frac{\partial Q_\beta}{\partial \underset{\sim}{r}_d} \bullet \frac{\partial Q_{\beta'}}{\partial \underset{\sim}{r}_d} \right]_{\sigma=\sigma_0(\rho_1)} \equiv \mu^{-1} \delta_{\beta\beta'}, \tag{8}$$

where μ is an arbitrary parameter having dimensionality of mass. A different, but still arbitrarily chosen effective mass m is introduced for large-amplitude internal motion via the relation:

$$g^{\rho\rho}(\rho_1, 0) = \left[\sum_d \frac{1}{m_d} \left| \frac{\partial \rho_1}{\partial \underset{\sim}{r}_d} \right|^2 \right]_{\sigma=\sigma_0(\rho_1)} \equiv m^{-1}. \tag{9}$$

There is no special advantage in choosing the latter mass to be different, except that it helps to identify sources responsible for various scaling factors in the equations of motion.

By representing the equations of motion in the Lagrangian form (see Equations (1)–(8) and accompanying comments in [23]) the classical force, generating deviations of the nuclei from the reference geometry $\sigma_0(\rho_1)$, can be then written as

$$\mu \ddot{Q}_\beta = -\frac{\partial V}{\partial Q_\beta} - \Gamma^\beta_{\rho\rho} \mu \dot{\rho}_1^2 - 2\mu \dot{\rho}_1 \sum_{\beta'} \Gamma^\beta_{\rho\beta'} \dot{Q}_{\beta'} - \mu \sum_{\beta', \beta''} \Gamma^\beta_{\beta' \beta''} \dot{Q}_{\beta'} \dot{Q}_{\beta''}, \tag{10}$$

where $\Gamma^\nu_{\nu'\nu''}$ are the so-called Cristoffel symbols of the second kind. (Note that the factor 1/2 must be moved from the right-hand side of Equation (7a) in [23] into the following definition (7b) for the Cristoffel symbols of the first kind $\Gamma_{\mu'\mu'',\mu}$.) As directly follows from Equation (40) below (see also Appendix A in [24] for an explicit expression of the kinetic energy expressed in terms of the coordinates in question) use of the Eckart–Sayvetz coordinate transformation makes it possible to eliminate the last term in the right-hand side of (10) at $Q = 0$ [9]. Based on results of the author's earlier paper [14], it will be shown below that other sum in the right-hand side of (10) can be also eliminated at $Q = 0$ by means of a ρ_1-dependent orthogonal transformation mixing normal vibrational coordinates. One thus comes to the equation:

$$\mu \ddot{Q}_\beta \approx -\frac{\partial V}{\partial Q_\beta} - \kappa_\beta(\rho_1) m \dot{\rho}_1^2, \tag{11}$$

where the curvature coefficients $\kappa_\beta(\rho_1)$ are defined via the relation [25]:

$$g^{\rho\rho}(\rho_1, Q) = m^{-1} \left[1 + 2\sum_\beta \kappa_\beta(\rho_1) Q_\beta \right] + 0(Q^2). \tag{12}$$

Note that the effective mass m appeared in the second term in the right-hand side of (11) after substituting the explicit expression for the Cristoffel symbol $\Gamma^\beta_{\rho\rho}$ in (10).

It was Marcus [26, 27] who first realized that, because of non-zero curvature of the intrinsic internal path, nuclei must vibrate near slightly different reference geometries. He found a very interesting solution for the tunnelling path of a collinear atom-diatom exchange reaction, and below we show how this solution can be extended to molecular systems with more than one vibrational degrees of freedom.

Let us define a "quasi-equilibrium" model by the requirement that the force acting on the nuclei vanishes in the reference geometries sought for, i. e., that

$$\left.\frac{\partial V}{\partial Q_\beta}\right|_{\sigma=\sigma_0(\rho_1)} = -\kappa_\beta(\rho_1)\mu\dot{\rho}_1^2. \tag{13}$$

The equation for the reference geometries thus takes the form:

$$\left.\frac{\partial V}{\partial Q_\beta}\right|_{\sigma=\sigma_0(\rho_1)} = -2\kappa_\beta(\rho_1)[E - E_{\text{vib}}(\rho_1) - V_0(\rho_1)], \tag{14}$$

where E is the total energy of the molecule, $E_{\text{vib}}(\rho_1)$ is the energy of nuclear harmonic vibrations at the point ρ_1 on the reference path $\sigma_0(\rho_1,)$ and $V_0(\rho_1) \equiv V(\rho_1, 0)$. For systems with a single vibrational degree of freedom, Equation (14) for a "quasi-equilibrium" reference path coincides with Equations (17), (18) of Marcus [26]. (The 'minus' in the right-hand side of Equation (14) appears due to our choice for vibrational coordinates Q to be positive on the concave side of the "quasi-equilibrium" path, as suggested by Skodje, Truhlar, and Garrett [25], whereas the vibrational coordinate x used by Marcus [26, 27] is positive away from the concave side of this path.)

As a matter of fact, Marcus [26] came to Equation (14) from a different prospective, namely, he introduced the vibrational potential $U_{\text{vib}}(\rho_1, Q)$ via the relation

$$V(\rho_1, Q) = \{g^{\rho\rho}(\rho_1, Q)[E_{\text{vib}}(\rho_1) - E + V_0(\rho_1)]\} + U_{\text{vib}}(\rho_1, Q), \tag{15}$$

while including the term in figure brackets into the one-dimensional Schrödinger equation governing large-amplitude internal motion. One can easily verify that the vibrational potential $U_{\text{vib}}(\rho_1, Q)$ has an extermum (expected to be a minimum) at each point of the quasi-equilibrium path. An essential advantage of Marcus's derivation of Equation (14) is that the adiabatic separation of variables suggested in [26] is also applicable to the classically forbidden region. On the other hand, arguments presented above provide a simple purely classical picture for Marcus's rather remarkable ideas which have attracted undeservedly little attention so far.

An analysis of the equation (14) governing the reference path shows that the definition of the "quasi-equilibrium" reference geometry depends on how the kinetic energy is divided between small- and large-amplitude internal degrees of freedom. In particular, if all the kinetic energy is given to vibrations [23] then the right-hand side of Equation (14) vanishes and one comes again to Equation (7) for "the intrinsic internal path". (Note that, in contrast with [9], we distinguish here between "intrinsic" and "quasi-equilibrium" models – see Equations (7) and (14), respectively.) As explained in [23], classical trajectories with all the kinetic energy given to vibrations yield a dominant contribution to reactive probabilities, so that they were used in [23] to construct the dividing surface sought for, which gave rise to a new definition for the large-amplitude internal variable ρ_1.

The derivation presented in [23] was essentially based on the assumption that nuclei must vibrate near "the intrinsic internal path", and the analysis performed above indeed supports such a choice of the reference path for the aforementioned family of trajectories. However, for non-rigid molecules the most effective separation of small- and large-amplitude degrees of freedom is expected in the ground vibrational state, so that the kinetic energy of large-amplitude internal motion cannot be neglected any more.

Note that the reference path defined according to Equation (14) is energy-dependent, in contrast with "the intrinsic internal path". Another non-trivial feature of the quasi-equilibrium models is that their definition depends on the frequencies of harmonic vibrations which may vary depending on the definition of the large-amplitude internal variable in the neighbourhood of the selected reference path. As a result the question of defining reference geometries turns out to be closely related to the even more difficult issue of how to choose the curvilinear axes of the coordinate system in question. It is not sufficient any more to simply declare that they must be orthogonal to the reference path. One needs also to specify the curvature of each curvilinear axis, coming up with different harmonic frequencies depending on the choice of the hypersurface that defines "pure vibrations" of nuclei.

Before going to the discussion of this much more complicated problem, let us make some comparative comments concerning the "generalized valley approximation" of Do Dang, Bulgac, and Klein [28] giving rise [15] to the already mentioned "gradient extremals" [17]. First, one can easily see that Equation (13) here is completely equivalent to Equation (2.21) of Do Dang, Bulgac, and Klein [28], since, taking into account (6), the velocity $\dot{\rho}$ of large-amplitude internal motion and the momentum p_ρ conjugated to the coordinate ρ are related via the simple relation:

$$p_\rho|_{Q=0} = m\dot{\rho} \tag{16}$$

at any point lying on the reference path. However, Do Dang, Bulgac, and Klein [28] require for their Equation (2.21) to be fulfilled for any value of the momentum p_ρ. The latter requirement led these authors to a rather restrictive constraint

$$\kappa_\beta(\rho_1) \equiv 0, \qquad \beta = 1, \ldots, \Gamma, \tag{17}$$

in addition to our conditions (6) and (7), used as a starting point in [9]. Since, as emphasized in [19], Equation (17) is usually inconsistent with (6) and (7), Do Dang, Bulgac, and Klein [28] seek an optimum way to satisfy all three conditions approximately. To do it, they introduce the invariant

$$U(\sigma) \equiv \frac{\partial V}{\partial \sigma^\mu} \mathbb{G}^{\mu\mu'}(\sigma) \frac{\partial V}{\partial \sigma^{\mu'}} \tag{18}$$

and then search for the direction of its slowest change, as is prescribed by "the mountaineer's algorithm" independently suggested by Basilevsky and Shamov

[18]. (Although the "generalized valley approximation" was formulated in [28] in more general terms applicable to collective motion having large amplitudes in several dimension, here I intentionally limited the discussion to a single large-amplitude internal degree of freedom to more clearly emphasize the difficulties which arise even in this "simplest" case.) The outlined variation procedure leads one [18, 28] to the equation for a "gradient extremal", in terms of Hoffman, Nord, and Ruedenberg [17]:

$$\frac{\partial U}{\partial \sigma^{\mu}} = \lambda(\sigma)\frac{\partial V}{\partial \sigma^{\mu}}\,, \tag{19}$$

where λ is the Lagrange multiplier, or in a more familiar form:

$$V_{,\mu\mu'}\mathbb{G}^{\mu',\mu''}(\sigma)V_{,\mu''} = \frac{1}{2}\,\lambda(\sigma)V_{,\mu} \tag{19'}$$

When deriving (19') from (19), we took into account that covariant derivatives $V_{,\mu}$ and $U_{,\mu}$ of the invariants $V(\sigma)$ and $U(\sigma)$ coincide with the corresponding partial derivatives, whereas covariant derivative of the inverse $\underset{\approx}{\mathbb{G}}(\sigma)$ of the metric tensor $\underset{\approx}{\mathbb{T}}(\sigma)$ is identically equal to zero, like the derivative of the metric tensor $\underset{\approx}{\mathbb{T}}(\sigma)$ itself.

To relate the condition (19') to the zero-curvature constraint (17), let us express the left-side of the condition (19') in the coordinate system ρ_1, Q under assumption that the reference path satisfies the extremum condition (7). In fact, in the latter case one finds that

$$\begin{aligned} V_{,\mu'\beta}\mathbb{G}^{\mu'\mu''}V_{,\mu''}\Big|_{Q=0} &= \frac{1}{m}\frac{dV_0}{d\rho_1}\left[\frac{\partial^2 V}{\partial\rho_1\partial Q_\beta} + \frac{dV_0}{d\rho_1}\,\Gamma^{\rho}_{\rho\beta}\right]_{Q=0} \\ &= -\frac{2}{m}\,\kappa_\beta(\rho_1)\left(\frac{dV_0}{d\rho_1}\right)^2 . \end{aligned} \tag{20}$$

where we took into account that the mixed derivative of the potential V with respect to ρ_1 and Q_β vanishes at $Q = 0$. If the curvature coefficients $\kappa_\beta(\rho_1)$ for the internal intrinsic path were identically equal to zero, then making use of (7) once again immediately leads one to Equation (19'), since only the ρ-components of both sides of the latter equation differ from zero.

However, the curvature coefficients $\kappa_\beta(\rho_1)$ generally differ from zero, and as a result Equations (7) and (19') describe two different curves. (The fact that deviations of the gradient extremal from the internal intrinsic path are proportional to the curvature of the latter path was first recognized by Basilevsky and Shamov [18].) The suggestion of Do Dang, Bulgac, and Klein [28] to use the gradient extremal, instead of the internal intrinsic path, does not seem to be properly justified. As explained above, the internal intrinsic path will necessarily provide a more effective adiabatic separation of variables at least for motion with a dominant part of the kinetic energy given to vibrational degrees of freedom.

On the other hand, as recently emphasized by Shida, Almöf, and Barbara [16], round-off errors induced during numerical calculations make the internal intrinsic path unstable and inaccurately determined, in contrast with gradient extremals which are defined via the local criterion [18]. In particular, Shida, Almöf, and Barbara [16] computed the "inversion" energy levels of a two-dimensional model Hamiltonian describing the inversion of the NH_3 and H_3O^+ molecules and compared the resultant energy spectrum with eigenvalues of the one-dimensional problems obtained by cutting the corresponding two-dimensional potential surface along the internal intrinsic path and gradient extremal, respectively. Their calculations showed that differences between eigenvalues of the one-dimensional problems are negligibly small compared with errors caused by omitting the coupling terms between the umbrella motion and the stretching vibration.

One can better understand the results of the cited work by considering a more familiar model [29] which describes the inversion of the ammonia molecule as motion of the nitrogen relative to the rigid equilateral triangle formed by the hydrogens. The remarkable feature of this model is that it is depicted as a straight line in the "plane" representing C_{3v} configurations in the space of internal variables. The fact that the inversion potentials associated with the internal intrinsic path and the gradient extremal give rise to practically the same energy spectra indicates that the simple "straight-line" model is expected to be sufficiently accurate for reproducing the shape of the inversion potential.

Note that the term "plane" was used above in a more or less usual sense, since the two-dimensional subspace in question is Euclidean, taking into account that the coefficients coupling vibrational and angular velocities vanish in any C_{3v} configuration. As a result, the corresponding Hamiltonian can be represented in exactly the same form as that used by Marcus [26, 27] for the atom-diatom collinear problem. The same is true for the subspace of geometries representing C_{2v} configurations of water in the space of internal variables. Recognizing this fact earlier would significantly simplify the analytical expansion used by the author [30] to construct the "intrinsic" semirigid model of the water molecule.

By comparing the inversion splittings reported by Shida, Almöf, and Barbara [16] for the exact two-dimensional potential surface and for its section along the internal intrinsic path, one finds that the one-dimensional double-well problem slightly overestimated the inversion splittings. On the contrary, the calculations earlier performed by Brown, Tucker, and Truhlar [31] for the cut of the multidimensional potential surface of ammonia [32] along the internal intrinsic path gave the inversion splittings which are always larger than the exact ones [33], in agreement with predictions of the small-curvature approximation of Skodje, Truhlar, and Garrett [25]. In the latter case the tunnelling path is drawn through the turning points of vibrations on the concave side of the internal intrinsic path, and as a result the splittings computed within this approximation are always larger than those obtained by cutting the potential surface along the internal intrinsic path. Brown, Tucker, and Truhlar [31] found that the exact inversion splittings lie between the values associated with the one-dimensional tunnelling along the both paths, so that the small-curvature approximation overestimates the splittings in this case.

It directly follows from Equation (14) that the quasi-equilibrium path always lies on the concave side of the intrinsic internal path within the classically forbidden region, so that its use, instead of the internal intrinsic path, would improve agreement between one- dimensional double-well problem and the exact results [33] for the multidimensional potential surface of ammonia [32], while leading to larger discrepancies for the model studied by Shida, Almöf, and Barbara [16].

To conclude this part of our discussion, let us point out that, due to the existence of a local criterion, the "generalized valley approximation" of Do Dang, Bulgac, and Klein [28] provides a very interesting extension of "the mountaineer's algorithm" [18], giving rise to a rather powerful numerical tool for constructing *ab initio* semirigid models with more than one internal degrees of freedom, *provided that the curvature of the corresponding "intrinsic" surface is sufficiently small.* Further details and corresponding bibliography can be found in a more recent review article by Klein, Walet, and Do Dang [34].

Suppose now that one has finally decided how to choose the reference path in the space of internal variables, though, as illuminated above, it is by no means an easy choice. Then we immediately run into the second problem of how to define the large-amplitude internal variable ρ_1 in the neighbourhood of the selected reference geometries. The reader may be surprised why the author focuses his attention only on the large-amplitude internal variable, since exactly the same problem should be apparently addressed for any axis of the curvilinear coordinate system in question. An essential difference, however, comes from the fact that any change in the definition of the large-amplitude internal variable will generally affect the frequencies of harmonic vibrations. On the other hand, these frequencies are independent of the particular choice of vibrational coordinates if the definition of the large-amplitude internal variable is kept the same [9]. In fact, the second term in the right-hand side of the relation

$$\left(\frac{\partial^2 V}{\partial Q_\beta \partial Q_{\bar\beta}}\right)_{\rho_1}\Bigg|_{Q=0} = \left(\frac{\partial^2 V}{\partial Q_\beta \partial Q_{\bar\beta}}\right)_{\tilde\rho_1}\Bigg|_{Q=0} + \frac{dV_0}{d\rho_1}\left(\frac{\partial^2 \tilde\rho_1}{\partial Q_\beta \partial Q_{\bar\beta}}\right)_{\rho_1}\Bigg|_{Q=0}, \tag{21}$$

would generally change eigenvalues of the $\Gamma \times \Gamma$ matrix $\underset{\approx}{G}(\rho_1)\,\underset{\approx}{F}(\rho_1)$ at points other than equilibrium configurations. On the contrary, if the large-amplitude internal variable remains unchanged, the $\underset{\approx}{G}(\rho_1)\,\underset{\approx}{F}(\rho_1)$ matrix undergoes a similarity transformation which cannot affect eigenvalues of the transformed matrix.

The conventional way of defining large-amplitude internal variables is to choose a semirigid model $\underset{\sim}{\bar a}(\rho_1)$ of the molecule and then to introduce the Eckart–Sayvetz coordinate system [1–3] via the relations:

$$\underset{\sim}{r}_d = \underset{\sim}{r}^{CM} + \underset{\approx}{S}(\Theta)\left[\underset{\sim}{a}_d(\rho_1) + \sum_t \underset{\sim}{c}^t_d(\rho_1) Q_t^{\mathrm{ES}}\right] \tag{22}$$

where $\underset{\sim}{r}^{CM}$ is the radius vector of the centre of nuclear mass; $\underset{\approx}{S}(\Theta)$ is the 3×3 orthogonal matrix with determinant equal to +1, and the coefficients $\underset{\sim}{c}^{t}_{d}(\rho_1)$ satisfy the following linear constraints:

$$\sum_d m_d \, \underset{\sim}{a}_d(\rho_1) \equiv \underset{\sim}{0} \, , \tag{23a}$$

$$\sum_d m_d \, \underset{\sim}{a}_d(\rho_1) \times \underset{\sim}{c}^{t}_{d}(\rho_1) \equiv \underset{\sim}{0} \, , \tag{23b}$$

and

$$\sum_d m_d \frac{d \, \underset{\sim}{a}_d}{d\rho_1} \bullet \underset{\sim}{c}^{t}_{d}(\rho_1) \equiv 0. \tag{23c}$$

It is also required that

$$V_{tt'}(\rho_1) \equiv \frac{1}{\mu} \, \omega_t^2(\rho_1)\delta_{tt'} \tag{24}$$

and

$$G^{tt'}(\rho_1) \equiv \frac{1}{\mu} \, \delta_{tt'}, \tag{25}$$

where we put

$$g^{tt'}(\rho_1, 0) \equiv G^{tt'}(\rho_1). \tag{26}$$

For the purpose of the present analysis it is, however, more convenient to define the appropriate large-amplitude variables as functions of nuclear Cartesian coordinates $\underset{\sim}{\bar{r}} \equiv \{ \underset{\sim}{r}_1, \underset{\sim}{r}_2, \ldots, \underset{\sim}{r}_N\}$ by minimizing the following least-squares form [35]:

$$L(\tilde{\Theta}, \tilde{\rho}_1; \, \underset{\sim}{\bar{r}} \,) = \sum_d m_d \left| \underset{\sim}{r}_d - \underset{\sim}{r}^{0}_{d}(\underset{\sim}{r}^{CM}, \tilde{\Theta}, \tilde{\rho}_1) \right|^2 \tag{27}$$

with respect to all possible orientations $\underset{\approx}{S}(\tilde{\Theta})$ and all allowed geometries $\tilde{\rho}_1$ of the dynamical molecular model

$$\underset{\sim}{r}^{0}_{d}(\underset{\sim}{r}^{CM}, \tilde{\Theta}, \tilde{\rho}_1) = \underset{\sim}{r}^{CM} + \underset{\approx}{S}(\tilde{\Theta}) \, \underset{\sim}{a}_d(\tilde{\rho}_1). \tag{28}$$

The large-amplitude variables Θ and ρ_1 are then chosen to coincide with the values of $\tilde{\Theta}$ and $\tilde{\rho}_1$ at which the form reaches its absolute minimum. It can be then easily shown [35] that the appropriate extremum conditions are equivalent to the Eckart–Sayvetz conditions. The author refers to the coordinate ρ_1 defined in such a way as "least-squares" large-amplitude internal variable to emphasize that it can be defined via the least-squares form (27) without any reference to the Eckart–Sayvetz transformation (22).

It is essential that the "least-squares" large-amplitude internal variable $\rho_1 \equiv \rho_1^{LS}$ defined as a function of the Cartesian coordinates $\underset{\sim}{\bar{r}}$ remains the same, regardless of how one orientates the semirigid model $\underset{\sim}{\bar{a}}(\rho_1)$ in space for different values of the variable ρ_1. In fact, let us consider a model

$$\underset{\sim}{\bar{a}}'(\rho_1) = \overline{\underset{\approx}{R}(\rho_1) \, \underset{\sim}{a}(\rho_1)} \tag{29}$$

which is obtained from the model $\underset{\sim}{\bar{a}}(\rho_1)$ by means of the rotation $\underset{\approx}{R}(\rho_1)$ dependent on the particular geometry ρ_1 of the model. By substituting $\underset{\sim}{\bar{a}}'(\rho_1)$ for $\underset{\sim}{\bar{a}}(\rho_1)$ in (28) and minimizing the least-squares form (27) with respect to $\tilde{\Theta}$ and $\tilde{\rho}_1$, one finds that the form has the minimum at exactly the same value $\tilde{\rho}_1 = \rho_1^{\text{LS}}(\underset{\sim}{r})$, as before. Therefore the only important point is how one chooses the reference geometries.

One thus comes to a new, rather broad family of mass-dependent variables. For example, substituting "the rigid bender" [3] for the static model $\underset{\sim}{\bar{a}}\,\rho_1)$ in (28) leads to the "least-squares" bending angle, the use of a model with internal rotation result in the "least-squares" torsional angle and so on.

The least-squares internal variables are a very important tool for freezing nuclear vibrations near slowly deforming geometries [36]. For example [36], by freezing the bending motion in a triatomic molecule one comes to the effective mass associated with the so-called "least-squares" bond lengths [36], rather than to the more familiar expression [37] obtained by differentiating a distance between two nuclei with respect to the nuclear Cartesian coordinates.

As pointed out by the author earlier [24], one can combine the least-squares internal variables with geometrical vibrational coordinates. The only requirement imposed on the choice of the latter coordinates is that they must vanish in any geometry allowed for the semirigid model $\underset{\sim}{\bar{a}}(\rho_1)$. For the rigid bender the possibility to use the symmetrized linear combinations of the deviations ΔR_1 and ΔR_2 of the bond lengths R_1 and R_2 from the equilibrium value R^e, instead of the rectilinear vibrational coordinates, has been already recognized by Hoy and Bunker [38] and more explicitly utilized by Jensen [39] to construct the so-called "Morse Oscillator-Rigid Bender Internal Dynamics (MORBID)" Hamiltonian.

Let us now consider an alternative definition of large-amplitude internal variable due to Quade [20]. (Further developments and applications of Quade's approach can be found in [40–43].) Again one starts from selecting geometries used as reference points for vibrations, so that $3N - 7$ vibrational coordinates vanish by definition at any point lying on the selected reference path. For example, the rigid bender can be again chosen as an approximate reference path for stretching vibrations in a triatomic molecule. The symmetric and antisymmetric combinations of the deviations ΔR_1 and ΔR_2 are then used as the corresponding normal coordinates.

Next step is to introduce an auxiliary geometrical variable τ used to construct the large-amplitude internal variable ρ_1 which satisfies the Hofacker–Marcus–Quade conditions (6), in contrast with the geometrical variable τ. In the case of large-amplitude bending motion in water, the notation τ may simply stand for the usual intrabond angle [43]. The new mass-dependent large-amplitude internal variable ρ_1 is then introduced via the relation:

$$\rho_1(\tau, Q) = \tau + \sum_{\beta} \Re^{\beta}(\tau) Q_{\beta}, \tag{30}$$

where the functions $\Re^{\beta}(\tau)$ are determined by the following linear equations:

$$g_{\beta}(\tau) + \sum_{\beta'} G^{\beta\beta'}(\tau)\Re^{\beta'}(\tau) = 0 \tag{31}$$

with

$$G^{\beta\beta'}(\tau) \equiv g^{\beta\beta'}(\tau, 0) \tag{32}$$

and

$$g_{\beta}(\tau) \equiv \sum_{d} \frac{1}{m_d} \left[\frac{\partial Q_{\beta}}{\partial \underset{\sim}{r}_d} \bullet \frac{\partial \tau}{\partial \underset{\sim}{r}_d} \right]_{\underset{\sim}{\bar{r}} = \underset{\sim}{\bar{a}}(\tau)}. \tag{33}$$

One can then easily verify that the variable ρ_1 indeed satisfies the Hofacker–Marcus–Quade conditions (6), as required.

For large-amplitude motions associated with a non-trivial molecular symmetry [44, 45] such as the inversion of the ammonia molecule [40] or internal rotation of a C_{3v} atomic group [20, 41] a special care should be also taken [24] to satisfy additional symmetry requirements which must be imposed on the coordinate τ to make the Hamiltonian invariant under the Feasible Permutation Inversion (FPI) group [44] (the molecular symmetry group in Bunker's terms [45]). However, it is relatively a side question in the present context.

The only essential point is that the use of the large-amplitude internal variable (30), instead of the least-squares coordinate ρ_1^{LS}, leads to a different set of harmonic frequencies at points other than equilibrium geometries; and the question is whether there is any additional criterion that allows one to prefer one harmonic approximation to another.

Recently the author [14] found a very remarkable feature of the least-squares large-amplitude internal variable, which may be used as an argument in support of a "privileged" status of the Eckart–Sayvetz coordinate system. To reveal the mentioned preferable feature of the least-squares large-amplitude internal variable, let us first point out that according to Equation (25b) in [9] the coefficients

$$\xi_{\beta'\beta''}(\rho_1) \equiv \left. \frac{\partial g^{\rho\beta'}}{\partial Q_{\beta''}} \right|_{Q=0} \qquad (\beta', \beta'' = 1, 2, \ldots, \Gamma) \tag{34}$$

evaluated in the Eckart–Sayvetz coordinate system form an antisymmetric matrix. Since any allowed change of variables has the form [9]

$$\rho_1 = \rho_1^{\mathrm{LS}} + \frac{1}{2} \sum_{\beta',\beta''=1}^{\Gamma} b_{\beta'\beta''}(\rho_1^{\mathrm{LS}}) + 0(Q^3), \tag{35a}$$

$$Q_{\beta} = \sum_{\beta'=1}^{\Gamma} C_{\beta\beta'}(\rho_1^{\mathrm{LS}}) Q_{\beta'}^{\mathrm{ES}} + 0(Q^2) \tag{35b}$$

one finds that the coefficients (34) evaluated in a ρ_1, Q coordinate system can be represented in general as

$$\sum_{\tilde{\beta}'=1}^{\Gamma} C_{\tilde{\beta}'\beta''}(\rho_1)\xi_{\tilde{\beta}\tilde{\beta}'}(\rho_1)$$

$$= \sum_{\beta=1}^{\Gamma} C_{\tilde{\beta}\beta}(\rho_1)\left(\xi^{\mathrm{ES}}_{\beta\beta''}(\rho_1) + \frac{1}{m}\, A_{\beta\beta''}(\rho_1) + \frac{1}{\mu}\, b_{\beta\beta''}(\rho_1)\right), \tag{36}$$

where the notation $\xi^{\mathrm{ES}}_{\beta\beta''}(\rho_1)$ is used for the coefficients (34) evaluated the Eckart–Sayvetz coordinate system, and

$$A_{\beta\beta''}(\rho_1) \equiv \sum_{\tilde{\beta}=1}^{\Gamma} C_{\tilde{\beta}\beta}(\rho_1)\frac{dC_{\tilde{\beta}\beta''}}{d\rho_1}\,. \tag{37}$$

(Note that the sum over γ' appears in error in the right-hand side of Equation (6) in [14].) Therefore, starting from the Eckart–Sayvetz coordinate system, one can always eliminate the coefficients (34) by making an orthogonal transformation from the rectilinear vibrational coordinates to the "adiabatic" vibrational coordinates:

$$Q^{\mathrm{ad}}_{\beta} = \sum_{\beta'=1}^{\Gamma} C_{\beta\beta'}(\rho_1^{\mathrm{LS}})Q^{\mathrm{ES}}_{\beta'}, \tag{38}$$

where the functions $C_{\beta\beta'}(\rho_1^{\mathrm{LS}})$ are solutions of the following differential equations [14]:

$$\sum_{\tilde{\beta}=1}^{\Gamma} C_{\tilde{\beta}\beta}(\rho_1)\frac{dC_{\tilde{\beta}\beta''}}{d\rho_1} + \xi^{\mathrm{ES}}_{\beta\beta''}(\rho_1) = 0. \tag{39}$$

The crucial point is that, due to existence of the symmetric term $\frac{1}{\mu}b_{\beta\beta''}(\rho_1)$ in the right-hand side of (36), the off-diagonal coupling term cannot be eliminated unless the variable ρ_1 coincides with the least-squares large-amplitude internal variable ρ_1^{LS} with the accuracy $0(Q^3)$.

In principle one can construct adiabatic vibrational coordinates by superimposing any set of normal vibrational coordinates, however, a definite advantage of using the Eckart–Sayvetz normal coordinates Q^{ES} comes from the fact that their use eliminates momentum coupling between normal vibrations in the vicinity of the reference path [9], namely,

$$\left.\frac{\partial g^{\beta\beta'}_{\mathrm{ES}}}{\partial Q^{\mathrm{ES}}_{\beta''}}\right|_{Q=0} \equiv 0, \tag{40}$$

where the symbols $g^{\beta\beta'}_{\mathrm{ES}}$ stand for the corresponding coefficients of momentum coupling evaluated in the Eckart–Sayvetz coordinate system. (Equation (40) has been used above to omit the last term in the right-hand side of the equation of motion (10).)

By combining the condition (8) with the requirement for the coefficient (34) to vanish at any value of the variable ρ_1 one finds that the Crystoffel symbols $\Gamma^{\beta}_{\rho\beta'}$ evaluated in the ρ_1^{LS}, Q^{ad} coordinate system are identically equal to zero at any point lying on the reference path. This remarkable feature of the ρ_1^{LS}, Q^{ad} coordinate system has been already used above to derive Equation (11) from the equation of motion (10).

A certain inconsistency [14] can be seen in the fact that, after the large- and small-amplitude internal variables have been separated using adiabatic vibrational coordinates Q^{ad}, the zeroth-order Schrödinger equation for vibrational degrees of freedom can be written in any other set of vibrational coordinates, including the normal vibrational coordinates Q^{ES}, provided that the large-amplitude internal variable remains unchanged. In other words the requirement for the coupling coefficients (34) to form an antisymmetric matrix turns out to be completely sufficient for the zeroth-order adiabatic separability, and the author simply does not know how to prove it.

As outlined above, a change in the definition of the large-amplitude internal variable generally affects both the frequencies and the shape of the normal vibrations near the given reference geometry, unless the selected geometry is a stationary point of the adiabatic potential governing nuclear motions. As a result, a change in the definition of the torsional angle for a vibrating molecule with internal rotation of atomic group may also change the symmetry of some normal modes with respect to the FPI group.

In fact, as initially recognized by Hougen [46] and by Bunker [47] for the dimethylacetylene molecule and then formulated in more general terms in our works [48–50], the symmetry classification of normal vibrations in molecules with internal rotations cannot be unambiguously performed until one diagonalizes the $\Gamma \times \Gamma$ matrix $\underset{\approx}{G}(\rho_1)\, \underset{\approx}{F}(\rho_1)$ at least at some values of the torsional angle ρ_1. This implies that a change in the definition of the torsional angle in the vicinity of the reference path will generally result in a change of the symmetry of normal vibrations. As explained below, the symmetry classification of normal vibrations becomes unique, if the reference path goes through at least two symmetrically non-equivalent stationary points of the potential which, in addition, have a higher point symmetry than that of other allowed configurations of the semirigid model in question. Since both the internal intrinsic path and the gradient extremal go through the equilibrium configurations of the molecules, the symmetry classification of normal vibrations becomes unambiguous for most molecules, if one of these two paths is used to construct the corresponding semirigid model (though one can still come with some counter-examples such as the CFHD-NO_2 molecule). On the other hand, the use of a quasi-equilibrium path may in principle lead to an unpredictable change in the symmetry of normal vibrations.

Let us start from a very brief introduction to the symmetry theory of normal vibrations developed in our works [24, 48–50]. First we require that any element of the FPI group convert the reference path in question onto itself. (Note that it is a priori unclear whether the reference paths discussed above satisfy this condition or not.) We refer to a feasible operation $\hat{g}'$ as "essentially trivial" if there is a "fixed" point $\rho_1 = \rho_1'$ which is kept unaffected by this operation. On the other hand, a feasible operation $\hat{g}'$ is referred as "non-trivial" if there is no fixed point and if this operation is represented by -1 in a unidimensional representation of the FPI group. The symmetry classification of normal vibrations may be ambiguous only if the FPI group has non-trivial elements.

Usually a non-trivial element of the FPI group can be represented as the product of two essentially trivial elements $\hat{g}'$ and $\hat{g}''$ with the fixed points ρ'_1 and ρ_1''. For example, the non-trivial element in the FPI group of the CF_3–NO_2 molecule with internal rotation of the nitro-group is the permutation (ab) of the oxygens. This permutation can be represented as the product of the permutation-inversions (12)*(ab) and (12)*, with (12) standing for the interchange of two fluorines. These permutation-inversions keep unchanged the configurations having the O–O bond either parallel or perpendicular to the bond between the fluorines 1 and 2.

To determine the symmetry of a normal vibration with respect to a non-trivial element it is thus sufficient to specify its symmetry with respect to both essentially trivial elements at the corresponding fixed points. Strictly speaking, one should first diagonalize the $\underset{\approx}{G}(\rho_1)\,\underset{\approx}{F}(\rho_1)$ matrix at intermediate points between ρ'_1 and ρ_1'' to correlate normal vibrations near a one higher-symmetry configuration to those near another. However, this step is usually avoided by assuming that there is no crossing between the curves $\omega_t^2(\rho_1)$ [46]. (If the semirigid model $\underset{\sim}{\bar{a}}(\rho_1)$ in question has a non-trivial point symmetry for any value of the variable ρ_1 then it is most often assumed that the curves $\omega_t^2(\rho_1)$ corresponding to normal vibrations of the same symmetry with respect to the point symmetry group of the model cannot cross each other [51].) With this assumption one only needs to diagonalize the $\underset{\approx}{G}(\rho_1)\,\underset{\approx}{F}(\rho_1)$ matrix at the higher-symmetry geometries ρ_1' and ρ_1''. (A change in the definition of the large-amplitude internal variable ρ_1 may generally lead to some crossings, however, the definition which makes it possible to avoid the crossings seems obviously preferable.)

Now we finally come to the point when we need to assume that the higher-symmetry geometries ρ_1' and ρ_1'' are the stationary points of the potential. In this case the shape of a normal vibration near geometries ρ_1' and ρ_1'' is independent of the choice of the large-amplitude internal variable ρ_1, so that the symmetry classification of normal vibrations with respect to the FPI group of *the selected isotopic modification* of a molecule with internal rotation becomes an intrinsic feature of the potential surface and nuclear masses. (As explicitly demonstrated by Groner [52] for the nitroethane molecule and its deutero-derivatives $CH_3CD_2NO_2$, $CD_3CH_2NO_2$, and $CD_3CD_2NO_2$ – see [24, 50] for a brief outline of this remarkable, never published work – isotopic substitution may change the symmetry of normal vibrations, even if the semirigid model retains its symmetry.) As mentioned above, both the intrinsic internal path and the gradient extremal must go through the corresponding equilibrium configurations, making the symmetry classification independent of the particular choice of the large-amplitude internal variable ρ_1. On the other hand, it is possible that the use of the quasi-equilibrium path (14), with the vibrational energy $E_{\text{vib}}(\rho_1)$ referred to the Eckart–Sayvetz coordinate system, would lead to different symmetry species, again raising the question whether there is any reason to prefer this coordinate system to all other possible choices.

Conclusions

As follows from the author's analysis, both questions, how to choose the reference path for vibrations and how to define the large-amplitude internal variable in the neighbourhood of the selected reference path, do not have a definite answer yet. The internal intrinsic path governed by Equation (3) seems to be the most natural solution so far, though its applicability to motions with the kinetic energy predominantly given to a large-amplitude internal degree of freedom remains questionable. The use of a quasi-equilibrium path determined by Equation (14) may be more appropriate in this case. However, such a choice of reference geometries makes them dependent on the definition of the large-amplitude internal variable since this definition affects the frequencies of normal vibrations. In addition, the quasi-equilibrium path does not go through equilibrium configurations, which may result in ambiguousness of the symmetry classification of normal vibrations with respect to the FPI group. As for the gradient extremals they should be considered solely as a useful numerical tool to approximate the intrinsic internal path provided that the latter has a relatively small curvature.

The possibility to eliminate the first-order momentum coupling between large- and small-amplitude internal motions by using the "adiabatic" vibrational coordinates [14] seems to be a very intriguing feature of the least-squares internal variable. However, some logical gaps in the cited arguments obviously require a further, more thorough investigation.

Coming back to the question formulated in the title of this paper, we obviously do not have a definite answer yet. However, the performed analysis throws some interesting new ideas which deserve to be tested from both logical and numerical points of view.

Acknowledgements

I am greatly indebted to Professor Yves G. Smeyers for his gracious encouragement, patience and constant assistance at different stages of my work on the manuscript.

References

[1] A. Sayvetz: "The kinetic energy of polyatomic molecules", *J. Chem. Phys.* **7** (1939) 383–389.
[2] C. Eckart: " Some studies concerning rotating axes and polyatomic molecules", *Phys. Rev.* **47** (1935) 552–558.
[3] J.T. Hougen, P.R. Bunker and J.W.C. Johns: "The vibration–rotation problem in triatomic molecules allowing for a large-amplitude bending vibration", *J. Molec. Spectrosc.* **34** (1970) 136–172.
[4] G.A. Natanson: *Modelling of nuclear motions in calculations of spectra of "non-rigid" molecules*, Ph.D. Thesis, Leningrad, U.S.S.R. (1974 in Russian).
[5] M.N. Adamov and G.A. Natanson: *Separation of variables describing motions of nuclei and electrons in "non-rigid" molecules*, Reprint of Inst. Theor. Phys. Acad. Sci. Ukr. SSR, ITF–76–82R, Kiev, USSR, (1976 in Russian).
[6] G.L. Hofacker: "Quantentheorie chemischer Reactionen", *Z. Naturforsch* **A18** (1963) 607–619.
[7] R.A. Marcus: "Local approximation of potential energy surfaces permitting separation of variables", *J. Chem. Phys.* **41** (1964) 610–616.
[8] K. Fukui: "The charge and spin transfer in chemical reaction paths", in *The World of Quantum Chemistry*, R. Dsudel, B. Pullman and D. Holland (Eds.), Reidel Publish. Co. (1974) p. 113.
[9] G.A. Natanson, "Internal motion of a non-rigid molecule and its relation to the reaction path", *Mol. Phys.* **46** (1982) 481–512.
[10] D.J. Rowe and A. Ryman: "Valleys and fall lines on a Riemannian manifold", *J. Math. Phys.* **23** (1982) 732–735.
[11] S. Kato, H. Kato and K. Fukui: "A theoretical treatment on the behavior of the hydrogen-bonded proton in malonaldehyde", *J. Amer. Chem. Soc.* **99** (1977) 684–691.
[12] M.V. Basilevsky: "Modern development of the reaction coordinate concept", *J. Molec. Struct. (Theochem)* **103** (1983) 138–152.
[13] H.M. Pickett: "Vibration–rotation interactions and the choice of rotating axes for polyatomic molecules", *J. Chem. Phys.* **56** (1972) 1715–1723.
[14] G.A. Natanson: "Optimum choice of the reaction coordinate for adiabatic calculations of the tunnelling probabilities", *Chem. Phys. Lett.* **190** (1992) 209–214.
[15] N.R. Walet, A. Klein and G. Do Dang: "Reaction path and generalized valley approximation", *J. Chem. Phys.* **91** (1989) 2848–2858.
[16] N. Shida, J.E. Almöf and P.F. Barbara: "Molecular vibrations in a gradient extremal path", *Theor. Chim. Acta* **76** (1989) 7–31.
[17] D.K. Hoffman, R.S. Nord and K. Ruedenberg: "Gradient extremals", *Theor. Chim. Acta* **69** (1986) 265–279.
[18] M.V. Basilevsky and A.G. Shamov: "The local definition of the optimum ascent path on a multi-dimensional potential energy surface and its practical application for the location of saddle points", *Chem. Phys.* **60** (1981) 347–358.
[19] G.A. Natanson: "A reduction of the reaction path formalism to the space of internal variables", *Chem. Phys. Lett.* **178** (1991) 49–54.
[20] C.R. Quade: "The interaction of a large amplitude internal motion with other vibrations in molecules. The effective Hamiltonian for the large amplitude motion", *J. Chem. Phys.* **65** (1976) 700–705.
[21] J.R. Villarreal, L.E. Bauman, and J. Laane: "Ring-puckering vibrational spectra of cyclopentene-d_1 and cyclopentene–1,2,3,3-d_4", *J. Phys. Chem.* **80** (1976) 1172–1177.
[22] L.E. Bauman, P.M. Killough, J.M. Cooke, J.R. Villarreal and J. Laane: "Two-dimensional potential energy surface for the ring puckering and ring twisting of cyclopentene-d_0, –1-d_1, –1,2,3,3-d_4, and -d_8", *J. Phys. Chem.* **86** (1982) 2000–2006.
[23] G.A. Natanson: "A new definition of the reaction coordinate via adiabatic dividing surfaces formed by classical trajectories", *Chem. Phys. Lett.* **190** (1992) 215–224.
[24] G.A. Natanson: "On invariance of localized Hamiltonians under feasible elements of the nuclear permutation-inversion group", *Adv. Chem. Phys.* **58** (1985) 55–126.
[25] R.T. Skodje, D.G. Truhlar and B.C. Garrett: "A general small-curvature approximation for

transition-state-theory transmission coefficients", *J. Phys. Chem.* **85** (1981) 3019–3023.
[26] R.A. Marcus: "On the analytical mechanics of chemical reactions. Quantum mechanics of linear collisions", *J. Chem. Phys.* **45** (1966) 4493–4499.
[27] R.A. Marcus: "On the analytical mechanics of chemical reactions. Classical mechanics of linear collisions", *J. Chem. Phys.* **45** (1966) 4500–4504.
[28] G. Do Dang, A. Bulgac and A. Klein: "Determination of the collective Hamiltonian in a self-consistent theory of large-amplitude adiabatic motion", *Phys. Rev.* **36** (1987) 2661–2671.
[29] R.R. Newton and L.H. Thomas: "Internal molecular motions of large amplitude illustrated by the symmetrical vibration of ammonia", *J. Chem. Phys.* **16** (1948) 310–319.
[30] G.A. Natanson: "A choice of the semirigid bender model of the water molecule", *J. Molec. Spectrosc.* **95** (1982) 63–67.
[31] F.B. Brown, S.C. Tucker and D.G. Truhlar: "Semiclassical reaction-path methods applied to calculate the tunnelling splitting in ammonia", *J. Chem. Phys.* **83** (1985) 4451–4455.
[32] P. Bopp, D.R. McLaughlin and M. Wolfsberg: "Variational calculations of the lower vibrational energy levels of the ammonia molecule", *Z. Naturforsch.* **37a** (1982) 398–400.
[33] B. Maessen, P. Bopp, D.R. McLaughlin and M. Wolfsberg: "An improved variational calculations of the lower vibrational energy levels of the ammonia molecule", *Z. Naturforsch.* **39a** (1984) 1005–1006.
[34] A. Klein, N.R. Walet and G. Do Dang: "Classical theory of collective motion in the large amplitude, small velocity regime", *Ann. Rev.* **208** (1991) 90–148.
[35] G.A. Natanson and M.N. Adamov: "Hamiltonian of a polyatomic molecule. II. Eckart–Sayvetz conditions", *Vestn. Leningr. Univ.* **10**, (1974, in Russian) 24–32.
[36] G.A. Natanson: "Comment on: Decoupling of the local mode stretching vibrations of water through rotational excitation. I. Quantum mechanics", *J. Chem. Phys.* **88** (1988) 7252–7253.
[37] E.B. Wilson Jr., J.C. Decius and P.C. Cross: *Molecular Vibrations*, McGraw-Hill, New York, (1955).
[38] A.R. Hoy and P.R. Bunker: "The effective rotation-bending Hamiltonian of a triatomic molecule, and its application to extreme centrifugal distortion in the water molecule", *J. Molec. Spectrosc.* **52** (1974) 439–456.
[39] P. Jensen: "A new morse oscillator-rigid bender internal dynamics (MORBID) Hamiltonian for triatomic molecules", *J. Molec. Spectrosc.* **128** (1988) 478–501.
[40] D.H. Cress and C.R.Quade: "The interaction of inversion with other vibrations in ammonia", *J. Chem. Phys.* **67** (1977) 5695–5701.
[41] C.R. Quade: "Contributions of the interaction of internal rotation with other vibrations in the effective potential energy for internal rotation in molecules with symmetric internal rotors", *J. Chem. Phys.* **73** (1980) 2107–2114.
[42] Y. Guan and C.R. Quade: "Curvilinear coordinate formulation for vibration–rotation large-amplitude internal motion interactions. I. The general theory", *J. Chem. Phys.* **84** (1986) 5624–5638.
[43] Y. Guan and C.R. Quade: "Curvilinear coordinate formulation for vibration–rotation large-amplitude internal motion interactions. II. Application to the water molecule", *J. Chem. Phys.* **86** (1987) 4808–4823.
[44] H.C. Longuet-Higgins: "The symmetry groups of non-rigid molecules", *Mol. Phys.* **15** (1963) 445–460.
[45] P.R. Bunker: *Molecular Symmetry and Spectroscopy*, Academic Press, New York (1979).
[46] J.T. Hougen: "Vibrational motions in dimethylacetylene", *Can. J. Phys.* **43** (1965) 935–954.
[47] P.R. Bunker: "Dimethylacetylene: An analysis of the theory required to interpret its vibrational spectrum", *J. Chem. Phys.* (1967) **47** 718–739.
[48] M.N. Adamov and G.A. Natanson: "New aspects in the symmetry theory of polyatomic molecules", *Fiz. Molek.* **2** (1976) 3–16 [see Los Alamos Translation LA-TR-77-32].
[49] G.A. Natanson: "Classification of normal vibrations in nonrigid molecules", *Fiz. Molek.* **6** (1978) 3–32.
[50] G.A. Natanson: "On symmetry classification of normal vibrations in molecules with internal rotation", in *Symmetries and Properties of Non-Rigid Molecules: A Comprehensive Survey*, by J. Maruani and J. Serre (Eds.), Studies in Phys. & Theor. Chem., Vol. 23 (1983), pp. 201–

218.

[51] A.J. Merer and J.K.G. Watson: "Symmetry considerations for internal rotation in ethylene-like molecules", *J. Molec. Spectrosc.* **47** (1973) 499–514.

[52] P. Groner: *Gas- und Matrixinfrarotspektren von Nitroethan-Isotopen und der Einfluss der Internen Rotation der Nitrogruppe*, Ph.D. Thesis, Diss. No. 5394, Swiss Fed. Inst. Technology, Zürich, Switzerland (1974).

GENERALIZING THE MOLECULAR SYMMETRY GROUP OF LONGUET-HIGGINS TO ASYMMETRIC TUNNELLING PROBLEMS

RICHARD G.A. BONE*
Department of Chemistry,
McMaster University,
Hamilton, Ontario,
Canada L8S 4M1

Abstract. This article describes the relationship of the Molecular Symmetry (MS) group of Longuet-Higgins and, ultimately, the Complete Nuclear Permutation-Inversion (CNPI) group to a molecular potential energy surface. The characterization of energy levels when "structural degeneracies" are lifted by finite tunnelling probabilities is outlined. Cosets of the MS group are used to obtain limiting numbers for the occurrence of structures of a given symmetry. It is shown how the conservation of point group symmetry elements along steepest descent pathways across the potential energy surface provides a useful framework for understanding isomerization. These ideas lead us to a generalization of the MS group to cases where a molecule samples more than one distinct structural form on the timescale of spectroscopic resolution. Tunnelling probabilities under these conditions are discussed and the circumstances under which the MS group could adequately describe "asymmetric" interconversions are explored. A number of simple examples is presented.

Outline

The basic tenet of this paper is one dating back to Curie [1], viz: "The characteristic symmetry of a phenomenon is the maximal symmetry compatible with the phenomenon". Although that might appear to be a straightforward concept, the steps that lead to its application in the context of non-rigid molecules are quite involved.

* Current address: Terrapin Technologies, Inc., 750-H Gateway Blvd., South San Francisco, CA 94080-7020, USA

Y.G. Smeyers (ed.), Structure and Dynamics of Non-Rigid Molecular Systems, 67–96.

In Section 1 of this paper, the origin of the permutational symmetry of the nuclear configurations of molecules is briefly discussed. In Section 2 the ramifications that this has for solutions to the Schrödinger equation are outlined. The apparent localization of wave functions and the communication between such wave functions manifested as observable tunnelling splittings are described [2]. In Section 3 the Molecular Symmetry group of Longuet-Higgins [3] is naturally introduced as the appropriate vehicle for classifying energy levels perturbed by tunnelling [4] and the subdivision of a potential energy surface into various domains [5], according to the criterion of feasibility [3], is outlined. The topology of a given domain is described, in a way which introduces such concepts as are necessary to predict the pattern of tunnelling splittings. This section is concluded with a mathematical derivation of the dimension of the MS group for a known interconversion. Section 4 demonstrates that the maximum level of point group symmetry present in any domain is well-defined and that, also, the number of structures of a given nature (geometry, Hessian index, etc.) is also limited by simple mathematical rules. The problems associated with deriving the MS group when geometrically distinct local minima are interconverting is highlighted.

Section 5 applies geometric symmetry concepts to paths of interconversion between structures and emphasizes the importance of the transition states for such processes. Section 6 describes the occurrence and characterization of symmetrically equivalent pathways and reinforces the notion that "symmetry-breaking" does not violate "symmetry-conservation" criteria. Section 7 discusses the possibility of tunnelling through asymmetric barriers and uses the foregoing discussions to demonstrate the applicability of the Longuet-Higgins Molecular Symmetry group [3]. A number of schematic cases are described, followed by a number of real molecular examples. The paper concludes with a summary in Section 8.

At this stage it is fair to note that since Longuet-Higgins's contribution there have been a number of other attempts to define and construct symmetry groups for non-rigid molecules. In particular, the real meaning of "feasibility" has been subjected to detailed analysis, mainly by Dalton [6]. In another respect the development of a group theory based on permutations and permutation-inversions has provoked many new definitions of symmetry operations and many reinterpretations thereof. Altmann was the first to do this [7], though based on what is now known to be a flawed assumption. Smeyers has devised a "Non-Rigid Group" (see, e.g., [8] and references therein) by synthesizing salient features of Altmann's and Longuet-Higgins's approaches. Smeyers and other workers [9, 10] have tried to keep the definition of "non-rigid" symmetry operations closely allied with the Hamiltonian for the problem, expressed in a coordinate system which is appropriate for the large-amplitude motions involved.

Nevertheless, for this present article, Longuet-Higgins's original formalism remains the simplest and most convenient framework for development and discussion.

1. Introduction

The indistinguishability of identical particles in quantum mechanics has profound implications for molecular behaviour [11]. The non-relativistic molecular Hamiltonian, $\mathcal{H}$, in configuration space for an isolated molecule, (comprising N nuclei and n electrons), is (in atomic units):

$$-\frac{1}{2}\sum_{i=1}^{n}\nabla_i^2 - \frac{1}{2}\sum_{a=1}^{N}\frac{\nabla_\alpha^2}{m_\alpha} - \sum_{i=1}^{n}\sum_{\alpha=1}^{N}\frac{Z_\alpha}{r_{i\alpha}} + \sum_{i<j}\frac{1}{r_{ij}} + \sum_{\alpha<\beta}^{N}\frac{Z_\alpha Z_\beta}{R_{\alpha\beta}}. \tag{1}$$

(Here the usual notation is employed: lower case Roman and Greek suffixes denote electronic and nuclear labels respectively; Z_α, is the charge on nucleus α; the inter-particle distances are lower case for electronic and upper case for nuclear coordinates.)

The fact that the molecular Hamiltonian is totally symmetric with respect to the permutation of the coordinates of identical particles and the inversion of all particle coordinates through some origin has led to the definition of the Complete Nuclear Permutation-Inversion (CNPI) group, G_{CNPI} [4] for the classification of molecular energy states. G_{CNPI} is the direct product group formed between the Complete Nuclear Permutation (CNP) group (the product of the symmetric groups for each set of identical particles present) and the inversion group (which consists of just two elements, the identity and the inversion operation, E^*). All of the operations in the CNPI group preserve inter-particle distances. It should be clear also that the CNPI group is independent of any reference structure.

It has been shown that an operation in the point group of any non-linear molecule may be represented by an element of G_{CNPI} [12, 13]. (Linear molecules are discussed elsewhere [14]). Specifically the relationship is such that proper rotations correspond to permutations and improper rotations correspond to permutation-inversions, in particular the principal plane of symmetry of a planar molecule becomes E^* [15]. Thus it is possible to construct a group, G_{PG}, which will be a subgroup of G_{CNPI} and which will be isomorphic to the point group of the structure under consideration [4, 13].

That we may attach labels to nuclei gives rise to the occurrence of "distinctly labelled forms" of a molecule. These have also been called variously "symmetrically equivalent nuclear equilibrium structures" [4], "configurations" [7], "permutational isomers" [16] and "versions" [17]. We will use the last of these and reserve the term "structure" to mean a particular nuclear geometry, i.e., each structure will have a number of distinct *versions*. For example, there are two *versions* of the pyramidal structure of ammonia. Therefore, different *versions* correspond to distinct points on the (3N–6)-dimensional potential energy surface of the system and are related to one another by elements of the CNPI group. Different "structures" are also distinct points on the potential energy surface but no symmetry operation may transform one structure into a different structure.

In general we will not be interested in all structures. Chemistry may largely be understood in terms of those structures for which gradients of the energy vanish with respect to all geometric distortions. These are stationary points on the potential energy surface. It is not our purpose here to review their occurrence or characterization (there are many comprehensive discussions in the literature [18]), except to say that at each we will assume that there are $(3N-6)$ non-vanishing eigenvalues of the second derivative matrix (or $(3N-5)$ if the structure is linear). At minima, all of those eigenvalues are positive; at transition states, one is negative [19]. We view chemical reactions, isomerizations, in fact any interconversion, as proceeding from an equilibrium structure (at a minimum), along a path to a transition state and thence to another minimum.

There are many molecules, however, for which the notion of a single "structure" is inappropriate, a fact which demands that we look for a special way of classifying their energy levels.

2. Stationary States of the Hamiltonian

The invariance of the molecular Hamiltonian to the operations of the CNP group requires that we should construct solutions which demonstrate that symmetry. Usually that is not reasonable. Similarly, the inversion-symmetry of the Hamiltonian suggests that its eigenfunctions should also be symmetric or anti-symmetric with respect to inversion of all particle coordinates. And again, it is common for molecular wave functions to be of mixed handedness. This is because we typically do not observe stationary states of the molecular Hamiltonian. Consequently, as remarked by Berry [2], any molecule which is observed to have a fixed geometry, for example one in which identical nuclei can be identified with symmetrically inequivalent sites in its structure, must be in a non-stationary state. Our ability to "trap" a species in such a state [20–22] depends on our timescale of measurement because there is an inherent uncertainty in energy resolution associated with the duration of the experiment. Accordingly, a sufficiently fast measurement may yield results pertinent to a mixture of stationary states [20]. The quantum mechanical picture [23] is that probability densities are localized in some regions of the nuclear configuration space.

Exact solutions, $\{\psi_i\}$, to the full molecular Hamiltonian, $\mathcal{H}$, transform as irreducible representations, Γ_i, of the CNPI group. We consider the operations of some group, G_{PG}, to pertain to the geometric symmetry of some equilibrium geometry. Suppose that a good approximation will be to represent the global potential as a sum of n "localized" potentials, each corresponding to a distinct *version* of an equilibrium structure [24]:

$$V_{\mathrm{global}} = \sum_{i=1}^{n} V_i^{\mathrm{local}}, \tag{2}$$

(a value for n will be derived later) and if a local Hamiltonian, h_i, contains a local potential, then

$$h_i\phi_i = \varepsilon_i\phi_i. \tag{3}$$

Here, the ϕ_i may be taken as ro-vibronic wave functions localized at each *version*. Eigenfunctions, $\{\phi_i\}$, of a local Hamiltonian whose potential energy is expanded about such a geometry, transform as irreducible representations, γ_i, of G_{PG}; the γ_i will not, in general, form a one-to-one correspondence with the Γ_i. Consequently, this (approximate) Hamiltonian will only be diagonal in a basis of functions which have the symmetry of the subgroup, i.e., the ϕ_i. Such considerations lead to a measure of the extent to which such a basis represents a set of non-stationary states [2]. Similarly the matrix of the global Hamiltonian in the basis of localized eigenfunctions is non-diagonal. The non-zero off-diagonal matrix elements, $\mathcal{H}_{ij}$, have the effect of mixing true stationary states, and their magnitude dictates the appropriate Hamiltonian for the problem.

For the majority of molecules under the usual conditions of observation, a description in terms of non-stationary states is both necessary and sufficient. The energy spectrum of each of the h_i is identical – that of any one of the *versions*. Therefore there is a "structural degeneracy" [4]. The degeneracy may manifest itself as a statistical factor (for example in a reaction probability [25–27]) but otherwise it is irrelevant: if we are to associate a given level with a particular state, (i.e., label it), it does not matter which of the several localized functions we choose. We cannot distinguish them; each represents the same limit of our need to approximate.

By associating the transition from a localized to a global description with the existence of off-diagonal terms in the Hamiltonian matrix, we find that otherwise degenerate levels become split by an amount which is precisely the $\mathcal{H}_{ij}$ (or some small multiple of them). By writing them in the localized basis (i.e., assuming that such a basis is a good first approximation),

$$\langle\phi_i \mid \mathcal{H} \mid \phi_j\rangle \tag{4}$$

we see that their magnitude is related to the overlap of the wave functions for different *versions*. If such *versions* are separated by some energy barrier, then the wave functions must penetrate that barrier in order to perturb one another. This is a non-classical effect, referred to as "tunnelling" [20].

There are many recorded attempts to simplify tunnelling-type problems in order to show a dependence of the splitting upon physical parameters such as mass. Most involve approximations to a one-dimensional form for the motion. Expressions for tunnelling matrix elements have been derived, the most famous of which is the Wentzel–Kramers–Brillouin–Jeffreys (WKBJ) approximation [28–31]. In that model, the magnitude of the tunnelling splitting is given by $h\nu_t/2$, with

$$\nu_t = \nu_0/A^2 \tag{5}$$

where ν_0 is the vibrational frequency along the tunnelling coordinate at each minimum, and

$$A = \exp\left\{\frac{1}{\hbar}\int_{s_1}^{s_0}[2\mu(V(s) - W)]\mathrm{d}s\right\} \tag{6}$$

with μ being the reduced mass for the motion, W the total vibrational energy in each well and s_0 the value at which the potential, V, has the energy of the ground-vibrational state. The critical feature of this description is the sensitivity to the height of the barrier. The "width" of the barrier has been perceived to be another important factor.

The tunnelling splitting has an associated frequency; this has been interpreted (classically) as the frequency of interconversion from one version to another. This is understood most clearly from the point of view of timescale of observation [2, 32]. In order to be able to resolve the splitting, we must make a measurement which is slow enough to allow the system to sample both (or all of the) *versions*. Alternatively, the precision in the energy measurement which allows us to see the effects of tunnelling demands a high uncertainty in the time taken to carry it out.

3. The Molecular Symmetry Group

3.1. THE CONCEPT OF FEASIBILITY

In a classic paper in 1963, Longuet-Higgins [3] identified the "feasibility" of a molecular rearrangement with the barrier to the change. Subsequently, Bunker [4, 33] associated it with the ability to resolve tunnelling splittings. (Natanson [24] has recently suggested that the "calculational accuracy" is an equally useful criterion.) The correct symmetry group to use for classifying molecular energy levels in all cases is the Molecular Symmetry (MS) Group, G_{MS}. It comprises just the "feasible" elements of G_{CNPI}, i.e., the collection of operations which relate those *versions* accessible to one another on the timescale of observation. Therefore, in the spirit of Bunker's remark [33], we will use "as much [symmetry] as necessary but not as much as possible".

It need not be the case that every *version* of an equilibrium structure will contribute to an observed splitting. Therefore, using a terminology introduced by Watson [5], we subdivide the global potential energy surface into a number of *domains*, each of which may contain a number of *versions*. There is no observable tunnelling between *versions* associated with different domains. (There may, of course, be a hierarchy of domains, or the complete potential energy surface itself may be a unique domain.) Accordingly, the Hamiltonian splits up into a sum of equivalent Hamiltonians, each localized in a separate domain [24]. The MS group is pertinent to a particular domain and elements of it relate *versions* within that domain to one another. The MS group for one domain will be isomorphic with the MS group for any other domain defined by the same criterion of feasibility. In the so-called "rigid-limit", where the tunnelling from any one version to any other is immeasurably small, the MS group is just G_{PG}. Therefore, in general, we have:

$$G_{PG} \subseteq G_{MS} \subseteq G_{CNPI}. \quad (7)$$

The correlation of energy levels (and their symmetries) from G_{PG}, through G_{MS} to G_{CNPI} has been outlined by Watson [5].

In the most widespread application of the MS group, the appropriate criterion of feasibility has been inferred intuitively from the "chemistry", i.e., a knowledge or good assumption of the relative magnitudes of the various barriers on the potential energy surface. Subsequent workers (in particular, Dalton [6], Pedersen [34], Natanson [35] and Watson [5, 36]) have attempted to address the physics which underlies the feasibility concept though Berry's discussion [2] (which predates Longuet-Higgins's paper) is probably the clearest. Dalton's approach is unique in that it seeks to identify significant off-diagonal matrix elements, (as discussed above) and therefore is an "ascending" process. Here, we have taken the opposite standpoint and look for reasons to exclude "unfeasible" operations.

3.1. TOPOLOGY OF THE POTENTIAL ENERGY SURFACE

We may assign any *version* of a given structure in a domain the status of "reference". Successive application of feasible operations to the reference generates a closed set of *versions* in that domain. Assuming that we understand the physical mechanism of interconversion, we may identify pairs of *versions* which are related by a single "step". The members of the pair are "adjacent" to one another in that domain, and usually may not be adjacent to one another under any other criterion of feasibility. But the interconversion between any pair of *versions* within one domain is represented by a feasible operation regardless of how many steps separate them – i.e., each *version* is not necessarily adjacent to every other in the domain. The "connectivity", C, of an interconversion type is defined to be the number of *versions* adjacent to any one *version*; it is the same for any *version* under a given interconversion type. A step will usually cross a single barrier on the potential energy surface; the path is its physical corollary. In this way, this definition of a "step" is closely allied to the definition of a "concerted" reaction suggested by Dewar [37].

It is possible to generate an "adjacency matrix" [38], $\mathbb{A}$, which is a symmetric square matrix whose dimension is the number of *versions* in the domain. Its diagonal elements are not defined; its other elements, $\mathbb{A}_{ij}$, are zero except when *versions* i and j are connected by a single step, in which case $\mathbb{A}_{ij} = 1$. The Hamiltonian matrix follows directly by setting the diagonal elements to be the zero of energy of any given *version* and replacing each other non-zero entry by the "tunnelling matrix element", $\mathcal{H}_{kl}$. Tunnelling state wave functions are the eigenvectors of the Hamiltonian matrix in the localised basis. These vectors span all the *versions* in the domain [6] if non-degenerate; they may necessarily have zero-amplitude in some regions of the potential energy surface if spanning degenerate energy levels. The pattern of energy levels depends upon the topology of the potential energy surface, but the symmetries (i.e., necessary degeneracies) of the tunnelling state wave functions depend only on G_{PG} and G_{MS} [5]; there may be additionally, accidental degeneracies arising from the Hamiltonian matrix. We could ultimately resolve all structural degeneracies by setting matrix elements between *versions* separated by n steps ($n > 1$) to a small n-dependent parameter.

3.2. COUNTING STRUCTURES IN A DOMAIN BY USING COSETS

One fact which it is important to know is the number of *versions* of any given structure found within a given domain. Suppose that there are n *versions* of a structure, **Q**, in a domain, **D**, and that the point group of **Q** has order h_{PG}; then the MS group for **D** contains nh_{PG} operations [5, 6].

The set of h_{PG} operations $\{r_m^1, m = 1, 2, \ldots, h_{\mathrm{PG}}\}$ of a "reference version", $\langle 1 \rangle$, of **Q** constitute G_{PG}^1, a subgroup of G_{CNPI}. We fix our attention on a point on the potential energy surface near $\langle 1 \rangle$ and note that any of the r_m^1 will induce a change in the internal coordinates that takes that point to another point near $\langle 1 \rangle$. (They may also cause a change in the 3 coordinates describing the overall orientation in space, but that is not relevant to the present discussion. We specify points *in the region* of $\langle 1 \rangle$ because we accept that the structure is vibrating, but although the motions may instantaneously destroy its point group symmetry, the operations in G_{PG}^1 remain useful symmetry operations [3, 4, 33].) Suppose that we find a feasible symmetry operation, $\hat{\mathcal{F}}_k$, which takes the reference point to a point near *version* k (also within the domain **D**). It follows that the operations $\{\hat{\mathcal{F}}_k r_m^1$ for m =1, 2, ..., $h_{\mathrm{PG}}\}$, all generate points near $\langle k \rangle$. There are h_{PG} of these, and they comprise a left coset of the MS group (and also the CNPI group) with respect to G_{PG}^1. They also generate all the h_{PG} symmetry-related points near *version* k so there can be no other operations that generate the *version* $\langle k \rangle$ of this structure.

In this way we can associate one of the cosets of G_{MS} with respect to G_{PG}^1 with each of the versions of **Q** in **D**. It follows that the number of elements, h_{MS}, in G_{MS} is given by

$$h_{\mathrm{MS}} = nh_{\mathrm{PG}}. \tag{8}$$

Formally, G_{MS} is the union of the left cosets of G_{PG}^1:

$$G_{\mathrm{MS}} = G_{\mathrm{PG}}^1 + \hat{\mathcal{F}}_2 G_{\mathrm{PG}}^1 + \ldots + \hat{\mathcal{F}}_n G_{PG}^1 = \sum_{k=1}^{n} \hat{\mathcal{F}}_k G_{\mathrm{PG}}^1 \tag{9}$$

with $\hat{\mathcal{F}}_1 = \hat{E}$. We note briefly that it is a fundamental property of cosets that they are disjoint [39, 40].

This discussion holds regardless of the height of the energy barriers, so n can be the total number of *versions* of a structure, in which case

$$h_{\mathrm{CNPI}} = n_{\mathrm{TOTAL}} h_{\mathrm{PG}}. \tag{10}$$

Similarly, the number of domains, n_D, is given by

$$h_{\text{CNPI}} = n_D h_{\text{MS}}. \tag{11}$$

Now, suppose that **Q** is an equilibrium structure and interconversion tunnelling between n of its *versions* is observed, through m *version(s)* of a transition state, **T**. Since the same MS group is used to describe the latter, the same argument can be used to deduce the order, h_{TS}, of its point group, i.e.,

$$h_{\text{MS}} = m h_{\text{TS}} = n h_{\text{PG}}. \tag{12}$$

This allows us to limit consideration of transition state structures to those with a precise level of symmetry and has formed the basis of discussion elsewhere [17]. Of course, analogous relationships hold for all structures under consideration.

The simple identities (11) and (12) are fundamental; they reflect the underlying symmetry of the potential energy surface. In the domain under consideration the number of feasible operations is a constant. Therefore the higher the local (point) symmetry of a given structure, the fewer *versions* of that structure there can be.

4. Structures Present in a Single Domain

A single domain contains an infinite number of different structures but a finite number of *versions* of each. The set of feasible operations (the MS group) define the most symmetry that needs to be used within the domain (e.g., to describe the approximate local Hamiltonian and its solutions); this total symmetry may be regarded as a conserved quantity within the domain. The domain-localized Hamiltonian itself may be written in any coordinate system and in general the potential energy term need not be expressed in terms of displacements from some equilibrium nuclear configuration. The MS group operations will commute with both the potential energy and the kinetic energy operator. Consequently, we appreciate that no individual structure within this region assumes any particular significance. This matter was probably first stressed by Natanson (see ref. [24] and references therein).

In a different (and more typical) approach, Bunker [4], factors the Hamiltonian into small- and large-amplitude displacement variables and defines a standard configuration to be one in which all the former are set to zero, but the latter can take any value. This is therefore assessing the feasibility of CNPI operations from their effect on an approximate Hamiltonian (and is therefore similar to the "isometric" group approach of Frei *et al.* [9], the semi-rigid model of Ezra [10] and the "Non-Rigid Group" of Smeyers [8]). A consideration of the associated potential shows immediately that there is no particular "structure" of importance and therefore that the feasible operations are pertinent to a region of the potential energy surface (defined by the range of values that the large-amplitude displacement variable(s) may take).

Therefore the MS group contains "hidden" information pertaining to any structure in the domain – often this might be a transition state. Such a relationship could be said to have been known intuitively for a few "classic" cases. The best example is that of ammonia inversion (see ref. [41] and ref. [42] and refs therein) in which the two versions of the pyramidal (C_{3v}) structure interconvert through a planar equilateral triangular (D_{3h}) transition state; it turns out that the MS group for the problem is isomorphic to the point group D_{3h}. Similarly, cyclic organic molecules which undergo facile ring-puckering (via planar transition states) have MS groups isomorphic to point groups D_{nh} [43]. It has also been found, (as stated by Jucks *et al.* [44]) that in spectra of some van der Waals molecules, the tunnelling splittings have nuclear spin weightings characteristic of the point group of the transition state for the interconversion. For example, the MS group for HF dimer is isomorphic to C_{2h} [45]; two *versions* of its C_s equilibrium structure interconvert through a C_{2h} transition state.

In these cases then, the MS group is found to be isomorphic to a point group and frequently the point group of a structure with some chemical significance. Nevertheless, it is erroneous to suppose that there will always be a structure in the relevant domain of the potential energy surface with that point symmetry. For example, in $(H_2O)_2$, whose MS group is isomorphic to D_{4h} [46–48], the potential energy surface has a cubic symmetry [49]. A square-planar structure of two water molecules plays no conceivable part in the interconversion processes. Furthermore there are other cases e.g., $(NH_3)_2$ [50] and one of the internal rotation mechanisms in $(C_2H_2)_3$ [17] – where the MS group is not isomorphic to any point group at all. Metiu *et al.* [51], and later Wales [52], suggested that symmetric structures only occurred at stationary points on the potential energy surface. Certainly, within the realm of exotic chemical architecture, where structures of high symmetry prevail, totally symmetric modes of distortion correspond to dissociative pathways which clearly may be ruled out if the structure is known to represent a stable chemical entity anyway. Consequently it is not surprising that one might try to search for some high-symmetry structure with which to associate the MS group.

At this stage it is useful to discuss the notion of the "highest symmetry structure on the potential energy surface" [53], an idea which probably dates back to Liehr [54]. Strictly this should be taken to be the structure whose point group is isomorphic with the CNPI group. Unfortunately the largest CNPI group for which this is possible is that formed from the inversion group and the symmetric group of 4 objects, S_4. (This group is isomorphic with point group O_h.) The principle could be reduced to locating the structure whose point group is isomorphic with the MS group, but, as discussed above, this is not always chemically meaningful. Certainly in the majority of cases known to this author, the most symmetric structure, if it exists, usually has no part to play in any observable interconversion process.

The use of the MS group and its irreducible representations to classify complicated molecular spectra is now well established. Bunker demonstrates its use for a number of simple molecules [4, 33]. The plethora of van der Waals molecules which undergo interconversion tunnelling (see for example [45, 46, 55, 56]) has reinforced the place of the MS group in current molecular spectroscopy.

The overwhelming majority of examples of the application of the MS group are to describe interconversion between a number of *versions* of equivalent structures. The matter of describing systems which are sampling more than one distinct type of minimum-energy structure would appear to be more difficult and certainly has received little explicit attention, except by Dalton and Nicholson [57]. There are three principal reasons for this.

Firstly, the number of systems which undergo such interconversion on the time-scale of observation and which have been studied spectroscopically is very small indeed. However, it is becoming more and more common to use isotopically substituted species in order to confirm spectroscopic assignments. The lowering of (nuclear permutational) symmetry which results is, when accompanied by differences in zero-point contributions at different points, often sufficient to transform a simple problem to one of this complexity. A number of workers have assessed the influence of isotope effects on studying reaction dynamics. The MS [58] and CNPI [27, 59] groups have found application here, in which case the application of feasible operations when more than one distinct structure is important has been tacitly assumed. The simplest envisageable circumstances in which two or more geometrically distinct local minima (and *versions* thereof) are sampled, occur when more than one distinct vibrational motion in a molecule is associated with a low barrier to change. Smeyers and co-workers have analysed a number of such systems [8, 60, 61].

Secondly there is the obvious difficulty, within the currently accepted framework, of envisaging a "symmetry operation" which could interconvert geometrically distinct structures. Certainly, feasible operations of the CNPI group do not appear to do this.

Thirdly, there is the physical aspect of the precise circumstances which can give rise to observable tunnelling (see, for example, [62]). As will be discussed later, asymmetric tunnelling probabilities are much smaller than those between potential wells which are degenerate. A particular feature of most tunnelling probabilities, though, is their extreme sensitivity to the form of the path chosen. Therefore we should have some idea of the path that the system will follow during an interconversion. It will become apparent that the changes in point group symmetry that come about as a molecule exhibits non-rigidity are also dictated by the path.

5. Symmetry and Chemical Change

In this section we show how the deformations that a molecule undergoes during the course of isomerization may be described and rationalized by group theory. Although we will use point groups, the arguments apply equally well to the isomorphic permutation-inversion group, G_{PG}.

5.1. THE SYMMETRY OF THE REACTION PATH

It is now an accepted fact that the chemical reaction path is "totally symmetric". By this it is meant that the displacement coordinates (which define the direction of motion along the path) transform as the totally symmetric irreducible representation of the point group of the structure at any intermediate point on the path – i.e., one encountered between a minimum and a transition state.

Pearson [63–66] was the first to state this explicitly, based on a treatment by Bader [67]. It is not the purpose of this article to describe the many attempts to calculate and define the reaction path in chemistry. Briefly, however, we note that a large amount of discussion has been given to the idea of expanding the potential energy surface in the region of a stationary point in terms of normal coordinates and thence to generalize this procedure to any point on the potential energy surface [51, 68]. But in general, approximate normal coordinate analyses are inappropriate and it is best to use the "intrinsic reaction coordinate" (IRC) of Fukui [69]. Since the IRC conforms to the direction of the gradient of the potential energy surface with respect to geometric displacements, it must itself be totally symmetric [70].

Whichever choice of path we rely upon, it is concluded that, at a stationary point, where the gradient vanishes, we may not say anything about the symmetry of the reaction path, but anywhere else it must be totally symmetric. There is an important corollary of this: that the point group symmetry may not change along the reaction path, except at a stationary point. Exceptions to this are branching and bifurcation points discussed elsewhere [51, 71].

5.2. THE PECHUKAS THEOREMS FOR TRANSITION STATES

The transition state plays an important role in understanding any chemical reaction and there have been many attempts to understand "symmetry rules" that pertain to it [68, 72–75]. Pechukas [76] showed that for paths of steepest descent from a transition state,

(i) a (point group) symmetry of the transition state is a symmetry of the entire path,
(ii) the transition state symmetry can be no greater than the symmetry of reactant or product, except when
(iii) "reactants" and "products" connected by such paths are physically indistinguishable, in which case, the transition state may contain symmetry which is neither present on the rest of the path nor present in the reactant (product). In such cases, the transition state symmetry may be greater than that of the reactant.

It follows from (i) and (ii) [76] that (except when reactant and product are related by symmetry) the point group of the transition state can be at most the largest common subgroup of the reactant and product point groups, i.e.,

$$\mathcal{G}_T \subseteq \mathcal{G}_R \cap \mathcal{G}_P \tag{13}$$

where $\mathcal{G}_T$, $\mathcal{G}_R$, and $\mathcal{G}_P$ are the point groups of transition state, reactant and product respectively. Such a reaction, in which reactant and product are distinct structures, is called a "non-degenerate" rearrangement (NDR). (Other terminology has been used – particularly, "polytopal isomerizations" [77] and "topomerizations" [78]. The former, however, appears to refer strictly to the rearrangement of ligands on a defined skeletal framework; the latter has not gained such wide acceptance.)

On the other hand, from (iii), if the transition state contains a symmetry not present on the rest of the path, then such symmetry must be "broken" by the reactive motion at the transition state and thus it is a symmetry which relates each point on the path on one side of the transition state to a geometrically and energetically equivalent point on the other side. Therefore the minima which interconvert through such a transition state are physically indistinguishable. This sort of reaction is called a "degenerate rearrangement" [79].

In the adiabatic model of chemical reactions, (where it is assumed that we can separate motion along the reaction path from motions along orthogonal coordinates) it has been shown that the steepest descent path in mass-weighted coordinates coincides with the normal mode whose eigenvalue is negative at the transition state [80]. Therefore the normal mode symmetries at the transition state are useful for identifying the conserved and "broken" symmetry elements at the transition state.

5.3. SYMMETRY PROPERTIES OF NORMAL COORDINATES

It has been assumed thus far that structures of chemical interest are stationary points on the potential energy surface, specifically minima and transition states. One important feature of stationary points is that normal modes are rigorously defined there, and normal modes can be shown [81] to transform as irreducible representations of the point group of the structure. Therefore we may describe any distortion of an equilibrium structure or transition state along a normal mode with an irreducible representation label, and usually from the character table of the original structure alone, find the point group of the distorted structure. These have been tabulated for every symmetry species in all point groups [82].

In summary, the following principles apply to any structure, where Γ_1 is the totally symmetric irreducible representation and $\mathcal{G}$ is the point group of the structure under consideration. Normal coordinates transforming as

1. Any non-degenerate irreducible representation (other than Γ_1) will destroy precisely half of the symmetry operations present, i.e., will lead to a "halving subgroup" [83] of $\mathcal{G}$;
2. A degenerate representation (or, in the absence of external fields, a component of a pair of separably degenerate irreducible representations) will destroy

more than half of the symmetry operations in $\mathcal{G}$.

The first of these arises as follows. Suppose that the normal coordinate, with symmetry Γ, takes us from structure $\mathbf{Q}$, to structure $\mathbf{Q}^{\Gamma}$; the symmetry operations, $\Re$ of $\mathbf{Q}$ are all such that

$$\Re\mathbf{Q} = \mathbf{Q} \tag{14}$$

i.e.,

$$\Re\mathbf{Q} = \chi^{\Gamma_1}(\Re)\mathbf{Q}. \tag{15}$$

Therefore the symmetry operations of $\mathbf{Q}^{\Gamma}$ are those for which $\chi^{\Gamma}(\Re) = 1$ in the point group of $\mathbf{Q}$. Because non-degenerate irreducible representations must have exactly the same number of +1 characters as -1 (by the Great Orthogonality Theorem [84]), the subgroup is half the order of the point group of $\mathbf{Q}$ and is obtained by deleting the elements of $\mathcal{G}$ which have character -1 in Γ. This applies regardless of whether the point group has rotational symmetry elements of order greater than 2, i.e., for any A or B representation.

The recipe for identifying the subgroup is not so straightforward for degenerate irreducible (and some reducible) representations, though the excellent discussion of "kernels" and "epikernels" by Ceulemans *et al.* [85] is an aid. These concepts have found great utility in describing Jahn–Teller systems [85–87]. Indeed, it is possible to associate the different regions of the potential energy surface around a high-symmetry stationary point with kernels and epikernels of its point group with respect to its various vibrational symmetries.

The matter of identifying the subgroup for an arbitrary displacement is similar, provided that its symmetry may be represented as a sum of irreducible representations.

It follows that any path which breaks symmetry at a given structure must be related to some other equivalent path or paths by the symmetry element(s) which were destroyed [88]. These relationships may be expressed in terms of permutation-inversion operations.

5.4. REPRESENTATIVE PATH SEGMENTS

Finally we conclude that there are only two distinct types of "path segment" which connect two minima via a transition state on a continuous region of the potential energy surface. These are illustrated as follows. Motions orthogonal to the path may break symmetry. There is no general relationship implied here between the point group orders of points $\mathbf{Q}$ and the minima, $\mathbf{M}$.

1. Interconversion between *versions* of the same structure (Figure 1a). The order of the point group of the transition state (T) must be exactly twice that of neighbouring points ($\mathbf{Q}$ and $\mathbf{Q}'$) on steepest descent paths. This is because the normal mode of imaginary frequency at the transition state is non-degenerate and non-totally symmetric.

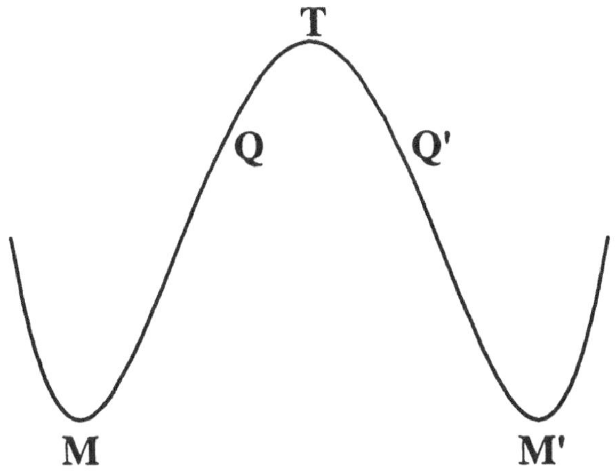

Figure 1a. Path segment for a degenerate isomerization. M and M′ are versions of the same structure.

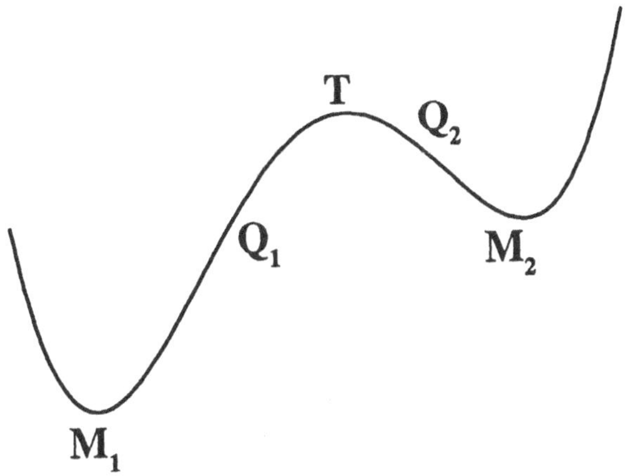

Figure 1b. Path segment for a non-degenerate isomerization. $\mathbf{M}_1$ and $\mathbf{M}_2$ are different structures in general but may be versions of the same structure. Motion at **T** has symmetry Γ_1.

2. For any isomerization (Figure 1b) though a non-degenerate one is assumed (i.e., one in which $\mathbf{M}_1$ $\mathbf{M}_2$ are different structures): from $\mathbf{Q}_1$ to **T** to $\mathbf{Q}_2$ all symmetry elements are preserved. ($\mathbf{M}_1$ and $\mathbf{M}_2$ may be *versions* of the same structure in which case the segment is asymmetric.)

It can be shown how *all* isomerizations may be analysed in terms of just these two path segments.

6. Symmetrically Equivalent Reaction Paths

Here we are able to relate the concept of the "connectivity" of an interconversion with the symmetry of the structure under consideration. The symmetry-breaking analysis applies to normal modes and to any displacement whose symmetry can be expressed as a reducible representation.

6.1. DEGENERACY-RELATED PATHWAYS

There may be displacements along several paths from a structure which are related by a rotational symmetry in its point group [88]. Such paths only derive from degenerate normal coordinates and the structures on each path lack the rotational symmetry in the parent structure. Displacements of opposite sign along each coordinate are only related by symmetry when some other symmetry element is destroyed, as below. These observations will apply separately for each component of a degenerate vibration because the components themselves are not necessarily related to one another by any symmetry operation.

6.2. PATHWAYS RELATED BY REFLECTION

Displacements along normal coordinates of planar structures either retain E^*, in which case the system remains planar, or do not. In the former case displacements of opposite sign are not necessarily related to one another by any (point) symmetry operation but in the latter case give rise to two symmetrically related pathways. These pathways are such that every point along one pathway is related to a point on the other by E^*. It is obvious that structures along the two paths have lower symmetry than the undistorted structure. If the coordinate transforms as a non-degenerate representation, i.e., preserves all rotational symmetries higher than 2-fold, then the distorted structure will have half of the symmetry of the parent structure.

In structures which are non-planar but have reflection symmetry, displacements in opposite directions along coordinates which are antisymmetric with respect to that plane of symmetry are related to one another by an improper rotation other than E^*.

Both of these behaviours may be a feature of degenerate as well as non-degenerate normal modes.

6.3. NON-DEGENERATE MODES WHICH DESTROY A 21-FOLD ROTATIONAL SYMMETRY

Suppose that a structure has a $2l$-fold rotation axis. If l is even, then normal coordinates which destroy the $2l$-fold rotation symmetry (but preserve the concurrent 2-fold symmetry for $l > 1$) [88] are related to one another by a pure permutation operation, P, (but not necessarily its corresponding permutation-inversion, P^*). If l is odd, then the normal mode is antisymmetric with respect to rotation about both the 2-fold and the $2l$-fold axis, but symmetric with respect to the concurrent l-fold rotations. These relationships must arise from non-degenerate modes (those of B symmetry).

It is therefore possible for pathways to simultaneously fall into more than one of the above categories. This is particularly the case in high symmetry. For example E'' modes of planar D_{3h} molecules can give rise to three rotationally equivalent distortions, each of which is further split into a pair of directions related by reflection in the σ_h plane.

6.4. PERMUTATIONALLY RELATED PATHWAYS

It can be possible for there to be n symmetrically equivalent paths from a structure without there being an element of n-fold rotation symmetry in the structure. This can usually be ascribed to the choice of coordinate system for the problem, e.g., it may be in internal coordinates which reflect a permutational symmetry.

Ethene provides a simple example. The four protons are, by a number of criteria, in geometrically equivalent positions in the planar (D_{2h}) equilibrium structure. But there is no axis of four-fold symmetry and any proton is only related to at most one other by a given point group symmetry operation. Therefore, although we might choose C–H bond stretches as internal coordinates and claim that the potential energy surface has a four-fold permutational symmetry with respect to dissociation to C_2H_3 + H, these channels are not degeneracy-related in the sense meant above. The collection of four equivalent C–H stretches transform as a reducible representation of G_{PG}.

7. Interconversions between Distinct Structures

It is now possible to deduce how one should describe non-degenerate isomerizations with the MS group. Dalton appears to be the only previous worker to have addressed this problem [57], but his result, although intrinsically the same as that given here, was arrived at by different means (detailed analysis of individual matrix elements) and there was no claim made that it was general. Furthermore, his result appears not to be widely known or appreciated [89].

We will show that the MS group is, in fact, no different from that used for degenerate isomerizations and that it is as large as the largest symmetry group required to describe a representative domain of the potential energy surface.

Bunker claimed [4] that since the G_{PG}'s for *cis-* and *trans*-dichloro-ethene contain the same operations, then G_{PG} itself was good for a 1,2-dichloro-ethene which underwent rapid torsional motion. The same author observed [4] (as did Liehr [54]), that the symmetry group for a Jahn–Teller system was that of the point group of the "most symmetrical geometry" on the surface. Therefore there is some precedent for our result.

7.1. ASYMMETRIC TUNNELLING PROBABILITIES

Although tunnelling, *per se*, will be a feature of all wave functions, the circumstances which permit an observable perturbation to result, are found to be quite restrictive (for a general discussion of tunnelling through an asymmetric barrier, see e.g., refs. [62, 91]). The "double-minimum" type potential, with which we deal here, i.e., in which both sides of the barrier represent bound states, is found to be a case which is especially sensitive to perturbations away from symmetry [62, 92–94]. In fact, a large number of 1-dimensional models show that an energy level in one well must be degenerate or nearly degenerate with that of the wave function in the other. This is a sufficiently severe restriction that tunnelling is only commonly observed between *versions* of the same structure. Here we outline why this should be so, for a model problem in which a single barrier separates two wells.

In order to define the problem, let us consider a hypothetical situation. In a symmetric double minimum potential, (Figure 1a), let us label the two halves "L" and "R", with respective localized wave functions ϕ_L and ϕ_R and unperturbed energies, $E_L = E_R = E$. If we allow the wave functions to tunnel through the barrier and mix with one another, in the usual way, we set up a 2×2 Hamiltonian matrix,

$$\begin{pmatrix} \langle \phi_L \mid \mathcal{H} \mid \phi_L \rangle & \langle \phi_L \mid \mathcal{H} \mid \phi_R \rangle \\ \langle \phi_R \mid \mathcal{H} \mid \phi_L \rangle & \langle \phi_R \mid \mathcal{H} \mid \phi_R \rangle \end{pmatrix} \tag{16}$$

and obtain two perturbed levels. The amplitudes of the delocalized wave functions,

$$\frac{1}{\sqrt{2}} (\phi_L \pm \phi_R) \tag{17}$$

are the same in both *versions* of the minimum. Both (17) are eigenfunctions of some operator, P_T, which interconverts the two *versions* and which is a symmetry operator of the transition state. Likewise, states which are above the barrier may be classified according to their symmetry about the transition state. The tunnelling splitting has been attributed to an "exact state of resonance" between the two wells [62].

If, now, we imagine that we may raise the right-hand well of the potential gradually, then the degeneracy of the unperturbed levels, E_L and E_R is lifted. Initially, the eigenfunctions of (16) are still substantially delocalized and we may retain P_T as a useful, if approximate, classifier of the levels. As the difference in the two wells increases, however, the extent of the mixing (the off-diagonal matrix element) diminishes and the eigenfunctions are, effectively localized in each of the two wells such that the lower energy solution is localized in the lower-energy well. If we may measure levels in one well, independent of the other, then we might as well use the full symmetry group of the structure to which that well pertains. If we can observe states of higher energy which are above the barrier, and, effectively delocalized, then the only useful symmetry operations are those which are common to the whole potential. The asymmetric potential is just that of Figure 1b and Pechukas's analysis [69] applies: the symmetry common to the whole path may be no more than that of the largest common subgroup of L and R.

It has been found that if one well is deepened by just 1% of the barrier height, the tunnelling splitting is decreased by a factor of 100 [62, 95]. The same analysis showed that the decrease in the tunnelling splitting was not due to any major change in the structure of the energy levels but by the ensuing localization of the wave functions in each well. That is to say, by becoming more localized, their propensity to penetrate the barrier and to overlap with one another is substantially reduced. Therefore the matter of whether *accidentally* degenerate levels in two wells of an asymmetric potential could interfere and give rise to a "spurious" tunnelling splitting is also rejected. Harmony has given an approximate formula for the tunnel-frequency, ν_t, in the limit of small asymmetry [95]:

$$\frac{\nu_t}{\nu_t^0} = \frac{2h\nu_t^0}{\sqrt{(h\nu_t^0)^2 + \Delta^2}} = \frac{2}{\sqrt{1 + (\Delta/h\nu_t^0)^2}} . \tag{18}$$

Here, ν_t^0 is the tunnelling frequency in the symmetric problem and Δ is the energy difference between the two minima.

We now attempt to put this reasoning in the wider context of the MS group, bearing in mind that there are three possible circumstances:

(i) When tunnelling itself is too small to be observable, physical interconversion is effectively "quenched" (signalled by the localization of probability densities), a state which is of no interest.

(ii) In order for interconversion between inequivalent structures to be "observed" the zero-point level must be very close to the top of the barrier, or above it. This may or may not lead to tunnelling splittings.

(iii) The asymmetric reaction profile may just be a segment of a longer chain of interconversion in which another *version* of one of the two distinct structures is encountered.

It will become clear how to describe the last two of these cases in the ensuing discussion.

7.2. A REDUCTIONIST VIEW OF THE POTENTIAL ENERGY SURFACE

It is clear that it is meaningless to attempt to find (or define) a feasible operation which relates a version of a particular structure with a version of another different structure. On the other hand, the operations in the MS (and hence CNPI) group are pertinent to many structures. There are now two perspectives. (Without loss of generality, we assume that the global minimum in each case plays some part in the observed process.)

1. Let the structure of the global minimum, **A**, in a particular domain, **D**, have a point group of order h_A with n_A *versions* in **D**. Let there also be some

local minimum, **B**, with h_B symmetry operations and n_B *versions* in **D**. The MS group for this domain has order $h_{MS} = n_A h_A = n_B h_B$, regardless of whether or not **A** and **B** may be feasibly interconverted. h_{MS} is regarded as the maximum amount of useful symmetry for the purpose of classifying the energy levels of the system.

2. Let us consider a representative path segment, as in Figure 1b. All point symmetry elements are preserved along the paths between the two minima. Therefore there are the same number of such path segments as there are transition states, **T**, in a given domain and we may look upon the feasible operations as interconverting path segments instead of individual (transition state) structures. This is tantamount to suggesting that the only points of the whole potential energy hypersurface that interest us are those points on steepest descent paths between second order stationary points.

From here we may show that the correct MS group is just the group of the domain, **D**, by consideration of a number of general cases. It should be noted that each general case is a model of typical cuts through the potential energy surface; it is quite possible that real systems may contain more than one of the general cases in juxtaposition. The extension to the more exotic occurrence of branching points should become clear.

Case 1: The Totally Symmetric Distortion
Suppose that the initial distortion of structure **A** is totally symmetric; it follows that B and **T** have the same level of symmetry as **A** and that only one *version* of each is involved. (It is also possible that a distortion of opposite sign from **A** may lead to a third structure **C** but the same arguments apply again. Note that distortions of opposite sign from **A** cannot be related to one another by a point group symmetry element of **A**.) Unless it is possible by some continued motion from **B** to reach another *version* of **A**, there is only one *version* of **A** and **B** per domain and elements of the CNPI group relate the various domains. Therefore the MS group has the same order of the point group of either of the structures. Although none of its operations may be interpreted as interconverting any of the structures, it is clear that all of them remain meaningful for classifying molecular energy levels because all elements of point group symmetry are maintained in the section of the potential energy surface under consideration. All wave functions transform as irreducible representations of the MS group.

Case 2: Other Distortions Which Belong to Non-Degenerate Irreducible Representations
If structure **A** undergoes a distortion of A (non-totally symmetric, e.g., A_2) or B symmetry, then the two signs of the motion are related by some other point group symmetry of A and the distorted structures have the same point group symmetry as each other – half of that of **A**. Any subsequent transition states, **T**, on the two

paths are different *versions* of one another (related themselves by symmetry), as are subsequent minima, **B**. Without loss of generality, we may assume that **B** has the same level of symmetry as **T**. This could be regarded as the "inverse" of the double-minimum curve (Figure 1a).

We may expand wave functions for this domain about its most symmetric point (**A**); alternatively we may localize wave functions in the two *versions* of **B** and construct sums and differences of them. All other point group symmetry operations of **A** are conserved along the entire path (which is a union of two *versions* of the basic path segment) and the wave functions we construct should reflect this.

Case 3: Distortions Belonging to Degenerate Irreducible Representations

If, without loss of generality, we consider **A** to have a 3-fold axis of symmetry, then, distortions (E-type) which break the three-fold symmetry lead to degeneracy-related pathways, on which structures have less than half of the symmetry of **A** (the distorted structures have point groups which are conjugate subgroups of the point group of **A**). Given that either of these motions is assumed feasible, it is meaningless to consider any of them in isolation; they will be related to one another by MS group operations which correspond to point group operations of **A**. It might be useful to expand wave functions about **A**; alternatively, linear combinations of wave functions localized at *versions* of minimum energy structures encountered further along the three path segments may be constructed which reflect the symmetry of the surface about **A**.

All conceivable examples of regular potential energy surfaces may be decomposed into one or more of the above cases. (It follows from this that, in a manner similar to that outlined elsewhere for a degenerate isomerization [17], the MS group can be of some assistance for locating transition states for non-degenerate isomerizations. We simply need to know the numbers of the two minima on the potential energy surface and a way of deducing the number of paths between them; there are as many transition states as there are paths. We must also know how many paths could arise from each minimum; this information may be obtained simply from the character table of its point group. Despite this, we still have no guarantee that the transition state will have any symmetry at all because there is no need for a symmetry operation to be generated at it.) It is the case that, even without assumptions as to the properties of a transition state [19], the same arguments will still apply because the coset rule with which we count *versions* of a structure applies regardless of the nature of the structure. Brief examples of this have been discussed elsewhere [17].

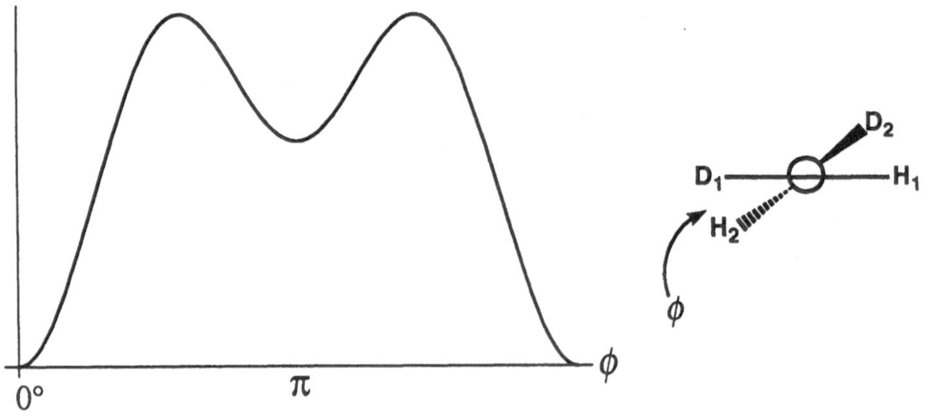

Figure 2. Torsional potential for $C_2H_2D_2$.

7.3. REPRESENTATIVE EXAMPLES

Here, we discuss a number of well-understood potential energy surfaces, each of which demonstrates the reasonability of the proposition.

Example 1: The HF ... DF van der Waals Complex
This would be an example of *Case 1*, above. The "Hydrogen-bonded" and "Deuterium-bonded" forms are at different energies due to their different zero point energy contributions [96–98]. Indeed, it is sufficient to mean that one form is observed in preference to the other [45, 99] and that tunnelling is effectively quenched between the two forms. (Certainly, no criteria of "exchange-symmetry" can now arise because the particles are different to one another.) Both forms have C_s point group, as does the transition state. Provided that the interconversion pathway in the isotopically substituted species is basically similar to that in the normal species, i.e., is an "in-plane" motion, then the system remains planar throughout. There is no reason to use any group other than the inversion group [4] to classify energy levels in this species. The potential energy curve looks like Figure 1b.

Example 2: Torsional Motion in 1,2-dideuterio-ethene
This corresponds to Bunker's discussion of $C_2H_2Cl_2$ [4], i.e., is an example of *Case 2*; here we see just why it works. We label the protons "1" and "2"; the deuterons "3" and "4" and the carbons "a" and "b". The collections of operations, G_{PG}, for structures on the path are as follows:

cis-planar; *trans*-planar	$\hat{E}$, $(12)(34)(ab)$, E^*, $(12)(34)(ab)^*$
all others	$\hat{E}$, $(12)(34)(ab)$

We assume that the two planar structures are both minima so that the potential has the form shown in Figure 2. It appears that the two-fold rotation symmetry is retained at every structure but because every point (except the two minima) is paired with its "labelled-atom enantiomer" [17], E^* remains feasible for the surface (as does, therefore, $(12)(34)(ab)^*$).

Example 3: Jahn–Teller Systems

As an example of *Case 3*, we note that it is common for Jahn–Teller systems to exhibit symmetry properties characteristic of the highest symmetry structure on the potential energy surface – a structure which frequently turns out to be that at which the conical intersection occurs. Clearly, as discussed, this will only necessarily be the case for systems with up to four identical nuclei. Where there is a possibility of permutational symmetry over and above maximal point group symmetry, as in octahedral or icosahedral systems, the potential energy surface is broken up into domains each of which contains at least one *version* of the structure with highest point group symmetry. Tunnelling between such domains need not be excluded.

We recall that in order for Jahn–Teller activity to be displayed, there must be a degenerate electronic state at a geometry at which a distortion exists which can lift that degeneracy. This will usually require a three-fold axis of symmetry or higher and the distortion itself must be degenerate. Therefore the point groups of the distorted geometries will be conjugate subgroups of the point group of the undistorted structure. Similarly, *versions* of the distorted structure will be interconverted by point group operations of the undistorted structure, i.e., MS groups of the domain. Good examples are provided by the $^2T_2$-state of CH_4^+ [100] and T_2 and E states of tetrahedral MX_4 metal complexes [86]. In the former, a T_d structure has two *versions*; assuming that they occupy non-interacting domains, there are six interconverting *versions* of the distorted C_{2v} geometry in each domain. (The topological representation is an octahedron; its twelve edges denote C_s symmetry structures.) The MS group is, of course, isomorphic to T_d. The MX_4 complexes have been discussed at some length in ref. [86]; suffice it to say that the analysis is consistent – for example an E-type distortion can lead to three *versions* of D_{2d} structures.

Thus, in general, no distortion exists in isolation, being always related to more than one another by a rotational symmetry in the point group of the undistorted structure. An appropriate MS group is just G_{PG} for the undistorted structure and its elements can be regarded as interconverting different versions of the lower-symmetry structures.

Example 4: Isomerization in H_2ABH Molecules

Molecules like hydroxylamine (A=N; B=O) may undergo internal rotation (about the AB bond) and inversion at the A-centre, and therefore may co-exist in *cis-* and *trans-* forms. This example therefore includes both *Case 1* and *Case 2*. The scheme

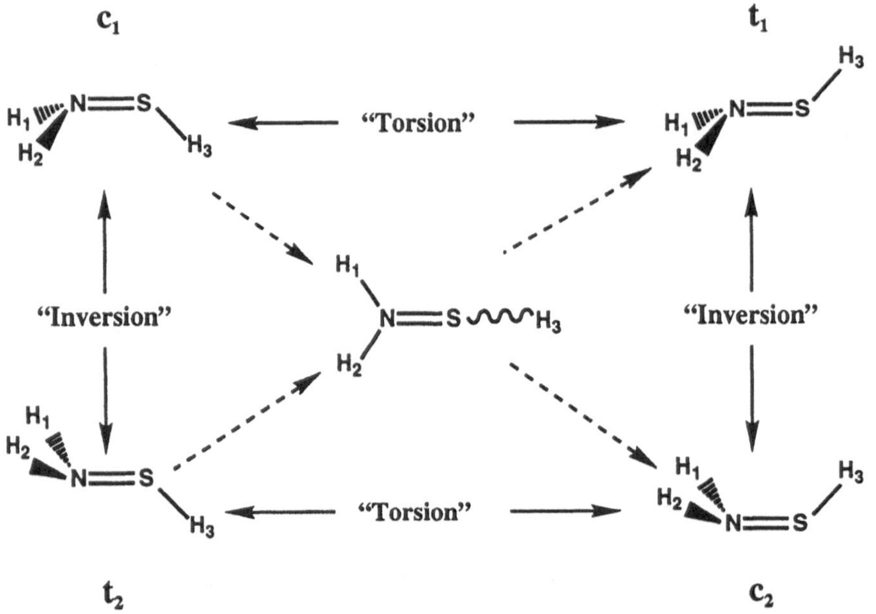

Figure 3. Internal motions of thio-hydoxylamine. The dashed lines denote high-energy paths through a hypothetical planar structure. The wavy line denotes an uncertain stereochemistry.

is as shown in Figure 3 for thio-hydroxylamine (in which both the barrier(s) and the energy difference are known to be very low ($\approx 50\ \mathrm{cm}^{-1}$), so that both forms have been observed [101]). Other molecules of this type which have been studied include the anion of methanol (A=C; B=O), $^{-}CH_2OH$, [102] and sulphilimine [103]. Note that there is, of course, pathway doubling of the torsional motion, i.e., equivalent motion in two opposite directions.

The group G_{PG} for either of the *versions* of both of the two (*cis* or *trans*) structures is $\{E, (12)^*\}$ (isomorphic to C_s). A cut through the potential energy surface is shown in Figure 4. We have assumed that the *trans*-structure is the lowest in energy and that the barrier to inversion is the lower of the two barriers. If the torsional barrier is too high to allow observable tunnelling between the two halves of the potential energy surface, then G_{PG} remains the most useful group and we have a path segment like Figure 1b representing the interconversion, via inversion, between one version of each of the *cis*- or *trans*- structures. If torsion becomes relevant, then (12) becomes a feasible operation. The correct group for the whole surface if both of the internal motions is feasible is thus i.e., $\{\hat{E}, (12), E^*, (12)^*\}$, i.e., $C_{2v}(M)$ in Bunker's notation [4]. Note that in this case there is no relevant C_{2v} symmetry structure for the interconversions described.

The salient features of this example may be transferred to other systems in which two distinct motions occur on the timescale of observation, see for example, the discussion in refs. [60, 61].

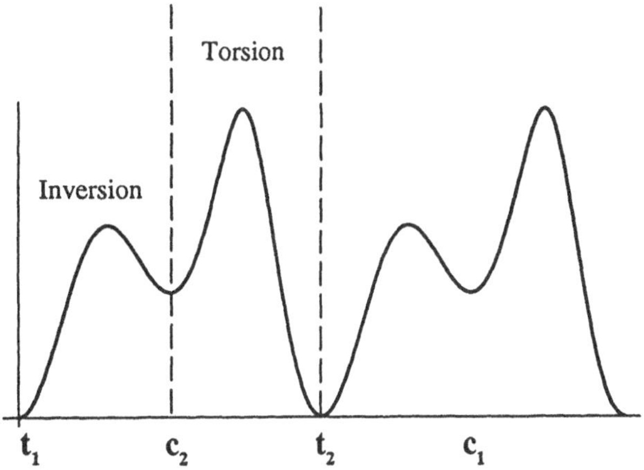

Figure 4. Inversion/torsion potential in H_2ABH molecules.

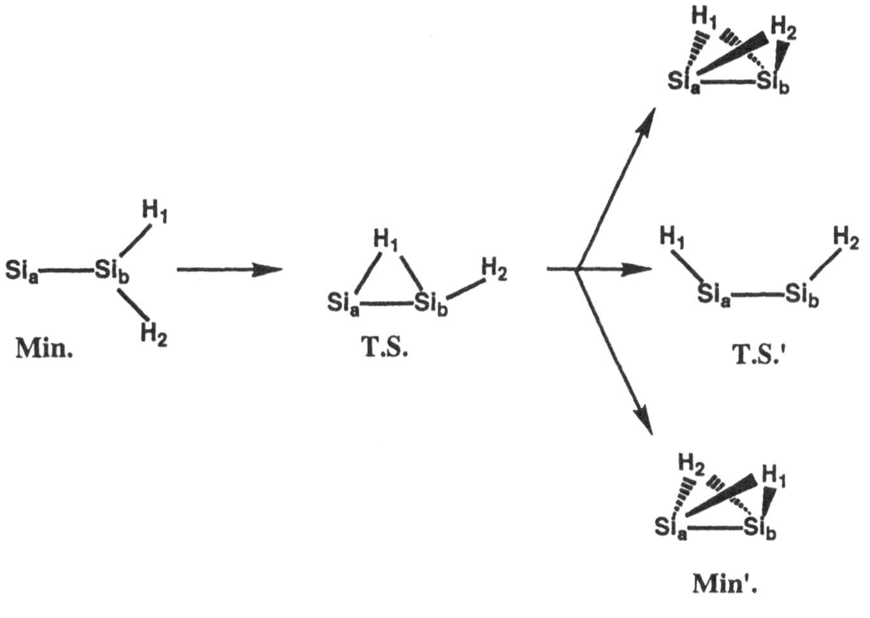

Figure 5. Simplified scheme for the potential energy surface of Si_2H_2. There is a bifurcation leading to two versions of a non-planar minimum energy structure. T.S. has an A′ imaginary frequency and T.S.′ has an A″ imaginary frequency.

Example 5: Isomerization of Si_2H_2 – Bifurcations

The potential energy surface [104] is schematically represented in Figure 5. The Global minimum is a "butterfly" structure. Only motion along one of the two symmetrically equivalent directions from the planar minimum is shown. It leads to subsequent planar structures which need not be related by any symmetry operation to the minimum itself. There is an overall E^* hyper-plane of symmetry

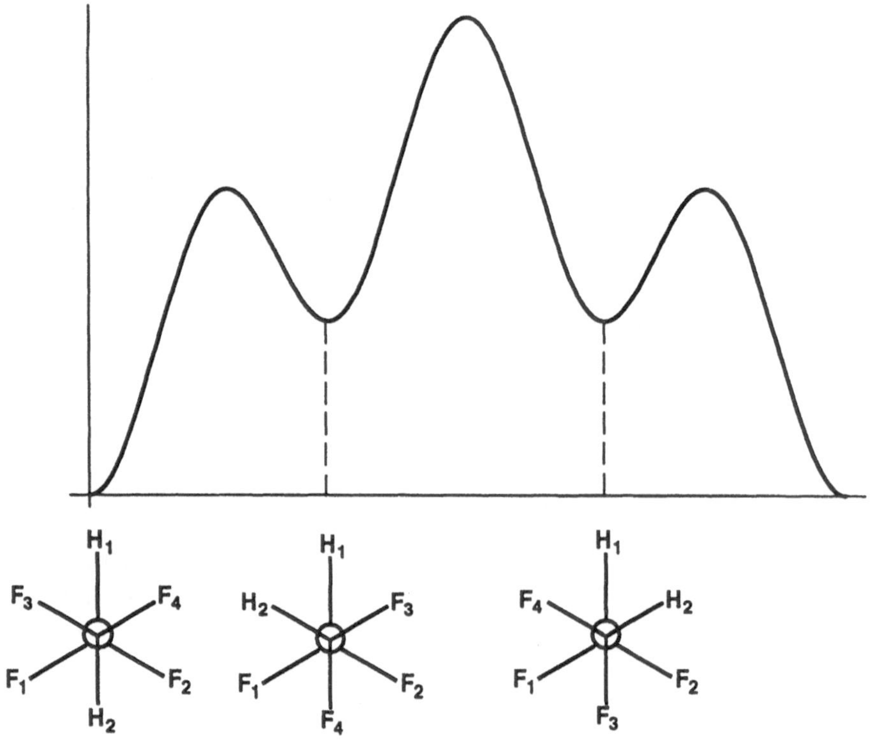

Figure 6. Internal rotation potential in H_2FCCFH_2 molecules.

in the surface but in addition the section shown is related by (12) to a equivalent portion. Subsequent out-of-plane motions from a point of bifurcation lead to structures related to one another by reflection (in this case minima). Symmetry group C_{2v}(M) is appropriate. Each symmetry operation is either conserved everywhere or corresponds to a feasible operation.

Example 6: Torsional Motion in HF_2CCF_2H

A final example suffices to illustrate the energy levels of a torsional problem with multiple minima (Figure 6). The occurrence of the two types of path-segments is again clear. In this example there will be two regimes of study but the same group, $C_{2h}(M)$, is pertinent to each because the global minimum happens to be, by assumption, the highest symmetry point on the surface. An analogous molecule, HD_2CCD_2H (deuterated ethane) has been observed in both "*trans*" and "*gauche*" forms; tunnelling splittings have been observed between the two forms of the latter [105, 106] but the spectrum is complicated by coupling between internal motion and overall rotation. There are similar measurable effects in 1,1-dideuterio-prop–3-ene (HD_2CCHCH_2) [107].

8. Summary

The earliest workers who attempted to obtain the correct symmetry group of non-rigid molecules, started from model Hamiltonians (see, e.g., [108]). It was customary to describe the large-amplitude motion by a single internal coordinate and identify its symmetries. Our standpoint here has been from a consideration of the complete potential energy surface.

In cases in which a non-rigid molecule samples distinct equilibrium structures on the timescale of spectroscopic resolution, we need to identify "conserved" symmetries, not just of the path followed but of the potential energy surface itself. In this way we are able to show how the non-rigid symmetry group may be constructed. An important ingredient of the discussion is a consideration of the point group symmetries of single structures and the changes in that symmetry which occur when following a path between two minima via a transition state. The barrier to change (i.e., the height of the transition state relative to the minima) determines the kinetic likelihood of the reaction; although the magnitude of that barrier cannot be predicted by group theory, it does tell us which symmetry operations are useful.

In order to be able to establish the applicability of the MS group to asymmetric tunnelling problems, we have needed to draw upon information from a number of sources. We have had to use a "counting rule" of molecular structure which demonstrates that we may always know how many *versions* of a particular structure are present in a given domain of the potential energy surface. This leads us to the notion that most reaction profiles for non-degenerate isomerizations are just part of a greater "whole" and that, if viewed in isolation, obscure the full symmetry that may be present.

It might be suggested that the role that a physical path plays in the definition is unsatisfactory. In general the path is the most difficult piece of information to obtain with any certainty, indeed it is probably only amenable to calculation, rather than measurement, and even then only in the smallest of systems. Even so, the properties of the path are well understood and appreciated. It is probably the notion of "adiabaticity" with respect to that path which is the most difficult to accept. Departures from the path may be real, but are usually taken to "average" to the assumed path.

The apparent contradiction between the discussion of a system evolving along a path and that of a fully delocalizable quantum state defined by amplitudes in the appropriate regions of the potential energy surface is quite straightforward to resolve. The stationary points themselves which dictate observable properties are always present and are invariant to the coordinate system. Amongst them the transition state(s) have a physical significance (defined by their energy relative to the minima) which control the level of observable tunnelling.

Acknowledgments

The author contributed this work whilst a SERC/NATO Postdoctoral Fellow. The bulk of this work formed a component of the author's PhD thesis [109]. SERC (U.K.) is acknowledged for a studentship held, 1988–91; Clare College, Cambridge and the University Chemical Laboratory, Cambridge are noted for financial support in the year 1991–2. Professor N.C. Handy is thanked for a number of constructive observations throughout the development of this work at Cambridge. The author is grateful to Dr. A.J. Stone and Dr. M.J. Bramley for making many detailed comments on terminology and self-consistency.

References

[1] P. Curie: *J. Physique* **5** (1894) 289.
[2] R.S. Berry: *Rev. Mod. Phys.* **32** (1960) 447.
[3] H.C. Longuet-Higgins: *Molec. Phys.* **6** (1963) 445.
[4] P.R. Bunker: *Molecular Symmetry and Spectroscopy*, Academic Press, New York (1979).
[5] J.K.G. Watson: *Can. J. Phys.* **43** (1965) 1996.
[6] B.J. Dalton: *Molec. Phys.* **11** (1966) 265.
[7] S.L. Altmann: *Proc. Roy. Soc.* **A298** (1967) 184.
[8] Y.G. Smeyers: *Adv. Quant. Chem.* **24** (1992) 1.
[9] H. Frei, A. Bauder and H.H. Günthard: "The isometric group of non-rigid molecules", in *Large Amplitude Motion in Molecules, I, Top. Curr. Chem.*, Vol. 81 (1979), p. 3.
[10] G.S. Ezra: *Molec. Phys.* **43** (1983) 771.
[11] P.A.M. Dirac: *The Principles of Quantum Mechanics*, 4th edition, Oxford University Press (1958).
[12] J.T. Hougen: *J. Chem. Phys.* **37** (1962) 1433.
[13] J.T. Hougen: *J. Chem. Phys.* **39** (1963) 358.
[14] P.R. Bunker and D. Papousek: *J. Molec. Spect.* **32** (1969) 419.
[15] J.T. Hougen: *J. Phys. Chem.* **90** (1986) 562.
[16] W.G. Klemperer: *J. Chem. Phys.* **56** (1972) 5478.
[17] R.G.A. Bone, T.W. Rowlands, N.C. Handy and A.J. Stone: *Molec. Phys.* **72** (1991) 33.
[18] P.G. Mezey: "Potential energy hypersurfaces", *Studies in Physical and Theoretical Chemistry* **53** (1987).
[19] J.N. Murrell and K.J. Laidler: *Trans. Farad. Soc.* **64** (1968) 371.
[20] F. Hund: *Z. Physik* **43** (1927) 805.
[21] E.B. Wilson Jr.: *J. Chem. Phys.* **3** (1935) 276.
[22] E.B. Wilson Jr.: *J. Chem. Phys.* **3** (1935) 818.
[23] R.G. Woolley: *Israel J. Chem.* **19** (1980) 30.
[24] G.A. Natanson: *Adv. Chem. Phys.* **58** (1985) 55.
[25] J.N. Murrell and G.L. Pratt: *Trans. Farad. Soc.* **66** (1970) 1680.
[26] R.G. Gilbert and S.C. Smith: *Theory of Unimolecular Recombination Reactions*, Oxford and Cambridge MA (1990).
[27] A.J. Karas, R.G. Gilbert and M.A. Collins: *Chem. Phys. Lett.* **193** (1992) 181.
[28] G. Wentzel: *Z. Phys.* **38** (1926) 518.
[29] H.A. Kramers: *Z. Phys.* **39** (1926) 828.
[30] L. Brillouin: *Compt. Rend.* **183** (1926) 24.
[31] H. Jeffreys: *Proc. Lond. Math. Soc.* **23** (1923) 428.
[32] R.S. Berry: "A general phenomenonology for small clusters, however floppy", in *Quantum Dynamics of Molecules: The New Experimental Challenge to Theorists*, R.G. Woolley (Ed.), NATO/ASI Series B, Vol. 57 (1979).
[33] P.R. Bunker: "Practically everything you ought to know about the molecular symmetry

group", *Vibrational Spectra and Structure*, Vol. 3, J. R. Durig (Ed.), Marcel-Dekker, New York (1976).
[34] T. Pedersen: *J. Chem. Phys.* **54** (1971) 4028.
[35] G.A. Natanson: *Adv. Chem. Phys.* **58** (1985) 55.
[36] J.K.G. Watson: *Molec. Phys.* **21** (1971) 577.
[37] M.J.S. Dewar: *Farad. Discuss.* **62** (1977) 197.
[38] N. Trinajstic (Ed.): *Mathematics and Computational Concepts in Chemistry*, Ellis-Horwood (1986).
[39] C.W. Curtis and I. Reiner: *Representation Theory of Finite Groups and Associative Algebras*, Wiley (1962).
[40] S. Fujita: *Symmetry and Combinatorial Enumeration in Chemistry*, Springer-Verlag, Berlin (1991).
[41] D.M. Dennison and G.E. Uhlenbeck: *Phys. Rev.* **41** (1932) 313.
[42] G. Herzberg: *Infrared and Raman Spectra of Polyatomic Molecules*, Van Nostrand (1945).
[43] J.M.R. Stone and I.M. Mills: *Molec. Phys.* **18** (1970) 631.
[44] K.W. Jucks, Z.S. Huang, R.E. Miller, G.T. Fraser, A.S. Pine, and W.J. Lafferty: *J. Chem. Phys.* **88** (1988) 2185.
[45] T.R. Dyke, B.J. Howard and W. Klemperer: *J. Chem. Phys.* **56** (1972) 2442.
[46] T.R. Dyke: *J. Chem. Phys.* **66** (1977) 492.
[47] G.T. Fraser: *Int. Rev. Phys. Chem.* **10** (1991) 189.
[48] L.H. Coudert and J.T. Hougen: *J. Mol. Spect.* **130** (1988) 86.
[49] L.L. Shipman, J.C. Owicki and H.A. Scheraga: *J. Phys. Chem.* **78** (1974) 2055.
[50] D. Nelson Jr. and W. Klemperer: *J. Chem. Phys.* **87** (1987) 139.
[51] H. Metiu, J. Ross, R. Silbey and T.F. George: *J. Chem. Phys.* **61** (1974) 3200.
[52] D.J. Wales: *J. Am. Chem. Soc.* **112** (1990) 7908.
[53] The concept of the "highest symmetry structure on the global potential energy surface" was suggested to me by Matthew J. Bramley.
[54] A.D. Liehr: *J. Phys. Chem.* **67** (1963) 471.
[55] G.T. Fraser, R.D. Suenram, F.J. Lovas, A.S. Pine, J.T. Hougen, W.J. Lafferty and J.S. Muenter: *J. Chem. Phys.* **89** (1988) 6028.
[56] D. Nelson Jr., G.T. Fraser and W. Klemperer: *J. Chem. Phys.* **83** (1985) 945.
[57] B.J. Dalton and P.D. Nicholson: *Int. J. Quant. Chem.* **9** (1975) 325.
[58] M. Quack: *Molec. Phys.* **34** (1977) 477.
[59] J. Ischtwan and M.A. Collins: *J. Chem. Phys.* **94** (1991) 7084.
[60] Y.G. Smeyers: *J. Mol. Struct.* **107** (1984) 3.
[61] Y.G. Smeyers, M.N. Bellido and A. Niño: *J. Mol. Struct.* **166** (1988) 1.
[62] R.P. Bell: *The Tunnel Effect In Chemistry*, Chapman and Hall (1978).
[63] R.G. Pearson: *Theor. Chim. Acta.* **16** (1970) 107.
[64] R.G. Pearson: *Pure Appl. Chem.* **27** (1971) 45.
[65] R.G. Pearson: *Acc. Chem. Res.* **4** (1971) 152.
[66] R.G. Pearson: *Symmetry Rules for Chemical Reactions*, Wiley-Interscience (1976).
[67] R.F.W. Bader: *Can. J. Chem.* **40** (1962) 1164.
[68] A. Rodger and P.E. Schipper: *Chem. Phys.* **107** (1986) 329.
[69] K. Fukui: *J. Phys. Chem.* **74** (1970) 4162.
[70] J.W. McIver Jr. and A. Komornicki: *Chem. Phys. Lett.* **10** (1971) 303.
[71] P. Valtazanos and K. Ruedenberg: *Theor. Chim. Acta.* **69** (1986) 281.
[72] J.W. McIver Jr. and R.E. Stanton: *J. Am. Chem. Soc.* **94** (1972) 8618.
[73] R.E. Stanton and J.W. McIver Jr.: *J. Am. Chem. Soc.* **97** (1975) 3632.
[74] J.W. McIver Jr.: *Acc. Chem. Res.* **7** (1974) 72.
[75] R.G.A. Bone: *Chem. Phys. Lett.* **193** (1992) 557.
[76] P. Pechukas: *J. Chem. Phys.* **64** (1976) 1516.
[77] E.L. Muetterties: *J. Am. Chem. Soc.* **91** (1969) 1636.
[78] G. Binsch, E.L. Eliel and H. Kessler: *Angew. Chem. Int. Ed.* **10** (1971) 570.
[79] R.E. Leone and P.v.R. Schleyer: *Angew. Chem. Int. Ed.* **9** (1970) 860.
[80] D.G. Truhlar and A. Kuppermann: *J. Am. Chem. Soc.* **93** (1971) 1840.

[81] E.B. Wilson, J.C. Decius and D.C. Cross: *Molecular Vibrations*, McGraw-Hill, New York (1955).
[82] A. Rodger and P.E. Schipper: *J. Phys. Chem.* **91** (1987) 189.
[83] S.L. Altmann: *Rev. Mod. Phys.* **35** (1963) 641.
[84] E.P. Wigner: *Group Theory and its Application to the Quantum Mechanics of Atomic Spectra*, Academic Press, New York (1959).
[85] A. Ceulemans, D. Beyens and L.G. Vanquickenbourne: *J. Am. Chem. Soc.* **106** (1984) 5824.
[86] A. Ceulemans and L.G. Vanquickenborne: *Structure and Bonding* **71** (1989) 126.
[87] E. Ascher: *J. Phys. C: Solid State Phys.* **10** (1977) 1365.
[88] A. Rodger and P.E. Schipper: *Inorg. Chem.* **27** (1988) 458.
[89] A survey of the "Science Citation Index" [90] has revealed that since its publication in 1975, Dalton's work has been referenced just 15 times in scientific papers and once in a book of conference proceedings. Approximately half of the references to ref. [57] are in connection with the non-rigidity of HF dimer, which the paper also addresses. One can only conclude that such an important result has been largely overlooked.
[90] SCI, the "Science Citation Index", Institute of Scientific Information, Philadelphia, U.S.A.
[91] P.-O. Löwdin: *Adv. Quant. Chem.* **2** (1965) 213.
[92] R.L. Sormorjai and D.F. Hornig: *J. Chem. Phys.* **36** (1962) 1981.
[93] J. Brickmann and H. Zimmermann: *J. Chem. Phys.* **50** (1968) 1609.
[94] J.H. Busch and J.R. de la Vega: *J. Am. Chem. Soc.* **99** (1977) 2397.
[95] M.D. Harmony: *Chem. Soc. Rev.* **1** (1972) 211.
[96] A.D. Buckingham and F. Liu: *Int. Rev. Phys. Chem.* **1** (1981) 253.
[97] S.A.C. MacDowell and A.D. Buckingham: *Chem. Phys. Lett.* **182** (1991) 551.
[98] M. Kofranek, H. Lischka and A. Karpfen: *Chem. Phys.* **121** (1988) 137.
[99] H.S. Gutowsky, C. Chang, J.D. Keen, T.D. Klots and T. Emilsson: *J. Chem. Phys.* **83** (1985) 2070.
[100] M.S. Reeves and E.R. Davidson: *J. Chem. Phys.* **95** (1991) 6551.
[101] F.J. Lovas, R.D. Suenram and W.J. Stevens: *J. Molec. Spect.* **100** (1983) 316.
[102] S. Wolfe, L.M. Tel and I.G. Csizmadia: *Can. J. Chem.* **51** (1973) 2423.
[103] P.G. Mezey: *Progr. Theor. Org. Chem.* **2** ("Applications of MO Theory in Organic Chemistry") (1977) 127.
[104] S. Koseki and M.S. Gordon: *J. Phys. Chem.* **93** (1989) 118.
[105] K.N. Rao, ed.: *Molecular Spectroscopy in Modern Research* (1986).
[106] E. Hirota, Y. Endo, S. Saito and J.L. Duncan: *J. Molec. Spect.* **89** (1981) 285.
[107] E. Hirota, T. Hirooka and Y. Morino: *J. Molec. Spect.* **26** (1968) 351.
[108] E.B. Wilson Jr.: *J. Chem. Phys.* **6** (1938) 740.
[109] R.G.A. Bone: *New Applications of the Molecular Symmetry Group*, Ph.D. Thesis, University of Cambridge (1992).

CHARACTERIZATION OF ROTATIONAL ISOMERIZATION PROCESSES IN MONOROTOR MOLECULES

GLORIA I. CÁRDENAS-JIRÓN, ALEJANDRO TORO-LABBÉ
Centro de Mecánica Cuántica Aplicada[†],
Departamento de Química,
Facultad de Ciencias, Universidad de Chile,
Casilla 653,
Santiago, Chile

CHARLES W. BOCK
Department of Physical Sciences,
Philadelphia College of Textiles and Science,
Philadelphia, Pennsylvania,
U.S.A.

AND

JEAN MARUANI
Laboratoire de Chimie Physique-Matière et Rayonnement,
CNRS and UPMC,
11, rue Pierre et Marie Curie,
75005 Paris, France

Abstract. A theoretical approach, in which the potential functions representing rotational isomerization processes are expressed in terms of linear combinations of local potentials, is presented. Partitioning the torsional potential function allows the identification of specific contributions that are at the origin of the shape of the potential curves at different regions along the torsional variable. Key properties, such as barrier heights, are then expressed parametrically in terms of properties associated to the stable conformations. Simple analytic expressions are formulated in order to explore, quantitatively and qualitatively, the main characteristics of intermediate conformers connecting the reference isomers. This procedure is used to analyse *ab initio* results concerning the *cis-trans* isomerization reaction of three series of molecules: XY–NY, OXC–CXO, and XS–SX (X =H, F, or Cl; Y = O or

[†]CMCA Contribution No. 9

Y.G. Smeyers (ed.), Structure and Dynamics of Non-Rigid Molecular Systems, 97–120.

S). We determine the relative stabilities of the different isomers and evaluate the associated potential barriers. It is shown that the mathematical procedure used to obtain potential functions is convenient enough to be applied to more complex isomerization reactions.

1. Introduction

Besides its importance in understanding fundamental molecular mechanisms, the modelization of isomerization processes is a problem of current relevance to various fields in molecular sciences, from biochemistry to engineering of molecular materials. Determination of the relative stabilities of different isomers, identification of the transition mechanisms, and evaluation of the potential barriers and transfer rates between various conformations are essential components of investigations aiming at the optimization of biological, chemical, or physical properties.

Analytic potential functions representing specific intramolecular interactions and, in particular, torsional motions are becoming widely used, especially because of the rapid development of Molecular Mechanics and Molecular Dynamics algorithms. Torsional potential functions and, in general, the conformational dependence upon internal rotation of any molecular property are usually characterized through fitting a few experimental data to a limited Fourier series. The expansion coefficients must comply to symmetry selection rules that come out from considering the composite symmetry of the rotors [1, 2, 3]. In particular, the potential energy function, $V(\alpha)$, describing a rotational isomerization process between two reference conformations in a single-top symmetric molecule, such that $V(-\alpha) = V(\alpha)$, is usually represented through the limited Fourier series expansion:

$$V(\alpha) = \frac{1}{2} \sum_{n=1}^{N} v'_n [1 - \cos n\alpha], \tag{1}$$

where the v'_n's are Fourier expansion coefficients.

In cases where the reference conformations represent stable isomers and the energy barrier separating them is sufficiently high, it is possible to assume that the molecule spends most of its time near the potential minima. Therefore, the hindered internal rotation can be treated semi-classically as a chemical exchange between the reference isomers, and the thermodynamic parameters characterizing such a process can be directly estimated from the potential function given in Equation (1). This will be, in the simplest case, a double-well potential function in which the coefficients v'_n, three in this case, are determined by fitting $V(\alpha)$ to three independent experimental data, namely, the torsional force constants (k_r and

k_p) of the reactive and product and the energy difference between them (ΔV°):

$$\Delta V^\circ \equiv V(\alpha_p) - V(\alpha_r) = \frac{1}{2}\sum_{n=1}^{N=3} v'_n[\cos n\alpha_r - \cos n\alpha_p]. \tag{2}$$

Experimentally, ΔV° can be estimated by measuring the relative intensities, $I_{p/r}(T)$, of the absorption bands associated to the stable isomers at various temperatures, thus giving the equilibrium constant $K_{\text{eq}}(T)$ [4]. Although the relative populations observed are determined by the free energy difference ($\Delta G^\circ = \Delta H^\circ - T\Delta S^\circ$), *ab initio* MO calculations can, at best, approach the enthalpy change for the intramolecular conversion and, therefore, we shall assume that $\Delta V^\circ \simeq \Delta H^\circ$. However, estimations of the entropic contribution to the free energy involved in a rotational isomerization through fitting the van't Hoff equation show that, in most cases, the population ratio is governed by ΔH° [5]. The equilibrium constant can therefore be written as:

$$K_{\text{eq}}(T) = \frac{I_p(T)}{I_r(T)} \approx \exp\left(\frac{-\Delta V^\circ}{RT}\right). \tag{3}$$

On the other hand, force constants are determined from the torsional frequencies associated to the reference isomers through the following equation:

$$k_{r/p} \equiv \left(\frac{d^2V}{d\alpha^2}\right)_{\alpha_{r/p}} = 4\pi^2\nu^2_{r/p}(G^{-1}_{tt})_{r/p}, \tag{4}$$

where (G^{-1}_{tt}) represents the torsional mode diagonal element of the inverse kinetic energy matrix. Putting Equations (2) and (4) in terms of the potential coefficients of Equation (1) leads to a system of equations whose solutions are the v'_n's. This provides a complete characterization of $V(\alpha)$ along the torsional variable. Torsional barriers ($\Delta V^{\neq}$) are then estimated by evaluating $V(\alpha)$ at the critical point located between the reference conformations. This is in fact an interpolation procedure, which has been successful in characterizing barrier heights for internal rotations.

The idea of interpolating data associated to reference conformations can be applied to theoretical calculations, where a number of calculated points can be fitted to a given analytic function, such as Equation (1). Advantages of using such a procedure are: (a) it may produce a quite accurate potential function from a small number of computed energy points; (b) the effects of changes in the molecular structure can be systematically included in the calculation of the parameters by allowing geometry relaxation along the torsional variable; (c) the procedure allows the iterative refinement of the interpolated function to correctly describe the energetic properties of any *gauche* conformation; and, perhaps the most interesting point: (d) the resulting potential function contains only a few parameters (normally three or four) which are functions of the input data and present a specific physical meaning [6, 7].

Moreover, since theoretical calculations allow one to investigate molecules and molecular conformations that are inaccessible to direct experimental study, it is possible to obtain double-barrier potential functions from the knowledge of properties associated with two unstable conformations. In these cases, the procedure is used to characterize a *gauche* stable conformation trapped between two potential barriers [8].

In this work we analyse potential functions describing the internal rotation of a few series of molecules, namely (a) XY–NY (X = H or F; Y = O or S); (b) OXC–CXO (X = H, F, or Cl); and (c) XS–SX (X = H, F, or Cl). These molecules have been studied through *ab initio* calculations at the HF level using extended basis sets. In series (a) and (b) most reference conformations correspond to stable isomers while in series (c) the starting reference conformations correspond to unstable isomers.

(A) THE XY–NY (X = H OR F; Y = O OR S) SERIES OF MOLECULES

Although nitrite compounds have been studied extensively, the corresponding sulphur analogues have been widely ignored. The importance of these latter compounds resides in the fact that they have been observed to take part in many photochemical reactions as reactive intermediates. The detailed mechanism of these photoreactions is far from being determined, but the relative stabilities and geometrical structures of the possible isomers may provide an understanding of these processes [9, 10]. The HY–NY type of molecules presents stable isomers at the *trans* and *cis* conformations. The two isomers are separated by a potential barrier located about midway between the reference conformations.

Compounds containing an S(O)–N bond may serve as a prototype for the S(O)–N linkage in some oxymes and inorganic compounds, and provide a starting point for understanding the structure and intramolecular dynamics of complex systems. The most studied molecule in this series is nitrous acid (HONO), whose *cis* and *trans* forms have been generated by UV photolysis of hydrazoic acid in oxygen-doped nitrogen matrices [11]. The characterization of these two stable structures has been performed through microwave and infrared spectroscopy [11, 12, 13, 14, 15, 16].

In the analogue fluorine series (FY–NY), very few informations have been reported. Although the FNO molecule has been studied in some detail [17, 18] and nitryl fluoride (FNO_2), with the fluorine atom bonded to the nitrogen, has been well characterized for many years [19, 20], relatively little is known about nitrosyl hypofluoride (FONO). This includes some theoretical studies [21, 22, 23] as well as some experimental works [24, 25]. The FONO molecule has been identified by Smardzewski *et al.* through infrared spectroscopy, as a product of the reaction between fluorine and nitrogen dioxide in matrix isolation experiments [24]. Normal coordinate analysis, performed by Sorenson *et al.* using Smardzewski's data, allowed determination of the bonding characteristics and definite assignment of the absorption bands [25]. From the theoretical viewpoint, FONO is considered as a *problem* molecule for conventional *ab initio* methods. To our knowledge, for FONS, FSNO and FSNS no theoretical or experimental studies have been reported.

(B) THE OXC–CXO (X = H, F, OR CL) SERIES OF MOLECULES

Glyoxal (HOHC–CHO) is the simplest molecule in this series that can undergo rotational isomerization. It has been shown, both spectroscopically [26, 27, 28] and computationally [29, 30, 31, 32, 33], that it exists in two planar, *trans* and *cis* forms. The experimental structure of *trans*-glyoxal was determined from electron diffraction experiments [34], and the *cis* structure by rotational spectroscopy [28]. Oxalyl fluoride (OFC–CFO) also exists in both planar forms [30, 35], while for oxalyl chloride (OClC–CClO) the spectroscopic evidence is not entirely definitive, indicating either a *trans* and a *cis* conformer [36] or a *trans* and a *gauche* conformer [37]. To our knowledge, no experimental data are available on either fluoroglyoxal or chloroglyoxal.

(C) THE XS–SX (X = H, F, OR CL) SERIES OF MOLECULES

Compounds containing S–S or S–X bonds serve as prototypes for the S–S linkage in proteins and provide a starting point for understanding these systems. For this reason, considerable experimental and theoretical efforts have been made to study molecules which contain the S–S unit. Knowledge of the factors that influence properties such as bond lengths, bond strengths, and conformations in disulfides is therefore relevant to several areas in chemistry and biochemistry. In the last few years, many theoretical and experimental papers aiming at elucidating the features of internal rotation in hydrogen persulfide and related molecules have appeared [38, 39, 40, 41, 42]. They report molecular structures, barrier heights, rotational spectra, and related matters concerning the dynamics of the internal rotation. However, in contrast to the large amount of data for the parent molecule XOOX, much less information is available for molecules of the XSSX type.

The structure of hydrogen persulfide has been investigated recently by microwave spectroscopy and electron diffraction methods [38, 39]. Many years ago Kuczkowski recorded the mass and microwave spectra of FSSF and determined the structure and other properties of interest [40]. More recently, Davis and Firth [41] recorded the microwave spectrum of FSSF and confirmed most of the results already obtained by Kuczkowski. The experimental structure of ClSSCl was determined from electron diffraction data by Hirota [42]. All these molecules present two maxima, at the *trans* and *cis* conformations (dihedral angle $\alpha = 0°$ and $180°$, respectively). Throughout this paper these will be considered as the reference isomers. A *gauche* stable isomer is also found around midway between the above reference conformations.

A few studies of the torsional potentials of disulfide molecules based on *ab initio* SCF–MO calculations have been published recently [8, 43, 44]. The most sophisticated treatments report *cis* and *trans* barrier heights in close agreement with experimental estimations.

In this chapter, we shall characterize the reference molecular structures and the transition states (or the *gauche* stable conformations) connecting them, as well as the torsional potential describing the isomerization reaction, in terms of a few *ab initio* computed molecular energies. Throughout the paper a transition state or a *gauche* stable conformation will be referred to as a *critical intermediate state* (CIS). For all molecules of the series, we have performed *ab initio* SCF–MO calculations using extended basis sets to determine the potential function hindering the isomerization process. The article is organized as follows: in the next section we present the theoretical treatment that allows deriving analytic forms for reaction parameters, and in Section 3 we present and discuss the numerical results. Section 4 gives our concluding remarks.

2. Theoretical Background

2.1. LOCAL POTENTIALS AND CONFORMATIONAL FUNCTIONS

As mentioned above, potential functions hindering the internal rotation of monorotor molecules are adequately represented by symmetry-adapted, limited Fourier series expansions [1, 2, 3]. We have shown in recent papers that a symmetry-adapted Fourier potential representing the internal conversion between two planar isomers, say *cis* and *trans*, along a torsional variable α can be conveniently expressed as [6, 33, 45, 46]:

$$V(\alpha) = V_0 + \sum_{n=1}^{3} V_n \cos(n\alpha), \tag{5}$$

where the coefficients are related to those of Equation (1) through the following relations:

$$V_0 = \frac{1}{2}\sum_{n=1}^{3} v'_n, \tag{6}$$

$$V_n = -\frac{1}{2}\, v'_n. \tag{7}$$

These parameters are determined from the energy difference between the reference isomers ($\Delta V^\circ \equiv V(\alpha_c) - V(\alpha_t)$) and the torsional force constants (k_c and k_t) associated with the *cis (c)* and *trans (t)* potential wells (or barriers). Choosing the origin of the energy at the *trans* conformation ($\alpha = 0^\circ$), the potential parameters are found to be given by:

$$V_0 = \frac{1}{2}\Delta V^\circ + \frac{1}{8}(k_t + k_c), \tag{8}$$

$$V_1 = -\frac{1}{2}\Delta V^\circ + \frac{1}{16}(k_t - k_c - \Delta V^\circ), \tag{9}$$

$$V_2 = -\frac{1}{8}\,(k_t + k_c), \tag{10}$$

$$V_3 = -\frac{1}{16}(k_t - k_c - \Delta V^\circ). \tag{11}$$

The torsional force constants must be determined previously by differentiating the associated local potentials, $V_{t/c}(\alpha)$, which are assumed to describe the function accurately in the vicinity of the reference conformations:

$$k_{t/c} = \left(\frac{d^2 V_{t/c}(\alpha)}{d\alpha^2}\right)_{\alpha_{t/c}} . \tag{12}$$

In previous studies we have used both, a simple cosine expansion and the harmonic oscillator approximation to represent $V_{t/c}(\alpha)$. As expected, both representations produce quite similar results concerning the $k_{t/c}$ numerical values [6, 7, 8, 33, 45]. In this work we choose to evaluate the k_t and k_c constants through the following expression, that comes out from expressing $V_{t/c}(\alpha)$ as a cosine expansion:

$$k_{t/c} = -\sum_{n=1}^{N}\sum_{i=1}^{N} n^2 C_{ni} V(\alpha_i)\cos(n\alpha_i), \tag{13}$$

where N is the number of computed energy points used to fit $V_{t/c}(\alpha)$, the coefficients C_{ni} are elements of the matrix $[\cos(n\alpha_i)]^{-1}$, and $V(\alpha_i)$ is the energy calculated at point α_i. Since it has been shown that the $k_{t/c}$ values are, to a good approximation, independent of N, we have used $N = 2$ in our calculations [6, 7]. In fact, k_t was determined from the energy points at $\alpha = 0°$ and $\alpha = 10°$, whereas k_c was calculated from the energies at $\alpha = 170°$ and $\alpha = 180°$. The resulting potential function may be written as [6, 33, 45, 46]:

$$\begin{aligned} V(\alpha) &= \frac{1}{2}\Delta V^\circ(1-\cos\alpha) + \frac{1}{4}\,(k_t + k_c)(1-\cos^2\alpha) \\ &\quad + \frac{1}{4}\,(k_t - k_c - \Delta V^\circ)(1-\cos^2\alpha)\cos\alpha, \end{aligned} \tag{14}$$

which can be conveniently expressed as a sum of two independent terms:

$$V(\alpha) = V_0(\alpha) + V_1(\alpha), \tag{15}$$

with

$$V_0(\alpha) = \frac{1}{2}\,\Delta V^\circ(1-\cos\alpha) + \frac{1}{4}\,(k_t + k_c)(1-\cos^2\alpha) \tag{16}$$

and

$$V_1(\alpha) = \frac{1}{4}\,(k_t - k_c - \Delta V^\circ)(1-\cos^2\alpha)\cos\alpha. \tag{17}$$

$V_0(\alpha)$ represents an approximated potential function that can be written in terms of two local potentials, $V_t(\alpha)$ and $V_c(\alpha)$, localized at the *trans* and *cis* wells, respectively:

$$V_0(\alpha) = \omega_t^0(\alpha)V_t(\alpha) + \omega_c^0(\alpha)V_c(\alpha), \tag{18}$$

where

$$V_t(\alpha) = \frac{1}{2}\,k_t(1 - \cos\alpha), \tag{19}$$

$$V_c(\alpha) = \frac{1}{2}\,k_c(1 + \cos\alpha) + \Delta V^\circ. \tag{20}$$

The functions $\omega_t^0(\alpha)$ and $\omega_c^0(\alpha)$ are the so-called *conformational functions* (see for example [46] and references therein) that give the statistical weights of the respective reference conformations when moving along the torsional variable. These functions are defined as:

$$\omega_c^0(\alpha) = \frac{\partial V_0}{\partial \Delta V^\circ} = \frac{1}{2}\,(1 - \cos\alpha) \equiv \omega_0(\alpha), \tag{21}$$

$$\omega_t^0(\alpha) = 1 - \omega_0(\alpha) = \frac{1}{2}\,(1 + \cos\alpha). \tag{22}$$

Since $(k_t - k_c - \Delta V^\circ)$ is generally small for molecules in which both reference conformations correspond to stable isomers, Equation (13) can be regarded as a correction to the shape of $V_0(\alpha)$ at the intermediate region located between two critical points, region in which $V_1(\alpha)$ has its highest amplitude. Now we let $\omega(\alpha) = \omega_0(\alpha) + \omega_1(\alpha)$ be the conformational function associated to the whole function $V(\alpha)$, treat $V_1(\alpha)$ as a small perturbation and, in analogy with Equation (21), define $\omega_1(\alpha)$ as:

$$\omega_1(\alpha) = \frac{\partial V_1(\alpha)}{\partial \Delta V^\circ} = -\frac{1}{4}\,(1 - \cos^2\alpha)\cos\alpha. \tag{23}$$

It is clear from Equation (23) that $\omega_1(\alpha)$ is not a distribution function by itself, but a function that slightly modulates $\omega_0(\alpha)$ at intermediate regions of α, between the critical points. Since in the region where the CIS is found $(\alpha \sim 90^\circ)$ $\omega_1(\alpha) \sim 0$, Equations (21) and (22) do not need to be corrected by the term $\omega_1(\alpha)$. Finally, we note that comparison of Equations (23) and (17) shows that $V_1(\alpha) = \lambda\omega_1(\alpha)$ with $\lambda = -(k_t - k_c - \Delta V^\circ)$ being a linear-response coefficient.

2.2. BARRIER HEIGHTS AND BRØNSTED COEFFICIENTS

The CIS associated to the function $V_0(\alpha)$ (or the *near*-CIS for the function $V(\alpha)$) is found to be located at an angle α_0' such that $(\frac{dV_0}{d\alpha})_{\alpha_0'} = 0$, this leading to:

$$\cos\alpha_0' = -\frac{\Delta V^\circ}{(k_t + k_c)}\,. \tag{24}$$

This expression is in agreement with the well known Hammond postulate [47], which states that activated complexes are located, along the reaction path, closer to the product(s) when $\Delta V^\circ > 0$, whereas for $\Delta V^\circ < 0$ they are located closer to the reactant(s).

The most important quantity to determine when characterizing a dynamical process described through a double-well potential function is the energy barrier separating the stable conformations. Since in most cases $(k_t + k_c) \gg \Delta V^\circ$ [33], it follows from Equation (24) that α'_0 is located around midway between the stable isomers. Because in that region $V_1(\alpha)$ is practically zero, according to Equation (17), the energy $V_0(\alpha'_0)$ should give a good estimate of the barrier height:

$$\Delta V_0^{\neq} \equiv V_0(\alpha'_0) = \frac{1}{4}(k_t + k_c) + \frac{\Delta V^\circ}{2} + \frac{(\Delta V^\circ)^2}{4(k_t + k_c)} . \tag{25}$$

The last two equations are quite important results. As already mentioned, Equation (24) is in fact a quantitative statement of the Hammond postulate [47]. It shows that the commonly used empirical concepts of *reactant-like* and *product-like* CIS are not completely described in terms of energy comparison alone; force constants describing the nature of the reference states are also relevant for such a description. On the other hand, Equation (25) is an analytic formula giving barrier heights of one-dimensional isomerization processes from the knowledge of characteristic properties associated to the reference conformations. It should be noted that in the case of double-barrier functions Equations (24) and (25) will give, respectively, the position and depth of the potential well trapping the stable CIS conformation.

To get more insight about an isomerization process described by a double-well potential, it is convenient to introduce the Brønsted coefficient [48]. This is a measure of the degree of resemblance of the transition state to the product(s) [7] and is used to quantify the empirical concepts of *reactant-like* and *product-like* transition states. Following the Leffler postulate [48], the Brønsted coefficient is defined as:

$$\beta_0 = \frac{\partial \Delta V_0^{\neq}}{\partial \Delta V^\circ} = \frac{1}{2} + \frac{\Delta V^\circ}{2(k_t + k_c)} \equiv \omega_0(\alpha'_0), \tag{26}$$

which is the same as the result obtained when replacing Equation (24) in Equation (21). The statistical weight of the product conformation at the CIS will then be greater than 0.5 when $\Delta V^\circ > 0$; for symmetric conversions $\Delta V^\circ = 0$ and then $\beta_0 = \frac{1}{2}$, otherwise β_0 will be smaller than 0.5, in agreement with our previous analysis of Equation (20). Substituting Equation (26) in Equation (25) now gives for $\Delta V_0^{\neq}$ the simple form:

$$\Delta V_0^{\neq} = (k_t + k_c)\beta_0^2, \tag{27}$$

which provides an alternative way to classifying barriers to internal rotation, in terms of the Brønsted coefficient rather than of the energy difference between the stable isomers.

2.3. REFINED POTENTIAL FUNCTIONS

Although the procedure described above allows one to get a rapid view of the potential curve and to estimate the position and relative energy of the CIS, one should not expect the interpolated potential function to be 100% reliable in all regions, because it was determined from data in the vicinity of the reference conformations. This interpolated function may be locally refined by introducing the energy of the optimized α_0-conformation, $E^{\neq} \equiv E(\alpha_0)$, as a fourth external input datum, to give $V^{\mathrm{ref}}(\alpha)$:

$$V^{\mathrm{ref}}(\alpha) = V_0(\alpha) + V_1(\alpha) + V_2(\alpha), \tag{28}$$

where

$$V_2(\alpha) = -\left[\frac{1}{4}\,(k_t + k_c) + 2f(\beta_0, \beta_1)\right] \sin^4 \alpha \tag{29}$$

and

$$\begin{aligned} f(\beta_0, \beta_1) &= \left[\frac{1}{2\sin^4 \alpha_0}\right] \\ &\times \{[(k_t + k_c)(2\beta_0 - 1) - (k_t - k_c)]\beta_1 - E^{\neq} + \beta\Delta V^{\circ}\}. \end{aligned} \tag{30}$$

As already said, the parameter β_0 is the Brønsted coefficient, whereas $\beta_1 \equiv \omega_1(\alpha_0)$. It should be noted that the correction to $V_0(\alpha) + V_1(\alpha)$, $V_2(\alpha)$, is proportional to $\sin^4 \alpha$, indicating that this term would be relatively important only near the CIS. Therefore, partitioning the torsional potential function enables the identification of specific contributions that are at the origin of the shape of the potential curve at different regions along α.

3. Results and Discussion

In this section, the discussion is focused on the torsional potentials we have obtained through *ab initio* calculations for the three series of molecules. We refer the reader to our previous papers for detailed discussions on the conformational dependences of other molecular properties [6, 7, 8, 33, 45, 46, 49, 50]. Calculations were performed at the SCF level, using extended basis sets including diffuse or polarization functions. It has been shown, in a variety of sulphur-containing compounds, that basis sets without diffuse or *d* polarization functions on the sulphur atom lead to quantitatively unreliable results. Structural optimizations were carried out employing numerical gradient techniques. We have used a combination of Newton–Raphson, steepest descent and Marquard's algorithms in the *Monster-gauss* code [51] in order to optimize critical points along the energy surface.

3.1. THE XY–NY SERIES OF MOLECULES

3.1.1. *Molecules of the type HY–NY (Y = O, S)*

The *ab initio* results were obtained using the split-valence 6–31G basis set in the case of HONO, and the doubly polarized split-valence 6–31G** basis set for HONS, HSNO and HSNS. Fully optimized molecular structures for *cis*, *trans* and CIS conformers were discussed in detail in previous papers [45, 46, 49, 50]. The experimental structure of HONO was determined for both isomers, *cis* and *trans*, from microwave and infrared spectroscopy [12, 13, 14, 15, 16]. It should be mentioned that, in general, a quite good agreement is found between our calculated molecular structures and the available experimental data.

The relative energies (ΔV°) for the HYNY series are included in Table I. These show that *trans* to *cis* isomerization, when occurring, is endoenergetic, except in HSNS, where the reaction is practically isoenergetic. The nitrogen non-bonding orbital containing the electronic lone pair lying in the molecular plane produces an attractive interaction that may explain, in part, the greater stabilization of the *trans* conformations of HONO, HONS and HSNO. In the case of HSNS, the intramolecular hydrogen bridge, H$\cdots$S, for the *cis* isomer seems to equilibrate the lone-pair$\cdots$H interaction. On the other hand, comparison between the ΔV° calculated value (0.83 kcal/mol) and experimental value (0.46 kcal/mol) for HONO [13] shows a satisfactory agreement. Unfortunately, experimental energies of the reference conformations for the remaining molecules are not available. In contrast, we have found in the literature theoretical works concerning HONS and HSNO in which results for ΔV° are in good agreement with ours [9]. For HSNS, Nakamura *et al.* [10] have reported a ΔV° value of -0.80 kcal/mol calculated using a 4–31G* basis set. Unsurprisingly, this result differs from ours (-0.15 kcal/mol), since we have used a more extended and flexible basis set with polarization orbitals on the hydrogen atom.

From the $k = k_t/k_c$ values given in Table I, we note that for all four molecules the relative symmetry of the local potential wells is nearly constant. From the experimental values of k_t and k_c for HONO reported by Deeley and Mills [13], a value of k = 0.76 is obtained, in good agreement with our value 0.82. It is also important to note that our *ab initio* force constants for this molecule are within 3% of the experimental data. For HSNO, Nonella *et al.* [9] reported k_t and k_c values from the *Transferable Valence Force Field* (TVFF) method. A value of k = 0.94 was obtained, higher than our value (k = 0.85) reported in Table I. When comparing the individual force constants, we found a very good agreement for k_t: our calculated value (18.19 kcal/mol rad^2) compares satisfactorily with the experimental one (17.67 kcal/mol rad^2). Our calculated value for k_c (21.49 kcal/mol rad^2) is however significantly higher than the experimental one (18.88 kcal/mol rad^2), this suggesting that the *cis* isomer may not be accurately described at the Hartree–Fock level. In HONS, Nonella *et al.* isolated experimentally only the *trans*

TABLE I. Input data ($k_t, k_c, \Delta V°$ and $E^{\neq}$) and resulting torsional potential properties for the internal rotation of the HY–NY series of molecules. Torsional force constants in kcal/mol · rad^2, energies in kcal/mol.

	HO–NO[a]	HO–NS[b]	HS–NO[b]	HS–NS[b]
k_t	19.03	23.37	18.19	26.00
k_c	23.17	29.13	21.49	27.83
$\Delta V°$	0.83	0.38	0.44	-0.15
k=k_t/k_c	0.82	0.80	0.85	0.93
α'_0	91°	90°	91°	90°
$\Delta V_0^{\neq}$	10.96	13.32	10.14	13.38
α_0	94°	94°	93°	91°
$\Delta V^{\neq}$	11.03	13.37	10.18	13.38
β	0.56	0.55	0.54	0.51
β_0	0.54	0.53	0.53	0.51
α_0^{opt}	95°	94°	94°	94°
$E^{\neq}$	9.91	14.44	10.60	12.63

[a] 6–31G
[b] 6–31G**

conformation. They determined a value for k_t of 18.42 kcal/mol rad^2, significantly smaller than our calculated value (23.37 kcal/mol rad^2). It is worth mentioning that the same authors have performed calculations for *trans*–HONS including the electron correlation through a CI scheme, this leading to $k_t = 25.10$ kcal/mol rad^2, even higher than the HF value. In HSNS, no experimental data are available. The *ab initio* values of k_t and k_c show a quite symmetric double-well potential with $k = 0.93$.

Near-transition states, defined by the angle α'_0, are found around midway between the reference isomers. A large value of $(k_t + k_c)$ compared to $\Delta V°$ confines the transition state to the $\alpha \sim \pi/2$ region, as can be inferred from Equation (20). The correct position of the transition state is obtained through full optimization of the structures. We found that α_0^{opt} is slightly shifted towards higher values with respect to α'_0. However, comparison of the predicted α'_0 with the optimized α_0^{opt} shows that Equation (20) approaches very well the actual position of the transition state. Finally, the values of the Brønsted coefficients show that the transition states resemble the *trans* and *cis* reference isomers in practically the same proportions.

Barrier heights calculated from $V_0(\alpha)$, denoted $\Delta V_0^{\neq}$, and from $V(\alpha)$, denoted $\Delta V^{\neq}$, are practically the same, this showing that the rotational isomerization process is correctly characterized by the function $V_0(\alpha)$. On the other hand, the optimized barriers $E^{\neq}$, also included in Table I, show a very good agreement with both $\Delta V_0^{\neq}$ and $\Delta V^{\neq}$. The barriers for HONS and HSNS are higher than those for HONO and HSNO, suggesting that in compounds with terminal sulphur atoms hyperconjugation is an important mechanism, making the rotation around the central bond more difficult. Figure 1 displays the potential curves, where these characteristic features are illustrated. From the above discussion we conclude that, in all cases, the potential $V_0(\alpha)$ describes correctly the isomerization process, and therefore no refinement to this function is necessary.

3.1.2. *Molecules of the type FY–NY (Y = O, S)*

The reference conformations and the transition states of the FYNY series were fully optimized by means of numerical gradient techniques, at the Hartree–Fock level, using a 6–31G basis set in FONO and a polarized 6–31G* basis set for the sulphur-containing compounds.

The experimental geometry of FONO, reported by Sorenson *et al.* [25], has been attributed to the *trans* isomer. Experiments suggest a planar conformation with a weak F–O bond, this meaning that the fluorine atom is loosely bonded to one of the oxygen atoms of the NO_2 radical. The highly electronegative fluorine atom acquires a significant amount of negative charge; the NO_2 portion of the molecule thus resembles an NO_2^+ ion, with a partial positive charge on the nitrogen atom. This charge separation explains why nitrosyl hypofluorite (FONO) can be stabilized only under matrix isolation, since attraction between the negative fluorine and positive nitrogen atoms should facilitate a rapid intermolecular reaction to form the more stable species nitryl fluoride (FNO_2). We are here in agreement with Dixon and Christe [22]: the reported experimental geometry for FONO [25], which had been attributed to the *trans* isomer, is actually compatible only with the *cis* isomer, since our theoretical results associated to the *cis* conformation are closer to the experimental data [52]. Finally, the structure of FONO (F···ONO), which has a non-typical F–O bond, explains why its geometry could not be duplicated correctly by traditional molecular orbital methods.

A detailed discussion of the molecular structures of these compounds can be found elsewhere [22, 25, 52]. We shall confine the present analysis to the description of the torsional potential function. However, results for $V(\alpha)$ in this series must be taken cautiously: the use of larger basis sets and CI to account for correlation effects seems to be in order to describe accurately the internal rotation.

For the FONS, FSNO, and FSNS molecules, no theoretical or experimental information is available. The input data k_t, k_c and ΔV°, together with the resulting potential parameters, are given in Table II. The potential functions hindering the internal rotation are displayed in Figure 2. As for the HYNY series, the

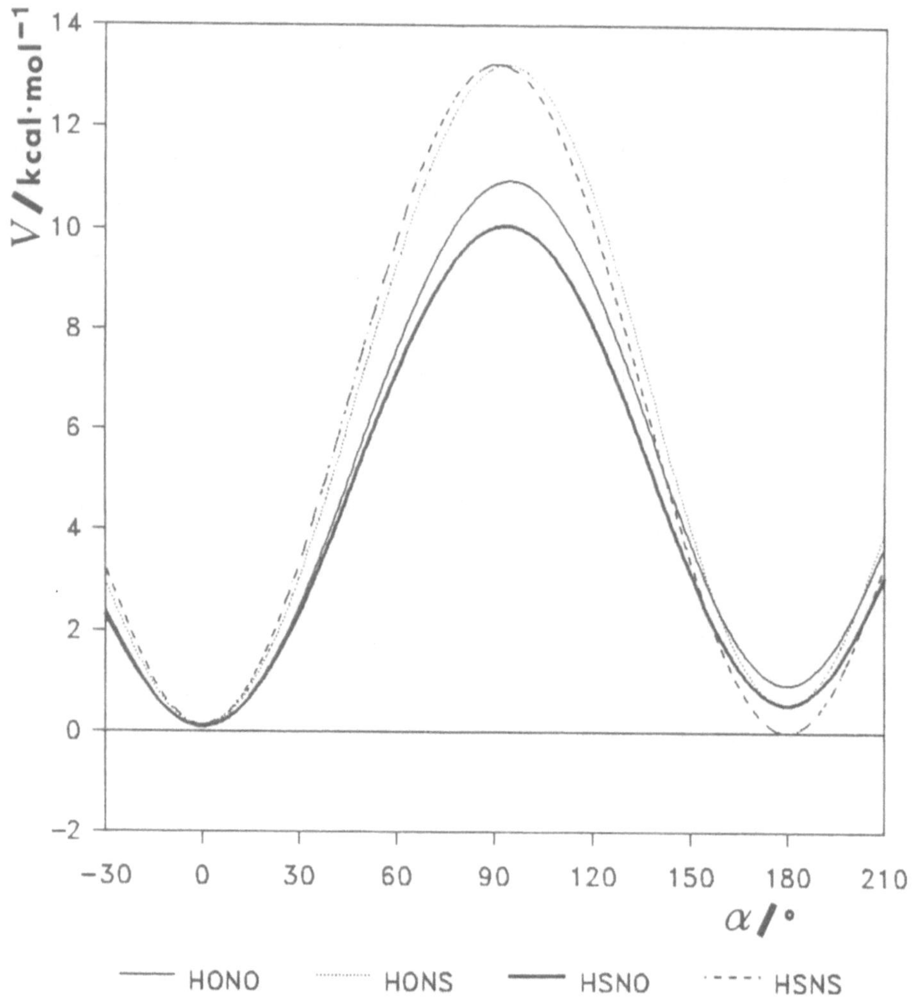

Figure 1. Torsional potential curves for the series HY–NY.

isomerization process can be characterized by the function $V_0(\alpha)$. The function $V_1(\alpha)$ represents a small correction to $V_0(\alpha)$ between the critical points.

A characteristic of this series is that, in all cases, the isomerization is exothermic. In Table II the relative stability of *cis*–FONO increases when substituting an oxygen by a sulphur in the torsional position, thus producing the FSNO molecule. This effect is still greater when substituting the second oxygen in terminal position, thus obtaining the FSNS molecule. This is mainly due to the fact that, in the *cis* isomer, the electron donation from the torsional sulphur to the fluorine atom is more important than that from the torsional oxygen. Then, the presence of torsional sulphur produces an increase of the positive partial charge in the terminal X (O,S) atom, this latter presenting a configuration near sp^3. According to

TABLE II. Input data ($k_t, k_c, \Delta V^\circ$ and $E^{\neq}$) and resulting torsional potential properties for the internal rotation of the FY–NY series of molecules. Torsional force constants in kcal/mol · rad^2, energies in kcal/mol.

	FO–NO[a]	FO–NS[b]	FS–NO[b]	FS–NS[b]
k_t	15.73	18.11	25.83	31.13
k_c	21.26	21.85	34.07	26.30
ΔV°	-1.88	-1.51	-2.21	-7.34
k=k_t/k_c	0.74	0.83	0.76	1.18
α'_0	87°	88°	88°	83°
$\Delta V_0^{\neq}$	8.33	9.25	13.89	10.92
α_0	90°	89°	91°	79°
$\Delta V^{\neq}$	8.31	9.24	13.87	11.42
β	0.50	0.49	0.51	0.36
β_0	0.50	0.49	0.51	0.40
α_0^{opt}	91°	91°	92°	90°
$E^{\neq}$	7.28	8.30	12.76	18.16

[a] 6–31G
[b] 6–31G*

this hypothesis, the fluorine atom should remain negatively charged. This should generate a strong intramolecular F···X interaction in both FSNO and FSNS and explain the greater stabilization of the *cis* conformation.

On the other hand, the force constant values displayed in Table II show that usually (except in FSNS) the *cis* potential well is sharper than the *trans* one. In FSNS, where $k_t > k_c$, the proposed hyperconjugative structure looses relevance, due to the presence of the torsional sulphur atom. The k parameters in this series indicate a relative symmetry of the local wells similar in FONO, FONS, and FSNO but quite different in FSNS.

The parameters characterizing the isomerization process indicate that it can be correctly represented by the function $V_0(\alpha)$. The optimized potential barriers for FONO, FONS, and FSNO differ only by $\sim$ 1.0 kcal/mol from the approximated values. Again, refinement of the potential functions is not necessary when dealing with a qualitative description of the isomerization process.

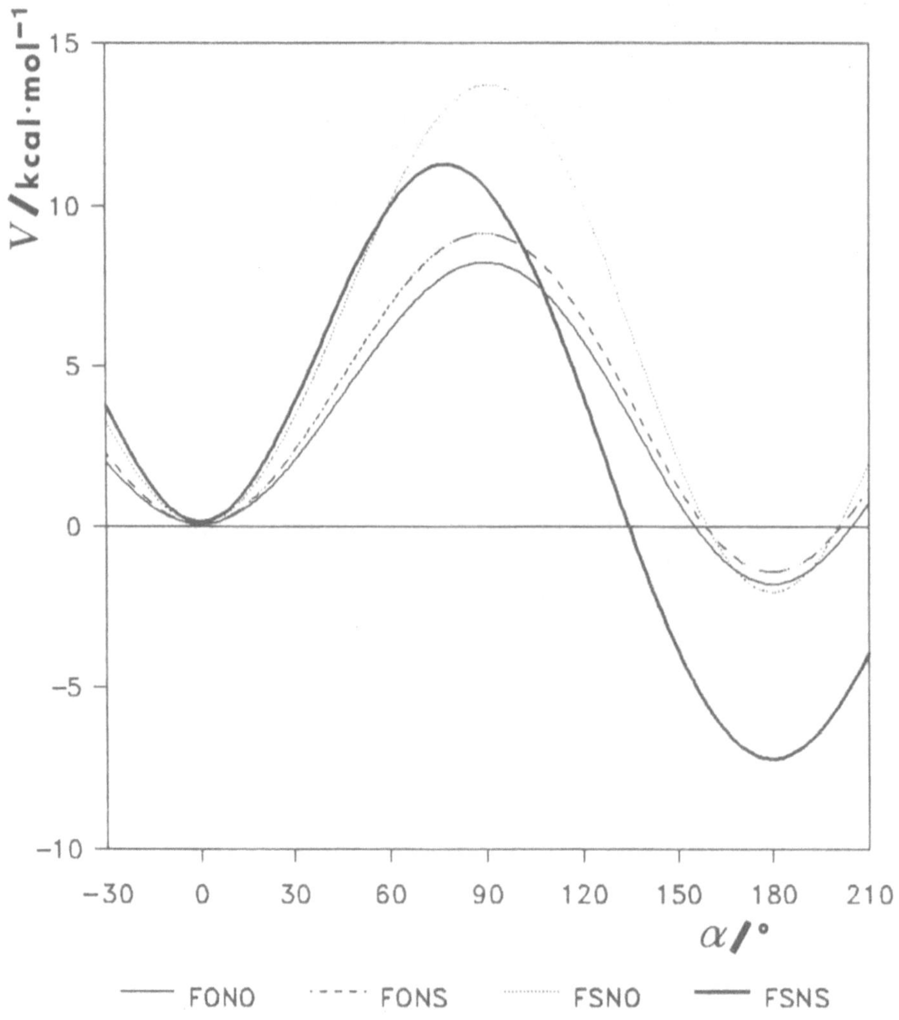

Figure 2. Torsional potential curves for the series FY–NY.

3.2. THE OXC–CXO (X = H, F, OR CL) SERIES OF MOLECULES

Ab initio results were obtained using the split-valence 6–31G basis set. The input data for the torsional potential function, as well as the resulting parameters characterizing the isomerization process, are given in Table III. ΔV° values show that *trans* to *cis* isomerization, when occurring, is endothermic. Years ago, Currie and Ramsay [26] reported a value of ΔV° for glyoxal of 3.21 ± 0.6 kcal/mol, and more recently a value of 3.85 ± 0.6 kcal/mol was reported by Parmenter *et al.* [53]. Our estimation of 5.69 kcal/mol [32] is above that experimental value by 1.84 kcal/mol. Unfortunately, experimental equilibrium thermodynamic quantities for this kind of molecules are difficult to obtain and, for the remaining halogen

TABLE III. Input data (k_t, k_c, ΔV° and $E^{\neq}$) and resulting torsional potential properties for the internal rotation of the OXC–CXO series of molecules. Torsional force constants in kcal/mol · rad^2, energies in kcal/mol.

Property	Glyoxal	F-Glyoxal	Cl-Glyoxal	Oxalyl-F	Oxalyl-Cl
k_t	11.74	9.25	9.86	6.62	6.30
k_c	3.66	6.90	4.06	5.03	-3.07
ΔV°	5.69	0.80	2.43	0.50	4.43
$k = k_c/k_t$	0.31	0.75	0.41	0.76	-0.49
α_0	108°	90°	93°	90°	
$\Delta V^{\neq}$	7.04	4.44	4.71	3.16	
β	0.73	0.50	0.54	0.50	
β_0	0.66	0.50	0.53	0.50	
α_0^{opt}	102°	90°	92°	90°	
$E^{\neq}$	7.81	5.29	5.33	4.30	
β^{opt}	0.55	0.50	0.51	0.50	

derivatives, are not yet available [35].

Comparison of torsional force constants demonstrates that *trans* wells are narrower than *cis* wells. The asymmetry shown by local wells, reflected in the values of k, is due to specific intramolecular forces that act in favour of a given conformation. On the other hand, it can be seen from the signs of k_t and k_c that *trans* conformations are associated with stable isomers, whereas the *cis* isomer appears to be unstable in oxalyl chloride. The interpolated potential function provides evidence for a *gauche* stable isomer for oxalyl chloride in the interval $[104°, 142°]$. This is based upon the fact that $V(\alpha)$ exhibits a change in curvature $[\frac{d^2V}{d\alpha^2} = 0]$. The same argument can be used to predict a barrier to internal rotation located in the interval $[42°, 104°]$. A more precise characterization of these critical points requires higher levels of approximation. However, it is worth mentioning that our theoretical results support the experimental evidence of Hagen and Hedberg [37], favouring the existence of *trans* and *gauche* conformers in oxalyl chloride. A recent theoretical investigation of the torsional potential in oxalyl chloride has been carried out by Hassett *et al.* [54]. Using different extended basis sets at both SCF and MP2 levels these authors were not able to find evidence for a second stable conformer.

Resulting torsional potential properties, α_0, β_0, β, and $\Delta V^{\neq} = V(\alpha_0)$, are displayed in Table III. Results show that, except for glyoxal, the potential curve is determined by the second term in Equation (16). It follows that the barrier to internal rotation in glyoxal is determined by both contributions to Equation (16). Looking at Equation (25), we see that for glyoxal all three terms are important in determining the barrier height, whereas for other molecules the barrier heights are essentially due to the first term.

Refinements of the potential functions that were necessary to obtain reliable descriptions of the isomerization processes were carried out following the methodology discussed in Section 2.3. Additional *ab-initio* calculations were performed at the predicted torsional transition state defined by α_0, and the relative energy $E^{\neq}$ was included in the determination of the potential parameters as a fourth external input datum. Resulting potential curves are shown in Figure 3.

In Table III we note a small shift of the potential barrier towards lower values of α for glyoxal. This appears through the changes occurring in the values of α_0 and the Brønsted coefficient. We note that α_0 goes from a first guess of 108°, with $\beta = 0.73$, down to an optimized set $\alpha_0^{\mathrm{opt}} = 102°$, $\beta^{\mathrm{opt}} = 0.55$. A similar but less marked effect occurs in chloroglyoxal, where our first guess of α_0 goes down by one degree only.

As mentioned above, the parameter β is a measure of the degree to which the transition state (TS) resembles the *cis* conformation. Values of β^{opt} collected in Table III indicate that the predicted TS resembles that isomer, in agreement with Hammond's postulate which is based upon an energetic criterium.

Comparative analysis of the potential curves and numerical results shows that most parameters associated with OHC–CXO have values intermediate between those corresponding to OHC–CHO and OXC–CXO. Barriers to internal rotation are ordered as follows: OHC–CHO > OHC–CClO > OHC–CFO > OFC–CFO. This ordering is due to the fact that substitution of some hydrogen by halogen atoms removes electronic charge from the central C–C bond region, producing lower barrier heights in the halogen-substituted molecules.

3.3. THE XS–SX (X = H, F, OR CL) SERIES OF MOLECULES

The Hartree–Fock, SCF–MO, *ab-initio* results were obtained using the split-valence 6–31G basis set with s and p diffuse functions and d polarization orbitals on the heavy atoms. Diffuse functions are used to get an accurate description of the SS bond. Also, diffuse s and p functions and d polarization orbitals were added to the halogen atoms in order to describe better the intramolecular interactions among these atoms. For the hydrogen atoms it was not necessary to include polarization or diffuse functions, since the SH bond is quite well described at a lower level: we have found an SH bond distance of 1.328 Å, which compares rather well with the experimental value of 1.327 Å.

Input data of the various properties under study are displayed in Table IV. The absolute reference energies for the *trans* conformations, determined through the

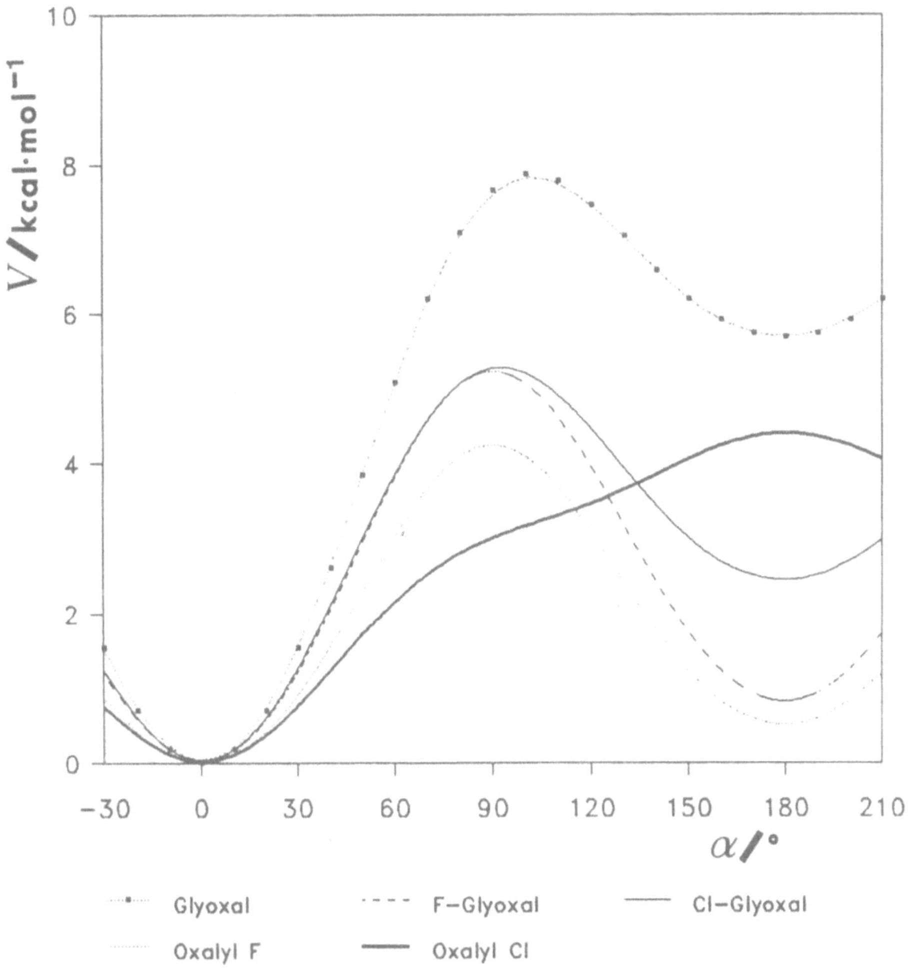

Figure 3. Torsional potential curves for the series OXC–CXO.

ab initio calculations, are -796.163232 *au*, -993.805549 *au*, and -1713.928904 *au* for HSSH, FSSF, and ClSSCl, respectively.

Many years ago Kuczkowski [40] has shown that the sulphur–sulphur bond distance in FS–SF was anomalously short, fact that was recently confirmed by Davis and Firth [41]. A value of r (S–S) = 1.889 Å was estimated from these experimental works. This should be compared to 2.029 Å and 2.055 Å in CH_3S–SCH_3 [42] and HS–SH [38, 39], respectively. Typical S–S single-bond lengths are around 2.05 Å, while the double bond-like distance obtained from S_2 is 1.89 Å. Both calculated and experimental S–S bonds in HSSH show a typical single bond distance of about 2.07 Å. In ClSSCl the calculated distance is found to be a little shorter, 2.01 Å (experimental, 1.97 Å), and lies between the typical single and double bond distances. In FSSF the double bond character is much clearer: its

TABLE IV. Parameters defining the analytic forms representing the conformational dependence, upon internal rotation, of the torsional potentials in the XSSX series of molecules. k_t and k_c in kcal/mol · rad^2; ΔV°, $\Delta V^{\neq}$ and $E_{\min}$ in kcal/mol.

Parameter	HS–SH	FS–SF	ClS–SCl
k_t	-9.47	-20.11	-8.26
k_c	-18.04	-59.11	-42.09
ΔV°	2.53	5.60	5.49
k=k_t/k_c	0.53	0.34	0.20
α_0^{opt}	90°	93°	95°
$E_{\min}$	-5.84	-19.71	-11.52
$\Delta V^{\neq}$	8.37	25.31	17.01

S–S distance is found to be 1.95 Å (experimental, 1.89 Å). This result is in close agreement with the infrared spectrum of FSSF, that unequivocally indicates that the SS bond has a definite double bond character, and therefore should be distinctly shorter than in XSSF (X = H, F), XSSCl (X = H, Cl), and HSSH [40, 41, 56].

Starting from the input data given in Table IV, we have determined the analytic functions representing the potential energy hindering the internal rotation of molecules in the XSSX series. The resulting functions were refined by introducing the energy of the fully optimized *gauche* stable conformations: $E_{\min} = E(\alpha_0^{\text{opt}})$. The potential curves obtained are displayed in Figure 4.

For HSSH our results are quite close to the experimental values measured by Herbs and Winnewisser [38]. For example, our ΔV° value of 2.53 kcal/mol is exactly the same as that obtained experimentally. Unfortunately, the experimental information concerning $V(\alpha)$ is meager. However, Herbs and Winnewisser have determined the *cis* and *trans* barrier heights for HSSH from a variety of milimeter-wave and far-infrared spectral data [38].

Our calculated *cis* barrier height is found to be 8.37 kcal/mol, very close to the experimental value 8.20 kcal/mol. There have been reports of *ab initio* studies of the barrier height in HSSH with bases of roughly double-ζ plus-polarization quality and full geometry optimization [44]. As expected, similar results obtained at the SCF level, without considering zero-point vibration effects. The torsional force constant for the *gauche* stable isomer (which is a measure of the bond strength), as determined from our potential function, is k_0 = 0.1018 mdyn Å/rad^2. This value compares fairly well with the 0.0992 mdyn Å/rad^2 derived from experimental

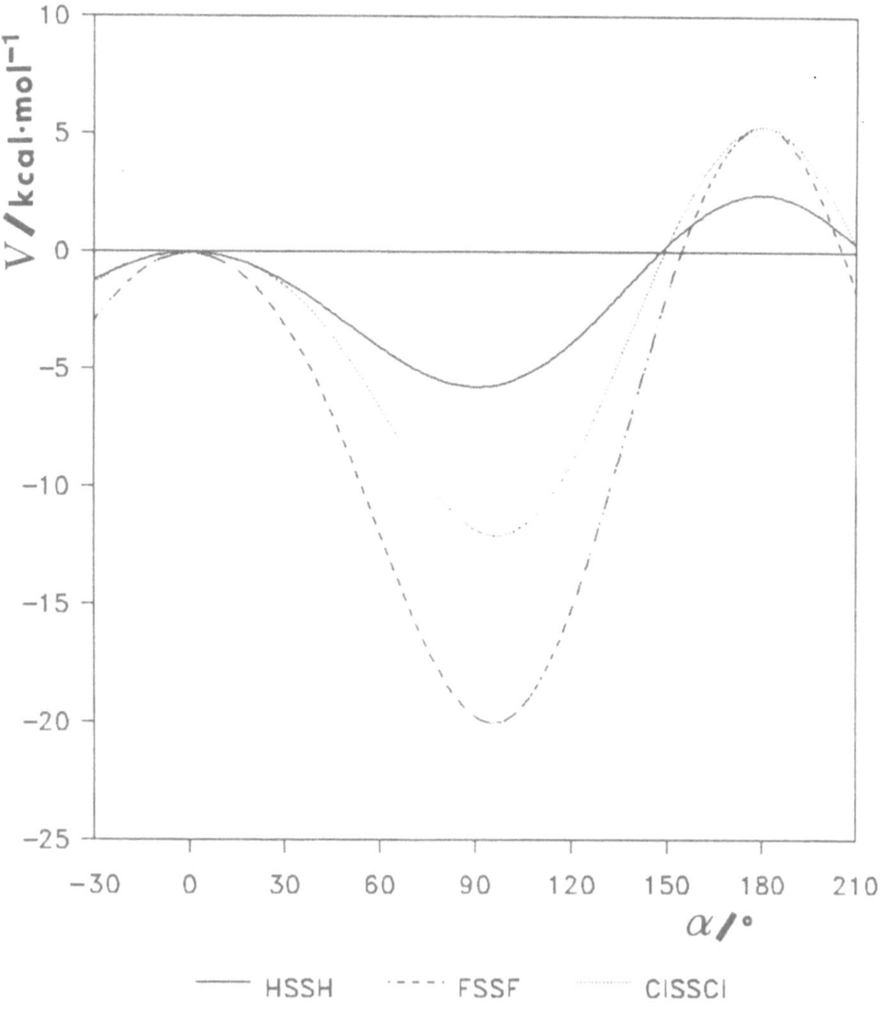

Figure 4. Torsional potential curves for the series XS–SX.

data.

Since the torsional barriers for all molecules are quite large, tunnelling under them should result in splitting of the lower energy levels. This effect has recently been observed in HSSH [39]. Large barrier heights are due to a hyperconjugative effect that results in a shortening of the SS bond length [56]. Comparison of the $\Delta V^{\neq}$ values for the substituted molecules shows that they strongly differ one from another, going from 25.31 kcal/mol for FSSF to 17.01 kcal/mol for ClSSCl. It is clear that the halogen atom favours a hyperconjugative interaction between the S–X σ-bond and the electron pair on the other sulphur. This hyperconjugative effect contributes to the π-bond character of the S–S bond and is favoured when the H atoms are replaced by the more electronegative halogens. To quantify roughly this effect, we consider the relative values of the XSSX potential barriers with respect to that of HSSH, given in Table IV. It is seen that $\Delta V^{\neq}(\text{FSSF}) \approx 3\Delta V^{\neq}(\text{HSSH})$ and $\Delta V^{\neq}(\text{ClSSCl}) \approx 2\Delta V^{\neq}(\text{HSSH})$.

The above issue concerning the hyperconjugative effect is confirmed by analysing the bond orders and torsional force constants. Calculated SS bond orders are ordered as: FSSF(1.3643) > ClSSCl(1.0940) > HSSH(0.9450). The numerical values of the torsional force constants at the minima (k_0) are 0.3610 mdyn Å/rad^2 and 0.1877 mdyn Å/rad^2 for FSSF and ClSSCl, respectively, to be compared to the 0.1019 mdyn Å/rad^2 of HSSH. In the light of the above results, one should expect a correlation between potential barriers and torsional force constants.

4. Concluding Remarks

We have presented a procedure for determining the analytic form of the conformational dependence of the potential energy of monorotor molecules. We have determined the relative stabilities of the various isomers, identified the *critical intermediate states* and evaluated the associated potential barriers. It has been shown that the conformational functions together with the Brønsted coefficients help determine the key properties of the CIS isomers, and provide a quantitative version of the Hammond postulate and the relative energy of any isomer along the torsional variable. The use of the conformational functions leads to an expression for the barrier heights in terms of properties associated to the reference isomers, as stated by Equations (25) and (27). This provides a basis for rationalizing experimental and theoretical data concerning the internal rotation of monorotor molecules.

Acknowledgements

The authors wish to thank Professors Yves G. Smeyers (CSIC, Madrid) and Ricardo Letelier (UCH, Santiago) for helpful discussions. This work was supported by FONDECYT through project No. 0835/1991. One of us (ATL) wishes to thank The Commission of European Communities for grants No. 44005 and 44083 for research visits in Spain and France, where part of this work was achieved.

References

[1] J. Maruani, A. Hernández-Laguna and Y.G. Smeyers: *J. Chem. Phys.* **63** (1975) 4515; *ibid.* **76** (1982) 3123 [Erratum **81** (1984) 1519].
[2] J. Maruani and A. Toro-Labbé: *Can. J. Chem.* **66** (1988) 1948.
[3] Y.G. Smeyers: *Introduction to Non-Rigid Molecule Theory*, Adv. Quantum Chem., Vol. 23, Academic Press, New York (1992), pp. 1–77.
[4] B. Pullman (ed.): *Quantum Mechanics of Molecular Conformations*, John Wiley & Sons, New York (1976).
[5] A. Toro-Labbé and C. Cárdenas-Lailhacar: *Int. J. Quantum Chem.* **32** (1987) 685.

[6] A. Toro-Labbé: *J. Mol. Struct. (Theochem)* **180** (1988) 209.
[7] A. Toro-Labbé: *J. Mol. Struct. (Theochem)* **207** (1990) 247.
[8] C. Cárdenas-Lailhacar and A. Toro-Labbé: *Theor. Chim. Acta* **76** (1990) 411.
[9] (a) M. Nonella, J.R. Huber and T.K. Ha: *J. Phys. Chem.* **91** (1987) 5203; (b) R.P. Müller, M. Nonella, P. Russegger and J.R. Huber: *Chem. Phys.* **87** (1984) 351.
[10] S. Nakamura, M. Takahashi, R. Okazaki and K. Morokuma, *J. Am. Chem. Soc.* **109** (1987) 4142.
[11] G.E. McGraw, D.L. Bernitt and I.C. Hisatsune: *J. Chem. Phys.* **45** (1966) 1392.
[12] H. Jones, R.M. Badger and G.E. Moore: *J. Chem. Phys.* **19** (1951) 1599.
[13] C.M. Deeley and I.M. Mills: *Mol. Phys.* **54** (1985) 23.
[14] R.T. Hall and G.C. Pimentel: *J. Chem. Phys.* **38** (1963) 1889.
[15] A.P. Cox, A.H. Brittain and D.J. Finnigan: *J. Chem. Soc. Faraday Trans.* **61** (1971) 2179.
[16] P.A. McDonald and J.S. Shirk: *J. Chem. Phys.* **77** (1982) 2355.
[17] L.A. Curtiss and V.A. Maroni: *J. Phys. Chem.* **90** (1986) 58.
[18] L.J. Lawlor, K. Vasudevan and F. Grein: *J. Am. Chem. Soc.* **100** (1978) 8062.
[19] R.E. Dodd, J.A. Rolfe and L.A. Woodward: *Trans. Faraday Soc.* **52** (1956) 145.
[20] A.C. Legon and D.J. Miller: *J. Chem. Soc.* **A** (1968) 1736.
[21] R. Vance and A.G. Turner: *Inorg. Chem. Acta* **149** (1988) 95.
[22] D.A. Dixon and K.O. Christe: *J. Phys. Chem.* **96** (1992) 1018.
[23] P.N. Noble: *J. Phys. Chem.* **95** (1991) 4695.
[24] R.R. Smardzewski and W.F. Fox: *J. Chem. Phys.* **60** (1974) 2980.
[25] S.A. Sorenson and P.N. Noble: *J. Chem. Phys.* **77** (1982) 2483.
[26] G.N. Currie and D.A. Ramsay: *Can. J. Phys.* **49** (1971) 317.
[27] J.R. Durig, C.C. Tong and Y.S. Li: *J. Chem. Phys.* **57** (1972) 4425.
[28] A.R.H. Cole, Y.S. Li and J.R. Durig: *J. Mol. Spectrosc.* **61** (1976) 346.
[29] C.E. Dykstra and H.F. Schaefer: *J. Am. Chem. Soc.* **97** (1975) 7210.
[30] J. Tyrrel: *J. Am. Chem. Soc.* **98** (1976) 5456.
[31] G.R. De Maré: *J. Mol. Struct. (Theochem)* **107** (1984) 127.
[32] Ch.W. Bock, Y.N. Panchenko and S.V. Krasnoshchiokov: *Chem. Phys.* **125** (1988) 63.
[33] Ch.W. Bock and A. Toro-Labbé: *J. Mol. Struct. (Theochem)* **232** (1991) 239.
[34] K. Kuchitsu, T. Fukuyama and Y. Morino: *J. Mol. Struct.* **4** (1969) 41.
[35] J.R. Durig, S.C. Brown and S.E. Hannum: *J. Chem. Phys.* **54** (1971) 4428.
[36] J.R. Durig and S.E. Hannum: *J. Chem. Phys.* **54** (1971) 4428.
[37] K. Hagen and K. Hedberg: *J. Am. Chem. Soc.* **95** (1973) 1003.
[38] E. Herbst and G. Winnewisser: *Chem. Phys. Lett.* **155** (1989) 572.
[39] E. Herbst, G. Winnewisser, K.M.T. Yamada, D.J. De Frees and A.D. McLean: *J. Chem. Phys.* **91** (1989) 5905.
[40] R.L. Kuczkowski: *J. Am. Chem. Soc.* **86** (1964) 3617.
[41] R.W. Davis and S. Firth: *J. Mol. Spectrosc.* **145** (1991) 225.
[42] E. Hirota: *Bull. Chem. Soc. Japan* **31** (1958) 130.
[43] A. Hinchliffe: *J. Mol. Struct.* **55** (1979) 127.
[44] D. Dixon, D. Zeroka, J. Wendoloski and Z. Wasserman: *J. Phys. Chem.* **89** (1985) 5334.
F. Grein: *Chem. Phys. Lett.* **116** (1985) 323.
T.K. Ha: *J. Mol. Struct.* **122** (1985) 225.
C.J. Marsden and B.J. Smith: J. Phys. Chem. **92** (1988) 347.
[45] G.I. Cárdenas-Jirón, C. Cárdenas-Lailhacar and A. Toro-Labbé: *J. Mol. Struct. (Theochem)* **210** (1990) 279.
[46] G.I. Cárdenas-Jirón, J.R. Letelier, J. Maruani and A. Toro-Labbé: *Molecular Engineering* **2** (1992) 17.
[47] G.S. Hammond: *J. Am. Chem. Soc.* **77** (1955) 334.
[48] J.E. Leffler: *Science* **117** (1953) 340.
[49] G.I. Cárdenas-Jirón and A. Toro-Labbé: *An. Quím.* **88** (1992) 43.
[50] G.I. Cárdenas-Jirón: M.Sc. Thesis, University of Chile, Santiago (1993).
[51] M.R. Peterson: *Program Monstergauss* (1977), Department of Chemistry, University of Toronto, Toronto, Ontario, Canada.

[52] G.I. Cárdenas-Jirón and A. Toro-Labbé: *Chem. Phys. Lett.* **222** (1994) 8.
[53] K.W. Butz, J.R. Johnson, D.J. Krajnovich and C.S. Parmenter: *J. Chem. Phys.* **86** (1987) 5923.
[54] D.M. Hassett, K. Hedberg and C. Marsden: *J. Phys. Chem.* **97** (1993) 4670.
[55] G. Winnewisser, M. Winnewisser and W. Gordy: *J. Chem. Phys.* **49** (1968) 3465.
[56] G.I. Cárdenas-Jirón, C. Cárdenas-Lailhacar and A. Toro-Labbé: *J. Mol. Struct. (Theochem)* **282** (1993) 113.

GROUP THEORY FOR THREE-DIMENSIONAL NON-RIGID MOLECULAR PROBLEMS. APPLICATION TO THE DOUBLE EQUIVALENT C_{3V} ROTATION PLUS BENDING, WAGGING OR TORSION MODE

YVES G. SMEYERS
Instituto de Estructura de la Materia,
C.S.I.C.
Serrano, 123,
E–28006 Madrid,
Spain

Abstract. In the present paper, the Non-Rigid Molecular Group (NRG) Theory is briefly presented, compared to the Longuet-Higgins one, and applied to some three-dimensional problems involving two equivalent C_S or C_{3v} rotations, plus another degree of freedom, of molecules having a C_{2v} frame in their most symmetric conformations. This third degree of freedom may be symmetric in-plane bending, out-of-plane wagging, asymmetric in-plane bending or torsion of the rotors with respect to a central axis. The respective Non-Rigid Molecular Groups are deduced, as well as their Character Tables and their Symmetry Eigenvectors developed on the basis of the solutions of the triple free rotor equation. The inner structures of the irreducible representations obtained in such a way is discussed in view of the variations of the electric dipole moments (or torque) introduced with the third degree of freedom in the molecules.

1. Introduction

A non-rigid molecule is a molecular system which presents large amplitude vibration modes. This kind of motion appears whenever the molecule possesses various isoenergetic forms separated by relatively low energy barriers. In such cases, intramolecular transformations occur.

This molecular plasticity confers interesting properties on the non-rigid mol-

Y.G. Smeyers (ed.), Structure and Dynamics of Non-Rigid Molecular Systems, 121–151.

ecules: such as a large specific heat capacity, etc. In particular, their Far Infrared spectrum presents a prolific band structure. On the other hand, small electronic excitations produce large structural changes in the excited states. As a result, because of the Franck-Condon factors, long band progressions are observed in their electronic spectra.

All these types of phenomena can be studied theoretically. For this purpose, the potential energy hypersurfaces on which the nuclei are moving have to be determined, the Schrödinger equations for the nuclear motions solved, and the level populations, band locations and band intensities deduced from the eigenvalues and eigenvectors.

Examples of such calculations can be found in the literature, such as the theoretical determination of band structures of the absorption spectra in acetaldehyde, thioacetaldehyde and thioacetone [1–4], as well as the far infrared torsional band structure in acetone [5]. In all these calculations, Group Theory for Non-Rigid Molecules have been largely used not only for simplifying the calculations, but also for classifying and labelling the transitions. Up to now, however, the calculations were limited to only two degrees of freedom. In the following, three degrees of freedom will be considered.

As is well known, the Non-Rigid Molecule Group is defined as the complete set of molecular conversion operations which commute with the nuclear motion Hamiltonian operator. The operations of such a set may be written either in terms of permutations and permutation-inversions of identical nuclei, either in terms of physical operations, such as rotations, inversions, etc. [6–8].

If the formalism of the permutations and permutation-inversions is adopted, we have the **Molecular Symmetry Group Theory** of Longuet-Higgins [9]. On the contrary, if the formalism of the rotations, inversions is retained, we have the **Non-Rigid Molecule Group Theory** such as developed in [6–8].

Now, it is interesting to remark that rotation in ethane and acetone presents the same Molecular Symmetry Group in the Longuet-Higgins Theory, while in the Non-Rigid Molecule Theory, it exhibits two isomorphic, but different groups.

In the same way, the double rotation plus a bending wagging or torsion mode, such as in acetone, dimethylamine or biacetyl, presents the same group as that of ethane and acetone in the Longuet-Higgins Theory, but it exhibits three new isomorphic different groups in the Non-Rigid Molecule Theory [6–8].

The reason for this difference stands in the fact that the Longuet-Higgins theory does not explicitly take into account the Hamiltonian operator, i.e., the potential energy hypersurface on which the nuclei are moving, but only permutations and permutation-inversions between identical nuclei [9]. For this reason, the Longuet-Higgins groups for different kinds of motions and/or different potential energy surfaces, but involving the same number of identical particles, can be the same.

In the present paper, we shall develop the **Non-Rigid Molecule Group Theory**, using physical operations, such as rotations and inversions [6–8], and we shall apply the theory to the double C_{3v} equivalent rotation plus another degree of freedom, such as bending, wagging or torsion. We shall obtain different isomorphic groups, which will contain more explicit information than the Longuet-Higgins groups. From these groups, the different sets of symmetry eigenvectors, which diagonalize the Hamiltonian matrix into boxes, will be more easily obtained.

2. The Non-Rigid Molecule Group

2.1. GENERAL THEORY

Before starting to study the Group Theory for Non-Rigid Molecules, it is convenient to outline that it is defined in the framework of the Born-Oppenheimer approximation, in which the movements of the electrons and nuclei can be considered separately [10]. The identical particles are regarded as distinguishable and labellable entities in order to establish the symmetry relation existing between them.

The Hamiltonian operator to be considered will be thus an effective nuclear Hamiltonian operator which describes the nuclear motions:

$$\hat{H}_N^n(X) = [\hat{T}(X) + V^n(X)]. \quad (1)$$

This Hamiltonian operator will depend on the electronic state n considered. Notice that the $V^n(X)$ is an effective potential which can exhibit eventually the same value for different nuclear coordinates Xs.

Group Theory for Non-Rigid Molecules considers the isoenergetic conformers, and the interconversion motions existing between them. Usually there are many possible interconversion movements between these isomers. These can be described by some operators, $\hat{M}_i(X)$ acting on X, and leaving the nuclear Hamiltonian operator (1) invariant.

2.2. THE NON-RIGID MOLECULE GROUP THEORY

As is well known, the complete set of such transformation operators which commute with the Hamiltonian operator [1] forms a group, called the "**Non-Rigid Molecule Group**"[8]. These operators can be expressed either in terms of permutations and permutation-inversions, as in the Longuet-Higgins Theory, either in terms of physical operations such as rotations, inversions, ring-puckering, etc. [6–8].

Notice that the second description explicitly considers specific physical operations, i.e., specific transformation coordinates on a well defined potential energy surface $V^n(X)$ in (1). Usually, the transformation coordinates approximately coincide with some vibration modes, so that they approximately coincide with those of the lowest transformation paths.

On the contrary, the Longuet-Higgins description does not specify either the dimension or the shape of this energy hypersurface, but essentially the number of identical particles to be permuted as well as the possible inversions between them.

Group Theory for Non-Rigid Molecules essentially considers large amplitude motions on $V(X)$, ignoring the small amplitude ones. So, the Non-Rigid Molecule Group (NRG) will be strictly defined as the complete set of physical conversion operations which commute with a given nuclear motion Hamiltonian operator (1), limited to large amplitude motions [8].

2.3. THE FULL NON-RIGID MOLECULE GROUP

To make this definition clear, let us consider a molecule in the absence of any external perturbation, so that its nuclear motion Hamiltonian operator can be written in terms of relative coordinates with respect to the center of mass, neglecting the translation coordinates. Under this condition, the nuclear Hamiltonian operator may be written as:

$$\hat{H} = \hat{T}(X_e) + \hat{T}(X_i) + \hat{T}(X_e, X_i) + V(X_i) \tag{2}$$

In this equation, $\hat{T}(X_e)$ is the kinetic operator corresponding to the overall rotation of the molecule expressed in terms of rotational coordinates around some orthogonal axes, X_e. $\hat{T}(X_i)$ is the kinetic operator corresponding to the intramolecular motions expressed in terms of internal coordinates, X_i. $\hat{T}(X_e, X_i)$ is the kinetic coupling operator between these two kinds of motion. Finally, $V(X_i)$ is the potential energy operator, which depends only on the intramolecular coordinates.

The complete set of the molecular conversion operations which commute with the nuclear motion operator (1) will contain overall rotation operations, describing the molecule rotating as a whole, and intramolecular motion operations, describing molecular moieties moving with respect to the rest of the molecule. Such a set forms a group, which we call the **Full Non-Rigid Group** (full NRG) [8].

Such a group is isomorphic to that of Longuet-Higgins' theory, but it contains additional information. Notice that it is because the NRG is expressed in terms of transformation variables that it contains more information. Conversely, the Longuet-Higgins Symmetry Molecular Group does not, and therefore remains to some extent more general.

2.4. THE RESTRICTED NON-RIGID MOLECULE GROUP

Let us now remark that, when the molecule possesses a two- or three-dimensional frame (organic molecules), the external rotation coordinates can be easily distinguished from the internal coordinates, i.e.: the coupling term $\hat{T}(X_e, X_i)$ in (2) is relatively insignificant. As a result, it can be neglected at least at a first approach:

$$\hat{H}^L = \hat{T}(X_e) + [\hat{T}(X_i) + V(X_i)] \tag{3}$$

At this level of approximation, the overall rotation coordinates and the internal motion coordinates become completely separable. Because of this separability, such an approximate Hamiltonian operator may be regarded as local.

When the frame is reduced to a single point (coordinate complexes) or a single line (ethane) the external rotation cannot be separated from the internal motion. Some intramolecular motion products could indeed give rise to some external movements [11].

Let us now remark that the complete set of the molecular conversion operations which commute with the approximate Hamiltonian operator (3) will define another group, which we call the local full NRG. This new group may be larger than the actual full NRG group.

Let us now consider separately the external rotation Hamiltonian operator :

$$\hat{H}_e = \hat{T}(X_e) \tag{4a}$$

as well as Hamiltonian operator restricted to the large intramolecular movements:

$$\hat{H}_i = \hat{T}(X_i) + V(X_i). \tag{4b}$$

The complete set of overall rotation operations which transform one rotational state into another one isoenergetically, i.e., which commute with the external rotation Hamiltonian operator (4a), will form a group called the **External Rotation Symmetry Group**. Notice that because the external rotation is relatively slower than the intramolecular motions, the external rotation symmetry group may be expected to be isomorphic to the symmetry point group of the molecule in its most probable configuration, which is usually the most symmetric one [8] and [12].

In the same way, the complete set of intramolecular operations which commute with the Hamiltonian operator restricted to the large intramolecular motions (4b), will define another group, which we call the **Restricted Non-Rigid Molecule Group** (rNRG).

Remark that the External Rotation Symmetry Group and the restricted Non-Rigid Molecule Group commute, *at this level of approximation*, since the external and internal coordinates are completely separable. As a result, the local full Non-Rigid Group, defined by operator (3) is expected to be isomorphic to the direct product of the restricted NRG by the symmetry point group of the molecule in its most probable conformation:

$$G^L_{\mathrm{fNRG}} \sim G_{\mathrm{rNRG}} \times G_{\mathrm{point\ group}} \tag{5}$$

This expression reminds one somewhat of that forwarded by Altmann for the Schrödinger Supergroup [13]. This similarity, however, is only apparent because of Altmann's particular definition of the symmetry operations, so that the Schrödinger subgroup does not commute necessarily with the isodynamic one [13].

In order to go farther into the Non-Rigid Molecule Theory, let us analyze expression (5), and compare the full and restricted NRGs. For this purpose, let us consider the following borderline cases:

(a) Non-rigid molecules without any symmetry in a random configuration
When a molecule does not possess any symmetry in a random configuration, the full and restricted NRGs are seen to be isomorphic. To verify this affirmation, let us turn for a moment to the permutation and permutation-inversion formalism and express both groups in that language. Since the number of identical particles are the same, as well as the possible inversions, both groups have to be necessarily the same in the L.H. formalism. As a result, both groups will have at least the same group structure in the physical operation formalism: i.e., they are isomorphic.

When this molecule has no symmetry in any configuration, the symmetry point group in (5) will be the identity, and the full NRG will coincide with the restricted one.

On the contrary, when the molecule retains some symmetry in some particular configurations, such as in acetone, the external rotation group will be the symmetry point group of the molecule in its most probable configurations (C_{2v}). In this special case, the full local NRG will be larger than the exact full one, since this last is isomorphic to the restricted one.

(b) Non-rigid molecules having some symmetry in a random configuration
When the molecule retains some symmetry in a random configuration, the external rotation group in (5) will be that of the system in its most probable configuration, and the restricted NRG group is expected to be smaller than the full NRG.

One of the difficulties of the NRG theory is to construct correctly the full or restricted Hamiltonian operators, as well as to deduce properly the physical operations which commute with these operators. In many cases, however, the interconversion operations may be described easily as rotation of molecular moieties around some axes supported by a solid frame. In such cases, the concept of restricted NRG recovers special relevance, and the restricted NRG is expected to be equivalent to the isometric group [14].

When the separation of external and internal motions is not possible, Equation (5) does not hold, and the concept of restricted NRG vanishes.

In the next two sections, we shall consider molecules of increasing complexity, the non-rigidity of which proceeds from internal rotations, inversion, etc. And we shall deduce their restricted NRG.

3. Non-Rigid Molecule Groups for Systems with One, Two and Three Degrees of Freedom

In the following, we shall introduce the Non-Rigid Molecule Group Theory in a gradual manner, considering non-rigid systems of increasing complexity with one, two or three degrees of freedom. In such a way, we will introduce the necessary mathematical machinery for applying the NRG theory to more involved problems.

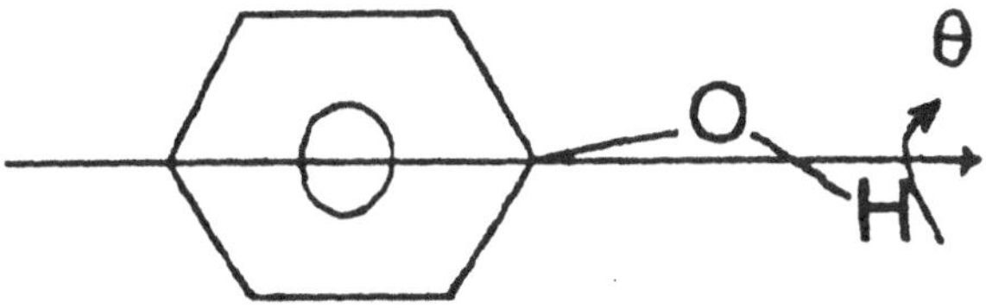

Figure 1. Internal rotation in phenol.

3.1. SINGLE ROTOR PROBLEM

Let us consider a molecule such as phenol which exhibits only one rotor: its hydroxyl moiety which is able to rotate around the C_2 symmetry axis of the phenyl ring.

The restricted Hamiltonian operator describing such a motion may be written as [15]:

$$\hat{H} = \left[-\frac{\partial}{\partial\theta} B(\theta) \frac{\partial}{\partial\theta} + V(\theta) \right] \tag{6}$$

where $B(\theta)$ is the internal rotation constant, which depends mildly on the rotation angle θ, and $V(\theta)$ is the potential energy operator according to which the rotor is moving.

$B(\theta)$ and $V(\theta)$ are structural parameters having the same symmetry. Therefore, in the following, we will regard $B(\theta)$ as a true constant and limit our considerations to the potential energy function. In order to deduce the symmetry properties of $V(\theta)$, let us note:

(a) The two-fold periodicity of the rotation;
(b) The invariance of the energy with respect to the sense of the rotation.

As a result, a two-fold potential energy function, which excludes any antisymmetrical terms with respect to the rotation coordinate may be written in (6):

$$\hat{H} = \left[-B \frac{\partial^2}{\partial\theta_2} + \sum_k A_K^C \cos 2K\theta \right] \tag{7}$$

where K is a positive integer.

The restricted NRG for the internal rotation in phenol may then be written as the set of the intramolecular operations which leaves (7) invariant. The following group is easily deduced:

$$G_{\mathrm{rNRG}} = [C_2^I \times U^I] \sim C_{2v} \tag{8}$$

which is a group of order four, isomorphic to the symmetry point group C_{2v}.

In expression (8), C_2^I is the subgroup of the two-fold rotation:

$$C_2^I \equiv [\hat{E} + \hat{C}_2] \tag{9}$$

where:

$$\hat{C}_2 f(\theta) = f(\theta + \pi). \tag{10}$$

TABLE I. Character table for the internal rotation in phenol.

	$\hat{E}$	$\hat{C}_2$	$\hat{U}$	$\hat{U}\hat{C}_2$
A_1	1	1	1	1
A_2	1	1	-1	-1
B_1	1	-1	1	-1
B_2	1	-1	-1	1

In the same way, U^I is the single switch subgroup:

$$U^I = [\hat{E} + \hat{V}] \tag{11}$$

defined by the single switch operation:

$$\hat{U} f\theta = f(-\theta) \tag{12}$$

The existence of this operation is conditioned by the presence of symmetry planes in the rotor and the frame [16].

The character table of such a group is written in Table I.

From this table, the symmetry eigenvectors, which factorize the matrix Hamiltonian corresponding to (7) into boxes, are easily deduced on the basis of the solutions of the planar free rotor equation. These are:

$$\begin{aligned} A_1 \rightarrow \chi_{A_1} &= \cos 2K\theta \quad B_1 \rightarrow \chi_{B_1} = \cos(2K+1)\theta \\ A_2 \rightarrow \chi_{A_2} &= \sin 2K\theta \quad B_2 \rightarrow \chi_{B_2} = \sin(2K+1)\theta \end{aligned} \tag{13}$$

Since phenol in a random conformation has no symmetry, its full NRG is expected to be isomorphic to the restricted one.

3.2. TWO ROTOR PROBLEMS

Let us now consider a molecule which has a second vibrational degree of freedom in addition to a rotor of order two. Such a molecule can be benzaldehyde, in which an out-of-plane wagging mode of the aldehydic hydrogen atom is allowed. See Figure 2.

In this figure, θ is the rotation angle of the phenyl moiety. α is the out-of-plane wagging angle, defined as a restricted rotation angle around a perpendicular axis to the CH bond projected in the plane of the molecule.

The restricted Hamiltonian operator corresponding to these motions may be easily deduced if we consider:

(a) The two-fold rotation of the phenyl group;
(b) The one-fold rotation of the aldehydic hydrogen atom;
(c) The invariance of the energy under the simultaneous change of the sense of the rotations. This last property is closely related with the existence of symmetry planes in both the rotor and the frame [17].

Figure 2. The benzaldehyde in standard and stereographic projections: torsion and wagging angles.

As a result, the restricted Hamiltonian operator is written as:

$$\hat{H} = \left[-B_{11}\frac{\partial^2}{\partial\theta^2} - 2B_{12}\frac{\partial^2}{\partial\theta\partial\alpha} - B_{22}\frac{\partial^2}{\partial\alpha^2}\right] + \sum_K \sum_L [A^{cc}_{KL}\cos 2K\theta\cos L\alpha + A^{ss}_{KL}\sin 2K\theta\sin L\alpha] \quad (14)$$

where the potential energy terms do not contain any antisymmetric products with respect to both coordinates, and K and L are positive integers.

The restricted NRG for the internal rotation and the wagging mode in benzaldehyde may then be written as the set of internal motion operators which leave the Hamiltonian operator (14) invariant:

$$G_{\mathrm{rNRG}} = [C_2^I \times V^I] \sim C_{2v} \quad (15)$$

which is another group of order four, isomorphic to the previous group of phenol (8) and to the symmetry point C_{2v}. In this new group, a new subgroup, V^I, appears:

$$V^I = [\hat{E} + \hat{V}]. \quad (16)$$

This is called the double switch subgroup because it contains the double switch operation:

$$\hat{V}f(\theta,\alpha) \equiv f(-\theta,-\alpha) \quad (17)$$

which is defined as a double simultaneous rotation from θ to $-\theta$ and from α to $-\alpha$. This operation was introduced by Altmann in his formalism [13].

As in the case of the simple switch operation, the existence of the double switch is conditioned by the presence of symmetry planes in the rotors and the frame [17]. The character group of such a group may be written as in Table II.

From this table, the symmetry eigenvectors are easily deduced on the basis of the solutions of the double free rotor equation. They factorize the Hamiltonian matrix, constructed with operator (14), into boxes.

These have the form:

$$A_1\begin{cases}\chi^{cc}_{A_1} = \cos 2K\theta\cos L\alpha\\ \chi^{ss}_{A_1} = \sin 2K\theta\sin L\alpha\end{cases} \qquad A_2\begin{cases}\chi^{cs}_{A_2} = \cos 2K\theta\sin L\alpha\\ \chi^{sc}_{A_2} = \sin 2K\theta\cos L\alpha\end{cases}$$

$$B_1\begin{cases}\chi^{cc}_{B_1} = \cos(2K+1)\theta\cos L\alpha\\ \chi^{ss}_{B_1} = \sin(2K+1)\theta\sin L\alpha\end{cases} \qquad B_2\begin{cases}\chi^{cs}_{B_2} = \cos(2K+1)\theta\sin L\alpha\\ \chi^{sc}_{B_2} = \sin(2K+1)\theta\cos L\alpha\end{cases} \quad (18)$$

TABLE II. Character table for the internal rotation and wagging vibration mode in benzaldehyde.

	$\hat{E}$	$\hat{C}_2$	$\hat{V}$	$\hat{V}\hat{C}_2$
A_1	1	1	1	1
A_2	1	1	-1	-1
B_1	1	-1	1	-1
B_2	1	-1	-1	1

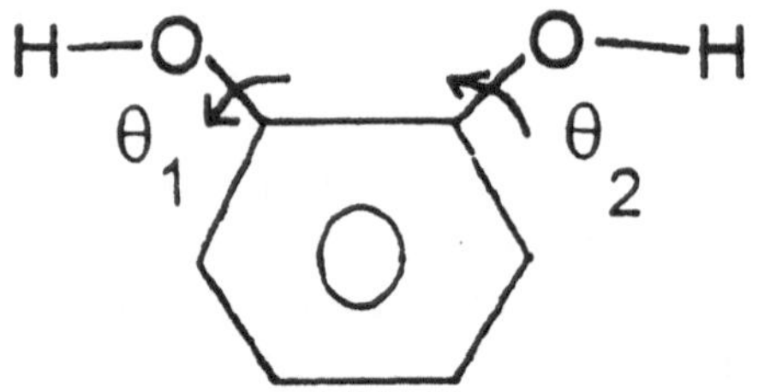

Figure 3. Structure of pyrocatechin, as well as the double equivalent rotation.

where the rotational functions of even and odd periodicity correspond to the A_n and B_n representations, respectively, as in phenol (Table I).

As a second example of a two-dimensional problem, let us consider a molecule such as pyrocatechin, in which there exist two equivalent one-fold rotors. The rotational coordinates for such a molecular system are θ_1 and θ_2 (see Figure 3).

The restricted Hamiltonian operator corresponding to these internal rotations may be easily deduced if we consider:

(a) The one-fold rotation of each rotor;
(b) The equivalence between the rotors;
(c) The invariance of the energy with respect to a simultaneous change of the sense of the rotations.

As a result, the Hamiltonian operator may be written as:

$$\begin{aligned}\hat{H} &= \left[-B_1\frac{\partial^2}{\partial\theta_1^2} - 2B_{12}\frac{\partial^2}{\partial\theta_1\partial\theta_2} - B_2\frac{\partial^2}{\partial\theta_2^2}\right] \\ &\quad + \sum_K\sum_L [A^{cc}_{KL}(\cos K\theta_1\cos L\theta_2 + \cos L\theta_1\cos K\theta_2) \\ &\quad + A^{ss}_{KL}(\sin K\theta_1\sin L\theta_2 + \sin L\theta_1\sin K\theta_2)] \qquad (19)\end{aligned}$$

where the potential energy terms are symmetric with respect to the exchange

TABLE III. Character table for the double equivalent internal rotation of pyrocatechin.

	$\hat{E}$	$\hat{W}\hat{V}$	$\hat{W}$	$\hat{V}$
A_1	1	1	1	1
A_2	1	1	-1	-1
A_3	1	-1	1	-1
A_4	1	-1	-1	1

and a simultaneous sign change of the rotation angles, and K and L are positive integers.

The restricted NRG for the double equivalent internal rotation in pyrocatechin may then be expressed as:

$$G_{\mathrm{rNRG}} = [C_1^I \times C_{1'}^I] \times [W^I \times V^I] \sim C_{2v} \tag{20}$$

which is again a group of order four isomorphic to the rNRGs (8) and (15), and the symmetry point group C_{2v}. In this group, V^I is the double switch subgroup, W^I is a new subgroup called the exchange subgroup, defined as:

$$W^I = [\hat{E} + \hat{W}]. \tag{21}$$

This subgroup is constructed by the identity and angle exchange operation, which reflects the indistinctiveness of the rotors. This is defined as:

$$\hat{W} f(\theta_1, \theta_2) \equiv f(\theta_2, \theta_1). \tag{22}$$

This operation was also introduced by Altmann [13]. The existence of this operation is conditioned by the equivalence of the rotors by superimposition [17]. The character table of this group is given in Table III.

Notice that in this table we retained the Longuet-Higgins' notation for the irreducible representations A_n of the G_{36} group of ethane [9], because the C_S rotors will be replaced next by C_{3v} ones.

From this character table, the symmetry eigenvectors for the double internal rotation in planar pyrocatechin are easily deduced on the basis of the double free rotor equation solutions:

$$\begin{aligned}
&A_1 \left\{ \begin{array}{l} \chi_{A_1}^{cc} = \cos K\theta_1 \cos L\theta_2 + \cos L\theta_1 \cos K\theta_2 \\ \chi_{A_1}^{ss} = \sin K\theta_1 \sin L\theta_2 + \sin L\theta_1 \sin K\theta_2 \end{array} \right. \\
&A_2 \rightarrow \chi_{A_2}^{cs} = \cos K\theta_1 \sin L\theta_2 - \sin L\theta_1 \cos K\theta_2 \\
&A_3 \rightarrow \chi_{A_3}^{cs} = \cos K\theta_1 \sin L\theta_2 + \sin L\theta_1 \cos K\theta_2 \\
&A_4 \left\{ \begin{array}{l} \chi_{A_4}^{cc} = \cos K\theta_1 \cos L\theta_2 - \cos L\theta_1 \cos K\theta_2 \\ \chi_{A_4}^{ss} = \sin K\theta_1 \sin L\theta_2 - \sin L\theta_1 \sin K\theta_2 \end{array} \right.
\end{aligned} \tag{23}$$

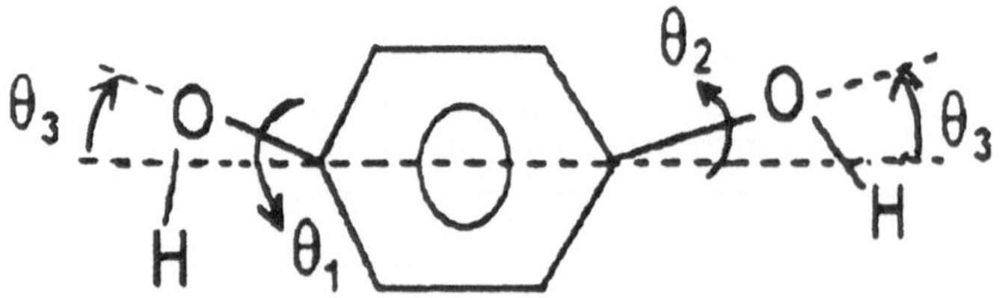

Figure 4. Para-hydroquinone structure, as well as the rotation angles, θ_1 and θ_2, and the bending angle θ_3.

where K and L are positive integers.

The equivalence of the rotors is seen to induce binomial forms, symmetric or antisymmetric with respect to the angle exchange, whereas the symmetry planes are seen to induce symmetric or antisymmetric functions with respect to the plane of the molecule, i.e., a simultaneous sign change of the rotation angles, A_2 being the completely antisymmetric representation.

Since the sine function depends on the sense of the rotations, the *sine* × *sine* symmetric binomial forms are considered to introduce a cog-wheel effect between the rotors into the Hamiltonian operator (19). These terms give rise to the elliptical form for the potential energy well [18].

As in phenol and benzaldehyde, pyrocatechin in a random configuration has no symmetry, and thus the full NRG is expected to be isomorphic with the restricted one.

3.3. THREE-DIMENSIONAL SIMPLE PROBLEMS

In the following, we shall consider some simple three-dimensional problems such as double one-fold rotation plus another third degree of freedom, and express their rNRGs using the operations defined in the previous sections for the one and two-dimensional problems.

The restricted Hamiltonian operator describing such a motion is a function of three variables: the two torsion angles θ_1 and θ_2, and a third degree of freedom, which may be expressed at least in principle as a rotation angle:

$$\hat{H} = \sum_i^3 \sum_j^3 \left[-\frac{\partial}{\partial \theta_i} B_{ij} \frac{\partial}{\partial \theta_j} \right] + V(\theta_1, \theta_2, \theta_3) \tag{24}$$

where B_{ij} are the so-called rotational constants which depend mildly on the torsion angles, and $V(\theta_1, \theta_2, \theta_3)$ the potential energy function.

Let us consider first a very symmetric molecule, such as para-hydroquinone (Figure 4), in which the two hydroxilic groups are allowed to rotate, as well as to oscillate symmetrically, into the molecular plane around the equilibrium position on the C_2 axis passing through the substituted carbon atoms.

Since the bending mode is symmetric, the restricted Hamiltonian operator will be invariant under the same operations as those of the two-dimensional problem (20) for pyrocatechin:

(a) The one-fold rotation of each rotor;
(b) The equivalence of the rotors;
(c) The invariance of the energy with respect to a simultaneous change of the rotations.
(d) In addition, the Hamiltonian operator has to be symmetric with respect to a symmetry plane perpendicular to the molecular plane along the para-phenylene C_2 axis: this operation can also be described by a 180° rotation of the benzene ring around this C_2 axis.

The restricted Hamiltonian operator describing such a nuclear motion, depending on three variables (Two torsions plus an additional large amplitude vibrational mode), is not so easy to transcribe in detail as in the previous cases. Therefore, the rNRG should be more conveniently deduced beforehand.

From the four conditions given earlier and the expression for the two-dimensional rNRG (20), the rNRG for the double rotation plus the 180° rotation of the benzene ring can be written as:

$$G_{\mathrm{rNRG}} = [C_1^I \times C_{1'}^I] \times [W^I \times V^I] \times U_b^I \tag{25}$$

where U_b^I is a switch operation subgroup for the 180° benzene rotation defined by the operation:

$$\hat{U}_b f(\theta_1, \theta_2, \theta_3) \equiv f(\pi - \theta_1, \pi - \theta_2, -\theta_3) \tag{26}$$

being the molecular plane the origin for the θ_1 and θ_2 rotations, and the symmetric equilibium position for bending angle θ_3.

This operation may be regarded as a reflection with respect to a plane perpendicular to the molecular plane and the C_2 rotation axis mentioned before (Figure 4).

The rNRG (25) for the double rotation and in-plane symmetric bending is now a group of order eight or a double group of order four, isomorphic to Longuet-Higgins' Complete Nuclear Permutation-Inversion Group for hydrogen peroxide [19].

The character table for such a group is easily deduced taking into account that the rNRG (25) can be written as a product of groups:

$$rNRG(25) = rNRG(20) \times U_b^I. \tag{27}$$

This can be found in [19]. From this table, the symmetry eigenvectors for the double internal rotation plus the in-plane symmetric bending mode, in planar para-hydroquinone, can be expressed as:

For the symmetric bending modes:

$$A_1 \begin{cases} \chi_{A_1}^{ccc} = (\cos K\theta_1 \cos L\theta_2 + \cos L\theta_1 \cos K\theta_2) \cos M\alpha \\ \chi_{A_1}^{csc} = (\sin K\theta_1 \sin L\theta_2 + \sin L\theta_1 \sin K\theta_2) \cos M\alpha \end{cases}$$

$$A_2 \rightarrow \chi_{A_2}^{csc} = (\cos K\theta_1 \sin L\theta_2 - \sin L\theta_1 \cos K\theta_2) \cos M\alpha$$

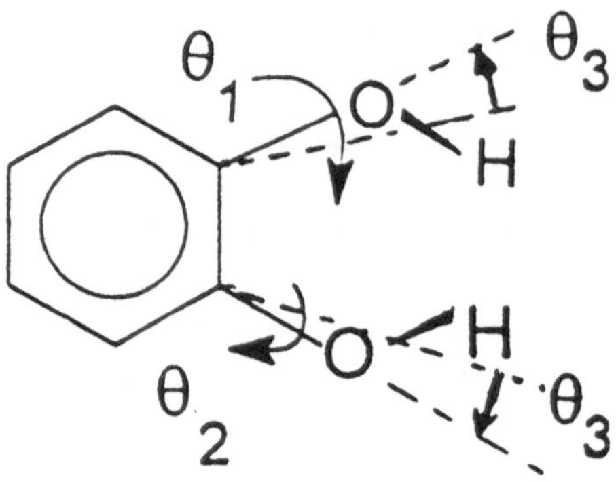

Figure 5. Double rotation and in-plane symmetric bending in pyrocatechin: θ_1 and θ_2 are the rotation angles, and θ_3 the bending angle.

$$A_3 \rightarrow \chi_{A_3}^{csc} = (\cos K\theta_1 \sin L\theta_2 + \sin L\theta_1 \cos K\theta_2) \cos M\alpha$$

$$A_4 \begin{cases} \chi_{A_4}^{ccc} = (\cos K\theta_1 \cos L\theta_2 - \cos L\theta_1 \cos K\theta_2) \cos M\alpha \\ \chi_{A_4}^{ssc} = (\sin K\theta_1 \sin L\theta_2 - \sin L\theta_1 \sin K\theta_2) \cos M\alpha \end{cases} \tag{28}$$

and for the anti-symmetric bending mode:

$$A_1' \begin{cases} \chi_{A_1}^{\prime ccs} = (\cos K\theta_1 \cos L\theta_2 + \cos L\theta_1 \cos K\theta_2) \sin M\alpha \\ \chi_{A_1}^{\prime sss} = (\sin K\theta_1 \sin L\theta_2 + \sin L\theta_1 \sin K\theta_2) \sin M\alpha \end{cases}$$

$$A_2' \rightarrow \chi_{A_2'}^{\prime css} = (\cos K\theta_1 \sin L\theta_2 - \sin L\theta_1 \cos K\theta_2) \sin M\alpha$$

$$A_3' \rightarrow \chi_{A_3'}^{\prime css} = (\cos K\theta_1 \sin L\theta_2 + \sin L\theta_1 \cos K\theta_2) \sin M\alpha$$

$$A_4' \begin{cases} \chi_{A_4}^{\prime ccs} = (\cos K\theta_1 \cos L\theta_2 - \cos L\theta_1 \cos K\theta_2) \sin M\alpha \\ \chi_{A_4'}^{\prime sss} = (\sin K\theta_1 \sin L\theta_2 - \sin L\theta_1 \sin K\theta_2) \sin M\alpha \end{cases} \tag{29}$$

where α states for θ_3, and K, L and M are positive integers.

As in the previous cases, the equivalence between the rotors, as well as the existence of symmetry planes in the frames and the rotors, induce binomial forms symmetric and anti-symmetric with respect to a simultaneous sign change of the rotation variables [8]. In addition, in the para-hydroquinone case, the symmetry eigenvectors have to be symmetric or anti-symmetric with respect to the bending angle α.

3.3.1. *Double Rotation plus Symmetric Bending Mode in Pyrocatechin*

Let us now again consider the pyrocatechin molecule, in which the two hydroxilic groups are able to rotate and simultaneously oscillate symmetrically with respect to the C_2 axis of the molecule into the molecular plane. Notice that this C_2 axis should not be confused with the above C_2 axis in para-hydroquinone (Figure 5).

The energy of pyrocatechin is not invariant anymore under the switch operation

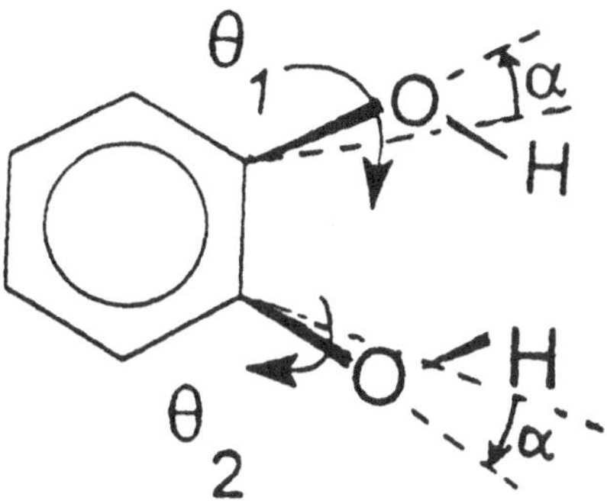

Figure 6. Double rotation and out-of-plane wagging in pyrocatechin: θ_1 and θ_2 are the rotation angles, and α the wagging angle.

(26), because the molecule lost its symmetry with respect to the molecular axis crossing the substituted carbon atoms.

The restricted Hamiltonian operator corresponding to the double internal rotation and the in-plane symmetric bending is easily deduced if we take into account:

(a) The one-fold rotation of each rotor;
(b) The equivalence of the rotors;
(c) The invariance of the energy with respect to a simultaneous change of the sense of the rotations.

From these conditions, the rNRG for the double rotation and in-plane symmetric bending is written as:

$$G_{\text{rNRG}} = [C_1^I \times C_{1'}^I] \times [W^I \times V^I] \sim C_{2v} \tag{30}$$

This expression coincides with the group of order four for the two-dimensional planar pyrocatechin (20).

The symmetry eigenvectors which transform according to the irreducible representations of this group are those given in (28 and 29), where α is the bending angle measured from the equilibrium position. However, because of the symmetry diminution of the Hamiltonian operator for the bending mode, the symmetry eigenvectors belonging to unprimed and primed representations combine between themselves for giving the four irreducible representations A_n.

3.3.2. *Double Rotation plus Out-of-Plane Wagging in Pyrocatechin*

Let us now consider once more pyrocatechin in which the two hydroxylic groups are rotating and wagging synchronously out-of-plane (see Figure 6). The restricted Hamiltonian operator describing such a motion (double rotation plus wagging) is again a function of three variables, the two rotation angles, θ_1 and θ_2, and the wagging angle α.

Due to the existence of the out-of-plane wagging, the two rotors are not equivalent any more by simple superimposition. They are, however, equivalent

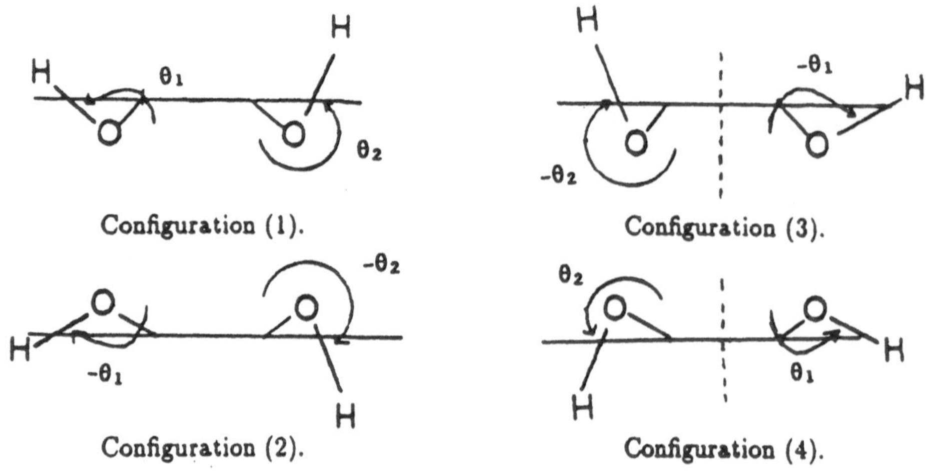

Figure 7. The four isoenergetic configurations of non-planar pyrocatechin in stereographic projection.

by reflection in a plane perpendicular to the molecular plane and containing the carbonyl bond. In fact it is easily seen that there are four isonergetic configurations as illustrated in Figure 7 [8].

In this figure, the non-planar pyrocatechin molecule has been presented in stereographic projection, with the hydroxylic groups rotated in a random way by angles θ_1 and θ_2, the molecular plane being the origin of the rotations. The two hydrogen atoms are deviating at an angle α from this plane, in conformity with (1).

The restricted NRG can be easily deduced by taking into account that the enantiomorphic configuration (2), obtained by reflection in the molecular plane, is isoenergetic. The corresponding operator is the triple switch operator:

$$(\hat{V}\hat{U})f(\theta_1, \theta_2, \alpha) \equiv f(-\theta_1, -\theta_2, -\alpha). \tag{31}$$

In the same way, it is clear that the configuration (3), obtained by reflection in a plane perpendicular to the molecular plane, is isoenergetic., i.e., the two rotors are equivalent by reflection. The operator corresponding to this internal motion is the double-switch-exchange operator:

$$(\hat{W}\hat{V})f(\theta_1, \theta_2, \alpha) \equiv f(-\theta_2, -\theta_1, \alpha). \tag{32}$$

Finally, the configuration (4) obtained by reflection in the molecular plane from the former (3) is also isoenergetic. The operator of such an internal motion is the single-switch-exchange operator:

$$(\hat{W}\hat{U})f(\theta_1, \theta_2, \alpha) \equiv f(\theta_2, \theta_1, -\alpha). \tag{33}$$

The non-rigid subgroup corresponding to these interconversion operators is:

$$G = [\hat{E} + \hat{U}\hat{V} + \hat{W}\hat{V} + \hat{W}\hat{U}] = [(WV)^I \times (VU)^I] \tag{34a}$$

TABLE IV. Character table for the double equivalent internal rotation and wagging in pyrocatechin.

	$\hat{E}$	$\hat{W}\hat{U}$	$\hat{W}\hat{V}$	$\hat{V}\hat{U}$
A_1	1	1	1	1
A_2	1	1	-1	-1
A_3	1	-1	1	-1
A_4	1	-1	-1	1

and the restricted non-rigid group:

$$G_{\mathrm{rNRG}} = [C_1^I \times C_{1'}^I] \times [(WV)^I \times (VU)^I] \sim C_{2v} \tag{34b}$$

which is again a group of order four isomorphic to the symmetry point group C_{2v}, as well as to the rNRGs for the single rotation in phenol (8), the double rotation in benzaldehyde (15) and in planar pyrocatechin (20), and the double rotation plus symmetric bending in pyrocatechin (30). The character table of such a group is given in Table IV.

From this character table, the symmetry eigenvectors are easily deduced on the basis of the solutions of the triple free rotor equation:

$$A_1 \begin{cases} \chi_{A_1}^{ccc} = (\cos K\theta_1 \cos L\theta_2 + \cos L\theta_1 \cos K\theta_2) \cos M\alpha \\ \chi_{A_1}^{ssc} = (\sin K\theta_1 \sin L\theta_2 + \sin L\theta_1 \sin K\theta_2) \cos M\alpha \\ \chi_{A_1}^{css} = (\cos K\theta_1 \sin L\theta_2 - \sin L\theta_1 \cos K\theta_2) \sin M\alpha \end{cases}$$

$$A_2 \begin{cases} \chi_{A_2}^{ccs} = (\cos K\theta_1 \cos L\theta_2 + \cos L\theta_1 \cos K\theta_2) \sin M\alpha \\ \chi_{A_2}^{sss} = (\sin K\theta_1 \sin L\theta_2 + \sin L\theta_1 \sin K\theta_2) \sin M\alpha \\ \chi_{A_2}^{csc} = (\cos K\theta_1 \sin L\theta_2 - \sin L\theta_1 \cos K\theta_2) \cos M\alpha \end{cases}$$

$$A_3 \begin{cases} \chi_{A_3}^{ccs} = (\cos K\theta_1 \cos L\theta_2 - \cos L\theta_1 \cos K\theta_2) \sin M\alpha \\ \chi_{A_3}^{sss} = (\sin K\theta_1 \sin L\theta_2 - \sin L\theta_1 \sin K\theta_2) \sin M\alpha \\ \chi_{A_3}^{csc} = (\cos K\theta_1 \sin L\theta_2 + \sin L\theta_1 \cos K\theta_2) \cos M\alpha \end{cases}$$

$$A_4 \begin{cases} \chi_{A_4}^{ccc} = (\cos K\theta_1 \cos L\theta_2 - \cos L\theta_1 \cos K\theta_2) \cos M\alpha \\ \chi_{A_4}^{ssc} = (\sin K\theta_1 \sin L\theta_2 - \sin L\theta_1 \sin K\theta_2) \cos M\alpha \\ \chi_{A_4}^{css} = (\cos K\theta_1 \sin L\theta_2 + \sin L\theta_1 \cos K\theta_2) \sin M\alpha \end{cases} \tag{35}$$

Here again the symmetry eigenvectors appear as different combinations of the unprimed and primed vectors of the more general case (30) for the symmetric in-plane bending of para-hydroquinone. These combinations are invariant under the operations of the rNRG (34).

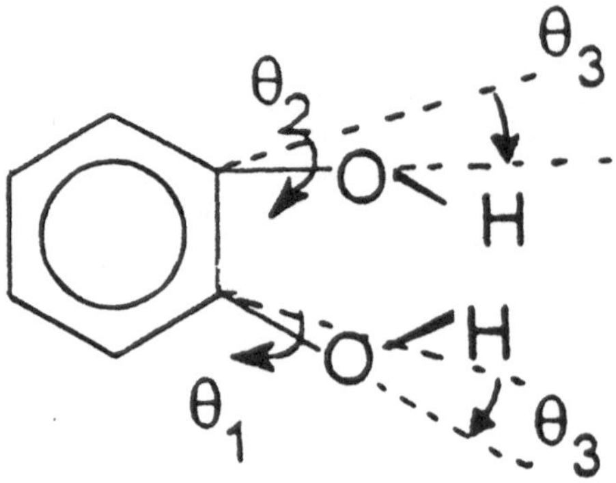

Figure 8. Double rotation and in-plane asymmetric bending in pyrocatechin: θ_1 and θ_2 are the rotation angles, and θ_3 the bending angle.

3.3.3. *Double Rotation plus Asymmetric Bending Mode in Pyrocatechin*

Let us consider next pyrocatechin in which the two hydroxylic groups are now rotating and oscillating asymmetrically with respect to the C_2 symmetry axis of the molecule into the molecular plane.

Since the bending mode is now asymmetric the equivalence of the rotors seems to be lost from a static point of view. However, taking into account the periodicity of the vibration, this equivalence is recovered dynamically: i.e., the Hamiltonian operator has to be invariant under an exchange-switch operator, $\hat{W}\hat{U}_b$, where the switch operator is defined as follows (Figure 8):

$$\hat{U}_b f(\theta_3) = f(-\theta_3) \qquad (36)$$

θ_3 being the in-plane bending angle measured from the equilibrium position.

The restricted Hamiltonian operator corresponding to the double internal rotation and in-plane asymmetric bending may be thus deduced if we consider:

(a) The one-fold rotation of each rotor;
(b) The dynamical equivalence of the rotors;
(c) The invariance of the energy with respect to a simultaneous change of the sense of the rotations.

From these conditions, the rNRG for the double rotation and in-plane asymmetric bending mode may be written as:

$$G_{\mathrm{rNRG}} = [C_I^I \times C_{I'}^I] \times [(WU_b)^I \times V^I] \sim C_{2v} \qquad (37)$$

where $(WU_b)^I$ is the subgroup of the dynamical exchange defined by the operation $\hat{W}\hat{U}_b$.

$$(WU_b)^I = [\hat{E} + \hat{W}\hat{U}_b]. \qquad (38)$$

The rNRG (37) is again a group of order four, isomorphic to the C_{2v} point group, and to the rNRGs for phenol (8), for benzaldehyde (15) and planar pyrocatechin (20), and the double rotation plus a symmetric bending or wagging mode in pyrocatechin (30) or (34). From the multiplication table, the character table is deduced. This table is identical to those of the mentioned group except for some operations. In particular, it is identical to that of planar pyrocatechin (Table III) in which the operation $\hat{W}$ is replaced by the $\hat{W}\hat{U}_b$ one.

From this character table, the following symmetry eigenvectors can be easily deduced:

$$A_1 \begin{cases} \chi_{A_1}^{ccc} = (\cos K\theta_1 \cos L\theta_2 + \cos L\theta_1 \cos K\theta_2) \cos M\alpha \\ \chi_{A_1}^{ssc} = (\sin K\theta_1 \sin L\theta_2 + \sin L\theta_1 \sin K\theta_2) \cos M\alpha \\ \chi_{A_1}^{ccs} = (\cos K\theta_1 \cos L\theta_2 - \cos L\theta_1 \cos K\theta_2) \sin M\alpha \\ \chi_{A_1}^{sss} = (\sin K\theta_1 \sin L\theta_2 - \sin L\theta_1 \sin K\theta_2) \sin M\alpha \end{cases}$$

$$A_2 \begin{cases} \chi_{A_2}^{csc} = (\cos K\theta_1 \sin L\theta_2 - \sin L\theta_1 \cos K\theta_2) \cos M\alpha \\ \chi_{A_2}^{css} = (\cos K\theta_1 \sin L\theta_2 + \sin L\theta_1 \cos K\theta_2) \sin M\alpha \end{cases}$$

$$A_3 \begin{cases} \chi_{A_3}^{csc} = (\cos K\theta_1 \sin L\theta_2 + \sin L\theta_1 \cos K\theta_2) \cos M\alpha \\ \chi_{A_3}^{css} = (\cos K\theta_1 \sin L\theta_2 - \sin L\theta_1 \cos K\theta_2) \sin M\alpha \end{cases}$$

$$A_4 \begin{cases} \chi_{A_4}^{ccc} = (\cos K\theta_1 \cos L\theta_2 - \cos L\theta_1 \cos K\theta_2) \cos M\alpha \\ \chi_{A_4}^{ssc} = (\sin K\theta_1 \sin L\theta_2 - \sin L\theta_1 \sin K\theta_2) \cos M\alpha \\ \chi_{A_4}^{ccs} = (\cos K\theta_1 \cos L\theta_2 + \cos L\theta_1 \cos K\theta_2) \sin M\alpha \\ \chi_{A_4}^{sss} = (\sin K\theta_1 \sin L\theta_2 + \sin L\theta_1 \sin K\theta_2) \sin M\alpha \end{cases} \quad (39)$$

As can be seen, these symmetry eigenvectors appear to be those of para-hydroquinone (28 and 29), or those of pyrocatechin for the in-plane symmetric bending, except that, in the present case, the unprimed and primed eigenvectors are combined in such a way that they are invariant under the operations of the rNRG (37). As before, the variable α stands for θ_3.

3.3.4. *Double Rotation plus Torsion Mode in Pyrocatechin*

Let us now consider again pyrocatechin, in which the two hydroxylic groups are allowed to rotate and simultaneously to twist up and down around the C–C bond, (see Figure 9).

The restricted Hamiltonian operator describing such a motion (double rotation plus torsion) is again a function of three variables: the two rotation angles, θ_1 and θ_2, and the torsion angle α. The two rotors lost the reflection symmetry in the molecular plane. But, because of the periodicity of the torsion, they possess a dynamic reflection symmetry: i.e., the Hamiltonian operator has to be invariant under the triple switch operation $\hat{V}\hat{U}_t$, where the switch operator, $\hat{U}_t$, is defined as follows:

$$\hat{U}_t f(\alpha) \equiv f(-\alpha) \quad (40)$$

where α is the torsion angle, the origin of which is the molecular plane.

The restricted Hamiltonian operator, corresponding to the double internal

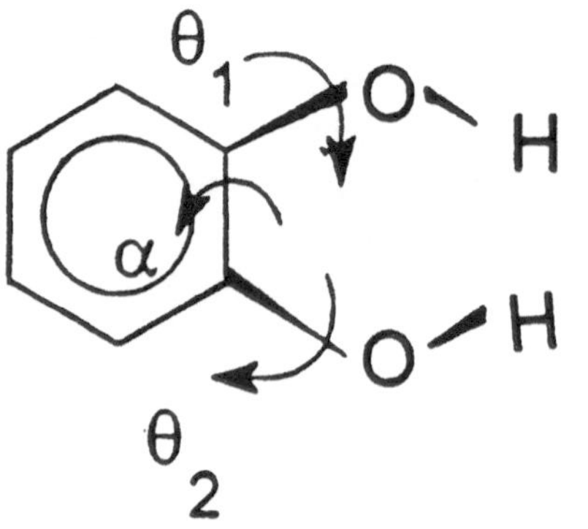

Figure 9. Double rotation and torsion mode in pyrocathchin: θ_1 and θ_2 are the rotation angles, and α the torsion angle.

rotation and torsion of the C–C bond of pyrocatechin, can be thus written, if we consider:

(a) The one-fold rotation of each rotor;
(b) The equivalence between the rotors;
(c) The dynamical invariance of the energy with repect to a simultaneous change of the sense of the rotations and the torsion.

From these conditions, the rNRG for the double rotation and torsion may be written as:

$$G_{\mathrm{rNRG}} = (C_1^I \times C_{1'}^I) \times [W^I \times (VU_t)^I]. \tag{41}$$

$(VU_t)^I$ is the subgroup of the triple switch defined by the operation $\hat{V}\hat{U}_t$:

$$(VU_t)^I = [\hat{E} + \hat{V}\hat{U}_t]. \tag{42}$$

The rNRG is once more a group of order four isomorphic to C_{2v} point group, as well as the rNRGs (8), (15), (20), (30), and (37). The character table of such a group is identical to that of planar pyrocatechin (Table III), except for the operation $\hat{V}$ which is replaced by $\hat{V}\hat{U}_t$.

From this character table, the symmetry eigenvectors are easily deduced on the basis of the solutions of the triple free rotor solutions:

$$A_1 \left\{ \begin{aligned} \chi_{A_1}^{ccc} &= (\cos K\theta_1 \cos L\theta_2 + \cos L\theta_1 \cos K\theta_2) \cos M\alpha \\ \chi_{A_1}^{ssc} &= (\sin K\theta_1 \sin L\theta_2 + \sin L\theta_1 \sin K\theta_2) \cos M\alpha \\ \chi_{A_1}^{css} &= (\cos K\theta_1 \sin L\theta_2 + \sin L\theta_1 \cos K\theta_2) \sin M\alpha \end{aligned} \right.$$

$$A_2 \left\{ \begin{aligned} \chi_{A_2}^{ccs} &= (\cos K\theta_1 \cos L\theta_2 - \cos L\theta_1 \cos K\theta_2) \sin M\alpha \\ \chi_{A_2}^{sss} &= (\sin K\theta_1 \sin L\theta_2 - \sin L\theta_1 \sin K\theta_2) \sin M\alpha \\ \chi_{A_2}^{csc} &= (\cos K\theta_1 \sin L\theta_2 - \sin L\theta_1 \cos K\theta_2) \cos M\alpha \end{aligned} \right.$$

$$A_3 \begin{cases} \chi_{A_3}^{ccs} = (\cos K\theta_1 \cos L\theta_2 + \cos L\theta_1 \cos K\theta_2) \sin M\alpha \\ \chi_{A_3}^{sss} = (\sin K\theta_1 \sin L\theta_2 + \sin L\theta_1 \sin K\theta_2) \sin M\alpha \\ \chi_{A_3}^{csc} = (\cos K\theta_1 \sin L\theta_2 + \sin L\theta_1 \cos K\theta_2) \cos M\alpha \end{cases}$$

$$A_4 \begin{cases} \chi_{A_4}^{ccc} = (\cos K\theta_1 \cos L\theta_2 - \cos L\theta_1 \cos K\theta_2) \cos M\alpha \\ \chi_{A_4}^{ssc} = (\sin K\theta_1 \sin L\theta_2 - \sin L\theta_1 \sin K\theta_2) \cos M\alpha \\ \chi_{A_4}^{css} = (\cos K\theta_1 \sin L\theta_2 - \sin L\theta_1 \cos K\theta_2) \sin M\alpha \end{cases} \quad (43)$$

These symmetry eigenvectors appear once more as combinations of the primed and unprimed ones of the more general cases for the symmetric in-plane bending of p-hydroquinone, and are invariant under the operations of the rNRG (41).

3.3.5. *Comparison between the Three-Dimensional Problems for the Symmetric Bending, Out-of-Plane Wagging, Asymmetric Bending and Torsion*

As may be seen, the irreducible representations, A_n and A'_n, of the most symmetric non-rigid molecule of para-hydroquinone, collapse in different ways into different A_n irreducible representations of rNRGs of less symmetry, according to some diminution of degrees of freedom.

When the irreducible representations of the most symmetric group are written as follows:

Symmetry eigenvectors for para-hydroquinone

Double rotation and in-plane symmetric bending:
$\{A_1\}; \{A_2\}; \{A_3\}; \{A_4\}$ and $\{A'_1\}; \{A'_2\}; \{A'_3\}; \{A'_4\}$.

The different irreducible representations for the symmetric bending, out-of-plane wagging, asymmetric bending, and torsion modes, can be expressed in terms of the A_n and A'_n representations as follows:

Symmetry eigenvectors for pyrocatechin

Double rotation and in-plane symmetric bending:
$\{A_1, A'_1\}; \{A_2, A'_2\}; \{A_3, A'_3\}; \{A_4, A'_4\}$.

Double rotation and out-of-plane wagging:
$\{A_1, A'_2\}; \{A_2, A'_1\}; \{A_3, A'_4\}; \{A_4, A'_3\}$.

Double rotation and in-plane asymmetric bending:
$\{A_1, A'_4\}; \{A_2, A'_3\}; \{A_3, A'_2\}; \{A_4, A'_1\}$.

Double rotation and torsion:
$\{A_1, A'_3\}; \{A_2, A'_4\}; \{A_3, A'_1\}; \{A_4, A'_2\}$.

where each term between brackets is an irreducible representation for the in-plane symmetric bending, out-of-plane wagging, in-plane asymmetric bending, and torsion rNRGs.

It is interesting to note that the different primed eigenvectors of the para-hydroquinone group combine with the unprimed ones according to the electric variations due to the third vibrational modes. So, the symmetric in-plane bending mode induces an electric dipole moment variation along the z axis of pyrocatechin, which can be classified according to the A_1 irreducible representation of the G_4 NRG [5]. In the same way, the out-of-plane wagging induces an electric dipole moment variation along the x axis, which can be classified according to the A_2 representation. The asymmetric in-plane bending mode induces an electric dipole moment variation along the y axis which can be classified according the A_4 representation. Finally, the torsion induces an electric torque variation around the z axis, which can be classified according to the A_3 representation.

4. Three-Dimensional Problems with Two Equivalent C_{3v} Rotors, plus Symmetric, Asymmetric Bending, Wagging or Torsion Mode

Next, we shall consider some more involved non-rigid systems which present two equivalent C_{3v} rotors, as well as an additional degree of freedom such as a bending, wagging or torsion mode. We shall deduce their restricted Non-Rigid Groups from those found in the previous sections for the one-fold rotors. From the rNRGs, the character tables and the symmetry eigenvectors will be deduced.

For these purposes, let us consider first the more symmetric case of para-xylene, in which the two methyl groups are allowed to rotate as well as oscillate symmetrically in the benzene ring plane, around their equilibrium position located on the phenylene C_2 axis.

According to expression (25) for para-hydroquinone, the rNRG for such a motion in para-xylene may be easily written if we substitute the one-fold rotation subgroups by three-fold ones:

$$G_{\mathrm{rNRG}} = (C_3^I \times C_{3'}^I) \wedge (W^I \times V^I) \times U^I \tag{44}$$

which is a group of order 72, or a double group of order 36, such as that of ethane [19].

The character table of such a group is easily deduced taking into account that the rNRG (44) may be written as a direct product of two subgroups: The G_{36} of acetone [8] or [20] and the U^I subgroup. The G_{36} character table can be found in [9] for ethane, or in [20] for acetone. From the character table the symmetry eigenvectors can be deduced as usual by using a projection operator [20].

In the following, we shall deduce them from the symmetry eigenvectors calculated for the double rotation in acetone [20]. As a result, the symmetry eigenvectors for the double rotation and symmetric bending in para-xylene can be written as products of those of acetone with those for the single switch subgroup, U^I.

For the non-degenerate representations, A_1, A_2, A_3, and A_4, we have:

$$A_1 \left\{ \begin{array}{l} \chi^{ccc}_{A_1} = (\cos 3K\theta_1 \cos 3L\theta_2 + \cos 3L\theta_1 \cos 3K\theta_2) \times \cos M\alpha \\ \chi^{ssc}_{A_1} = (\sin 3K\theta_1 \sin 3L\theta_2 + \sin 3L\theta_1 \sin 3K\theta_2) \times \cos M\alpha \end{array} \right. \tag{45}$$

where $L \geq K$ in order to avoid repetition.

$$A_2 \rightarrow \chi^{csc}_{A_2} = (\cos 3K\theta_1 \sin 3L\theta_2 - \sin 3L\theta_1 \cos 3K\theta_2) \times \cos M\alpha$$

$$A_3 \rightarrow \chi^{csc}_{A_3} = (\cos 3K\theta_1 \sin 3L\theta_2 + \sin 3L\theta_1 \cos 3K\theta_2) \times \cos M\alpha$$

where L and K may take all integer values.

$$A_4 \left\{ \begin{array}{l} \chi^{ccc}_{A_4} = (\cos 3K\theta_1 \cos 3L\theta_2 - \cos 3L\theta_1 \cos 3K\theta_2) \times \cos M\alpha \\ \chi^{ssc}_{A_4} = (\sin 3K\theta_1 \sin 3L\theta_2 - \sin 3L\theta_1 \sin 3K\theta_2) \times \cos M\alpha \end{array} \right.$$

where $L > K$. As well as the A_1', A_2', A_3' and A_4' representations in which the $\cos M\alpha$ factor has been replaced by the $\sin M\alpha$ one.

For the two-fold degenerate representations, E_1, E_2, E_3, and E_4, we have:

$$E_1 \left\{ \begin{array}{ll} \chi^{ccc}_{E_1^a} = & \{[\cos(3K \pm \delta)\theta_1 \cos(3L + \delta)\theta_2 \pm \sin(3K \pm \delta)\theta_1 \\ & \times \sin(3L + \delta)\theta_2 + \cos(3L + \delta)\theta_1 \cos(3K \pm \delta)\theta_2 \\ & \pm \sin(3L + \delta)\theta_1 \sin(3K \pm \delta)\theta_2]\} \times \cos M\alpha \\ \chi^{csc}_{E_1^b} = & \{[\cos(3K \pm \delta)\theta_1 \sin(3L + \delta)\theta_2 \mp \sin(3K \pm \delta)\theta_1 \\ & \times \cos(3L + \delta)\theta_2 - \sin(3L + \delta)\theta_1 \cos(3K \pm \delta)\theta_2 \\ & \pm \cos(3L + \delta)\theta_1 \sin(3K \pm \delta)\theta_2]\} \times \cos M\alpha \end{array} \right.$$

$$E_2 \left\{ \begin{array}{ll} \chi^{ccc}_{E_2^a} = & \{[\cos(3K \pm \delta)\theta_1 \cos(3L + \delta)\theta_2 \pm \sin(3K \pm \delta)\theta_1 \\ & \times \sin(3L + \delta)\theta_2 - \cos(3L + \delta)\theta_1 \cos(3K \pm \delta)\theta_2 \\ & \mp \sin(3L + \delta)\theta_1 \sin(3K \pm \delta)\theta_2]\} \times \cos M\alpha \\ \chi^{csc}_{E_2^b} = & \{[\cos(3K \pm \delta)\theta_1 \sin(3L + \delta)\theta_2 \mp \sin(3K \pm \delta)\theta_1 \\ & \times \cos(3L + \delta)\theta_2 + \sin(3L + \delta)\theta_1 \cos(3K \pm \delta)\theta_2 \\ & \mp \cos(3L + \delta)\theta_1 \sin(3K \pm \delta)\theta_2]\} \times \cos M\alpha \end{array} \right.$$

$$E_3 \left\{ \begin{array}{ll} \chi^{ccc}_{E_3^a} = & \{[\cos(3K \pm \delta)\theta_1 \cos(3L + \delta)\theta_2 \mp \sin(3K \pm \delta)\theta_1 \\ & \times \sin(3L + \delta)\theta_2 + \cos(3L + \delta)\theta_1 \cos(3K \pm \delta)\theta_2 \\ & \mp \sin(3L + \delta)\theta_1 \sin(3K \pm \delta)\theta_2]\} \times \cos M\alpha \\ \chi^{csc}_{E_3^b} = & \{[\cos(3K \pm \delta)\theta_1 \sin(3L + \delta)\theta_2 \pm \sin(3K \pm \delta)\theta_1 \\ & \times \cos(3L + \delta)\theta_2 + \sin(3L + \delta)\theta_1 \cos(3K \pm \delta)\theta_2 \\ & \pm \cos(3L + \delta)\theta_1 \sin(3K \pm \delta)\theta_2]\} \times \cos M\alpha \end{array} \right.$$

$$E_4 \begin{cases} \chi^{ccc}_{E^a_4} = \{[\cos(3K \pm \delta)\theta_1 \cos(3L+\delta)\theta_2 \mp \sin(3K \pm \delta)\theta_1 \\ \qquad \times \sin(3L+\delta)\theta_2 - \cos(3L+\delta)\theta_1 \cos(3K \pm \delta)\theta_2 \\ \qquad \pm \sin(3L+\delta)\theta_1 \sin(3K \pm \delta)\theta_2]\} \times \cos M\alpha \\ \chi^{csc}_{E^b_4} = \{[\cos(3K \pm \delta)\theta_1 \sin(3L+\delta)\theta_2 \pm \sin(3K \pm \delta)\theta_1 \\ \qquad \times \cos(3L+\delta)\theta_2 - \sin(3L+\delta)\theta_1 \cos(3K \pm \delta)\theta_2 \\ \qquad \mp \cos(3L+\delta)\theta_1 \sin(3K \pm \delta)\theta_2]\} \times \cos M\alpha \end{cases} \quad (46)$$

In all these expressions $\delta = \pm 1$ and $(3K \pm \delta)$ and $(3L + \delta)$ are always positive integers, $L \geq K$ for E_1 and E_3, and $L > K$ for E_2 and E_4. The superindexes a and b stand to distinguish the two E_n components. In the same way, the primed representations E'_1, E'_2, E'_3 and E'_4 are obtained replacing the $\cos M\alpha$ factor by the $\sin M\alpha$ ones.

For the four components of the G representation, G^a, G^b, G^c, and G^d, we have:

$$G \begin{cases} \chi^{ccc}_{G^a} = [\cos 3K\theta_1 \cos(3L+\delta)\theta_2 + \cos(3K+\delta)\theta_1 \cos 3L\theta_2] \\ \qquad \times \cos M\alpha \\ \chi^{ssc}_{G^a} = [\sin 3K\theta_1 \ \sin(3L+\delta)\theta_2 + \sin(3K+\delta)\theta_1 \ \sin 3L\theta_2] \\ \qquad \times \cos M\alpha \\ \chi^{csc}_{G^b} = [\cos 3K\theta_1 \sin(3L+\delta)\theta_2 + \cos(3K+\delta)\theta_1 \sin 3L\theta_2] \\ \qquad \times \cos M\alpha \\ \chi^{scc}_{G^b} = [\sin 3K\theta_1 \cos(3L+\delta)\theta_2 + \sin(3K+\delta)\theta_1 \cos 3L\theta_2] \\ \qquad \times \cos M\alpha \\ \chi^{ccc}_{G^c} = [\cos 3K\theta_1 \cos(3L+\delta)\theta_2 - \cos(3K+\delta)\theta_1 \cos 3L\theta_2] \\ \qquad \times \cos M\alpha \\ \chi^{ssc}_{G^c} = [\sin 3K\theta_1 \ \sin(3L+\delta)\theta_2 - \sin(3K+\delta)\theta_1 \ \sin 3L\theta_2] \\ \qquad \times \cos M\alpha \\ \chi^{csc}_{G^d} = [\cos 3K\theta_1 \sin(3L+\delta)\theta_2 - \cos(3K+\delta)\theta_1 \sin 3L\theta_2] \\ \qquad \times \cos M\alpha \\ \chi^{scc}_{G^d} = [\sin 3K\theta_1 \cos(3L+\delta)\theta_2 - \sin(3K+\delta)\theta_1 \cos 3L\theta_2] \\ \qquad \times \cos M\alpha \end{cases} \quad (47)$$

as well as the four components of the four-fold primed representation, G'^a, G'^b, G'^c and G'^d, in which the $cos M\alpha$ factor has been replaced by the $\sin M\alpha$ one.

Notice that we have used in (45) the symmetrized (or antisymmetrized) G eigenvectors with respect to the $\hat{W}$ operation in order to classify them in an univocal way.

4.1. DOUBLE C_{3V} ROTATION PLUS SYMMETRIC IN-PLANE BENDING

Let us consider the acetone (or the cis-dimethylglyoxal) molecule in which the two methyl moieties are allowed to rotate and simultaneously oscillate symmetrically into the molecular plane. Notice that the C_2 symmetry axis of the para-phenylene radical does not exist anymore in acetone, but a C_2 axis remains along (or between) the carbonyl bond(s).

The rNRG for such a motion is given by the equation (30) in which the one-fold rotation subgroups are replaced by the three-fold ones:

$$G_{\text{rNRG}} = (C_3^I \times C_{3'}^I) \wedge (W^I \times V^I) \tag{48}$$

which is a group of order 36.

This group coincides with the well known group for the double rotation in planar acetone, the character table of which is given in [3] and [20] in the Non-Rigid Molecular formalism, and in [9] in the Longuet-Higgins formalism. In Table V, we give this character table together with other information.

From this character table, the symmetry eigenvectors can be deduced, but they can be more easily obtained from those of para-xylene: Since the new Hamiltonian operator is not symmetric anymore with respect to the para-phenylene C_2 axis, the unprimed and primed irreducible representations (45), (46) and (47) collapse: i.e., the unprimed and primed symmetry eigenvectors have to be combined between themselves so that each new representation be symmetric or antisymmetric with respect to the operations of the new group (48). This combination is given in Section 4.5.

4.2. DOUBLE C_{3V} ROTATION PLUS OUT-OF-PLANE WAGGING

Let us consider again acetone in which the two methyl groups are rotating and the carbonyl rest wagging vertically out-of-plane. This situation is especially relevant in dimethylamine or the first excited states of acetone, which are no longer planar but pyramidal.

The rNRG for such a motion is given by Equation (34), in which the one-fold rotation subgroups are replaced by three-fold ones:

$$G_{\text{rNRG}} = (C_3^I \times C_{3'}^I) \wedge [(WV)^I \times (VU)^I]. \tag{49}$$

This group is isomorphic with those of acetone (30) or (48), but different because the operations of the subgroup $[(WV)^I \times (VU)^I]$ do not coincide with the operations of the subgroup $[W^I \times V^I]$.

The Hamiltonian operator now is not symmetric any more with respect to the double switch operation, $\hat{V}$, nor the exchange operation, $\hat{W}$; and the unprimed and primed irreducible representations, A_n, E_n, and G, have to be combined in such a way that the symmetry eigenvectors are symmetric or antisymmetric with respect to operations of the group (49).

For the non-degenerate representations, the combinations are given by ex-

TABLE V. Character table for the double internal rotation, plus symmetric in-plane bending, out-of-plane wagging, asymmetric in-plane bending or torsion mode in acetone or cis-dimethylglyoxal, with interaction between the moving parts.

Sym. bending	$\hat{E}$	$2\hat{C}_3, 2\hat{C}_{3'}$	$2\hat{C}_3\hat{C}_{3'}$	$2\hat{C}_3\hat{C}_{3'}^2$	$\hat{W}, 2\hat{W} \times\hat{C}_3\hat{C}_{3'}^2$	$6\hat{W} \times \hat{C}_3\hat{C}_{3'}$	$\hat{W}\hat{V}, 2\hat{W}\hat{V} \times\hat{C}_3\hat{C}_{3'}$	$6\hat{W}\hat{V} \times \hat{C}_3\hat{C}_{3'}$	$\hat{V}, 8\hat{V} \times\hat{C}_3\hat{C}_{3'}$
Wagging	$\hat{E}$	$2\hat{C}_3, 2\hat{C}_{3'}$	$2\hat{C}_3\hat{C}_{3'}$	$2\hat{C}_3\hat{C}_{3'}^2$	$\hat{W}\hat{U}, 2\hat{W}\hat{U} \times\hat{C}_3\hat{C}_{3'}^2$	$6\hat{W}\hat{U} \times \hat{C}_3\hat{C}_{3'}$	$\hat{W}\hat{V}, 2\hat{W}\hat{V} \times\hat{C}_3\hat{C}_{3'}$	$6\hat{W}\hat{V} \times \hat{C}_3\hat{C}_{3'}$	$\hat{V}\hat{U}, 8\hat{V}\hat{U} \times\hat{C}_3\hat{C}_{3'}$
Asym. bending	$\hat{E}$	$2\hat{C}_3, 2\hat{C}_{3'}$	$2\hat{C}_3\hat{C}_{3'}$	$2\hat{C}_3\hat{C}_{3'}^2$	$\hat{W}\hat{U}_b, 2\hat{W}\hat{U}_b \times\hat{C}_3\hat{C}_{3'}^2$	$6\hat{W}\hat{U}_b \times \hat{C}_3\hat{C}_{3'}$	$\hat{W}\hat{U}_b V, 2\hat{W}\hat{U}_b\hat{V} \times\hat{C}_3\hat{C}_{3'}$	$6\hat{W}\hat{U}_b\hat{V} \times \hat{C}_3\hat{C}_{3'}$	$\hat{V}, 8\hat{V} \times\hat{C}_3\hat{C}_{3'}$
Torsion	$\hat{E}$	$2\hat{C}_3, 2\hat{C}_{3'}$	$2\hat{C}_3\hat{C}_{3'}$	$2\hat{C}_3\hat{C}_{3'}^2$	$\hat{W}, 2\hat{W} \times\hat{C}_3\hat{C}_{3'}^2$	$6\hat{W} \times \hat{C}_3\hat{C}_{3'}$	$\hat{W}\hat{V}\hat{U}_t, 2\hat{W}\hat{V}\hat{U}_t \times\hat{C}_3\hat{C}_{3'}$	$6\hat{W}\hat{V}\hat{U}_t \times \hat{C}_3\hat{C}_{3'}$	$\hat{V}\hat{U}_t, 8\hat{V}\hat{U}_t \times\hat{C}_3\hat{C}_{3'}$
A_1	1	1	1	1	1	1	1	1	1
A_2	1	1	1	1	-1	-1	1	1	-1
A_3	1	1	1	1	1	1	-1	-1	-1
A_4	1	1	1	1	-1	-1	-1	-1	1
E_1	2	-1	2	-1	0	0	2	-1	0
E_2	2	-1	2	-1	0	0	-2	1	0
E_3	2	-1	-1	2	2	-1	0	0	0
E_4	2	-1	-1	2	-2	1	0	0	0
G	4	1	-2	-2	0	0	0	0	0

pressions (35) except for the periodicity of order three. For the two- and four-fold representations, they can be deduced in a similar way, and will be given hereafter in Section 4.5.

4.3. DOUBLE ROTATION C_{3V} PLUS ASYMMETRIC IN-PLANE BENDING

Let us now consider again the acetone (or cis-dimethylglyoxal) molecule in which the two methyl moieties are rotating and simultaneously bending asymmetrically in the molecular plane, with respect to the C_2 symmetry axis.

The rNRG of such a motion is given by Equation (37), in which the one-fold rotation subgroups are replaced by three-fold ones:

$$G_{\mathrm{rNRG}} = (C_3^I \times C_{3'}^I) \wedge [(WU_b)^I \times V^I]. \tag{50}$$

This expression is isomorphic to those for the double rotation in acetone (30), the double rotation plus symmetric bending (48), and the double rotation plus the out-of-plane wagging (49). It does not coincide with them, however, because of the operation $\hat{W}\hat{U}_b$.

Now, the Hamiltonian operator is not anymore symmetric with respect to the simple exchange operation $\hat{W}$, and the unprimed and primed irreducible representations (28) and (29) have to be combined in such a way that they are symmetric or antisymmetric with respect to the operations of the group (50), in particular to the dynamical exchange operation, $\hat{W}\hat{U}_b$.

For the non-degenerate representations, the combinations are given by expressions (39) except for the periodicity of order three. For the two- and four-fold representations, they can be deduced in a similar way, and will be given hereafter in Section 4.5.

4.4. DOUBLE C_{3V} ROTATION PLUS TORSION MODE IN CIS-DIMETHYLGLYOXAL

Let us now consider a molecule like cis-dimethylglyoxal which possesses a torsionable central bond, in which the two methyl groups are allowed to rotate and simultaneously to twist around the central bond.

The rNRG of such a motion is given by Equation (41), in which the one-fold rotation subgroups are replaced by three-fold ones:

$$G_{\mathrm{rNRG}} = (C_3^I \times C_{3'}^I) \wedge [W^I \times VU_t{}^I]. \tag{51}$$

This expression is isomorphic to those for the double rotation in acetone (30), the double rotation plus symmetric bending (48), and the double rotation plus the out-of-plane wagging (49), and double rotation plus the asymmetric bending mode (50). It does not coincide with them, however, because of the operation $\hat{V}\hat{U}_t$.

Now, the Hamiltonian operator is no longer symmetric with respect to the simple double switch operation $\hat{V}$; and the unprimed and primed irreducible representations (28) and (29) have to be combined in such a way that they are symmetric or antisymmetric with respect to the operations of the rNRG (51), in particular to the dynamical triple switch operation, $\hat{V}\hat{U}_t$.

For the non-degenerate representations, the combinations are given by expressions (43) except for the periodicity of order three. For the two- and four-fold representations, they can be deduced in a similar way, and will be given hereafter in Section 4.5.

4.5. COMPARISON BETWEEN THE THREE-DIMENSIONAL C_{3V} ROTOR PROBLEMS FOR SYMMETRIC BENDING, OUT-OF-PLANE WAGGING, ASYMMETRIC BENDING AND TORSION

As may be seen, the irreducible representations, A_n and A'_n, E_n and E'_n, and G and G' of the most symmetric case of para-xylene, collapse in different ways to yield different A_n, E_m, and G irreducible representations of rNRGs of less symmetry, according to some rules depending on the kind of diminution of degrees of freedom.

When the irreducible representations of the most symmetric group are represented as follows:

Symmetry eigenvectors for para-xylene
Double rotation and in-plane symmetric bending:
$\{A_1\}; \{A_2\}; \{A_3\}; \{A_4\}$ and $\{A'_1\}; \{A'_2\}; \{A'_3\}; \{A'_4\}$.
$\{E_1\}; \{E_2\}; \{E_3\}; \{E_4\}$ and $\{E'_1\}; \{E'_2\}; \{E'_3\}; \{E'_4\}$.
$\{G\}$ and $\{G'\}$.

Then, the different irreducible representations for the symmetric in-plane bending, out-of-plane wagging, asymmetric in-plane bending, and torsion modes, can be expressed in terms of the unprimed and primed representations of the para-xylene, as follows:

Symmetry eigenvectors for acetone or dimethylglyoxal
Double C_{3v} rotation and in-plane symmetric bending:
$\{A_1, A'_1\}; \{A_2, A'_2\}; \{A_3, A'_3\}; \{A_4, A'_4\}$.
$\{E^a_1, E'^a_1\}; \{E^a_2, E'^a_2\}; \{E^a_3, E'^a_3\}; \{E^a_4, E'^a_4\}$;
$\{E^b_1, E'^b_1\}; \{E^b_2, E'^b_2\}; \{E^b_3, E'^b_3\}; \{E^b_4, E'^b_4\}$.
$\{G^a, G'^a\}; \{G^b, G'^b\}; \{G^c, G'^c\}; \{G^d, G'^d\}$.

Double C_{3v} rotation and out-of-plane wagging:
$\{A_1, A'_2\}; \{A_2, A'_1\}; \{A_3, A'_4\}; \{A_4, A'_3\}$.
$\{E^a_1, E'^b_1\}; \{E^a_2, E'^b_2\}; \{E^a_3, E'^b_4\}; \{E^a_4, E'^b_3\}$;
$\{E^b_1, E'^a_1\}; \{E^b_2, E'^a_2\}; \{E^b_3, E'^a_4\}; \{E^b_4, E'^a_3\}$.

$\{G^a, G'^d\}; \{G^b, G'^c\}; \{G^c, G'^b\}; \{G^d, G'^a\}.$

Double C_{3v} rotation and in-plane asymmetric bending:
$\{A_1, A'_4\}; \{A_2, A'_3\}; \{A_3, A'_2\}; \{A_4, A'_1\}.$
$\{E_1^a, E_2'^a\}; \{E_2^a, E_1'^a\}; \{E_3^a, E_4'^a\}; \{E_4^a, E_3'^a\};$
$\{E_1^b, E_2'^b\}; \{E_2^b, E_1'^b\}; \{E_3^b, E_4'^b\}; \{E_4^b, E_3'^b\}.$
$\{G^a, G'^c\}; \{G^b, G'^d\}; \{G^c, G'^a\}; \{G^d, G'^b\}.$

Double C_{3v} rotation and torsion :
$\{A_1, A'_3\}; \{A_2, A'_4\}; \{A_3, A'_1\}; \{A_4, A'_2\}.$
$\{E_1^a, E_2'^b\}; \{E_2^a, E_1'^b\}; \{E_3^a, E_3'^b\}; \{E_4^a, E_4'^b\};$
$\{E_1^b, E_2'^a\}; \{E_2^b, E_1'^a\}; \{E_3^b, E_3'^a\}; \{E_4^b, E_4'^a\}.$
$\{G^a, G'^b\}; \{G^b, G'^a\}; \{G^c, G'^d\}; \{G^d, G'^c\}$.

Where each term between brackets is an irreducible representation for the in-plane symmetric bending, out-of-plane wagging, in-plane asymmetric bending, or torsion rNRGs, respectively.

In the present case of the double C_{3v} rotation, the different primed representations of the most symmetric case of para-xylene, collapse also with some of the unprimed ones according to the electric field variation induced by the third vibrational mode.

As a result, the in-plane symmetric bending mode induces an electric dipole moment variation along the z axis of the acetone molecule, which can be classified according to the A_1 irreducible representation of the G_{36} rNRG. In the same way, the out-of-plane wagging induces an electric dipole moment variation along the x axis, which can be classified according to the A_2 representation of the same group. The in-plane asymmetric bending induces an electric dipole moment variation along the y axis, which can be classified according to the A_4 representation. Finally, the torsion mode around the z axis induces an electronic torque arround the same axis, which can be classified according to the A_3 representation of the same group.

The selection rules for the vibrational transitions along the x, y and z axis, active in the FIR, are gathered in Table VI, [5, 8]. In the same table, the selection rules for polarized transition around the z axis, active in Raman, are given [21].

A direct relationship between the initial and final states, and the unprimed and primed irreducible representations of the G_{72} rNRG, can be seen for each type of electric variation. This relationship can also be generalized to the components of each representation.

TABLE VI. Selection rules for the electric dipole transitions in the FIR spectrum of acetone, as well as the electric torque around the z axis.

$\mu_z(A_1)$	$\mu_x(A_2)$	$\mu_y(A_4)$	R_z
$A_1 \leftrightarrow A_1$	$A_1 \leftrightarrow A_2$	$A_1 \leftrightarrow A_4$	$A_1 \leftrightarrow A_3$
$A_2 \leftrightarrow A_2$	$A_3 \leftrightarrow A_4$	$A_2 \leftrightarrow A_3$	
$A_3 \leftrightarrow A_3$			
$A_4 \leftrightarrow A_4$			
$E_1 \leftrightarrow E_1$	$E_1 \leftrightarrow E_1$	$E_1 \leftrightarrow E_2$	$E_1 \leftrightarrow E_2$
$E_2 \leftrightarrow E_2$	$E_2 \leftrightarrow E_2$		
$E_3 \leftrightarrow E_3$	$E_3 \leftrightarrow E_4$	$E_3 \leftrightarrow E_4$	$E_3 \leftrightarrow E_3$
$E_4 \leftrightarrow E_4$			$E_4 \leftrightarrow E_4$
$G \leftrightarrow G$	$G \leftrightarrow G$	$G \leftrightarrow G$	$G \leftrightarrow G$

5. Conclusions

In this paper, Longuet-Higgins' Symmetry Molecular Groups [9] for the single rotation in phenol, the double rotation in benzaldehyde or pyrocatechin, the double rotation plus another degree of freedom in pyrocatechin, etc., are seen to be the same. On the contrary, the Non-Rigid Molecular Groups, as developed in [3], appear to be different, although with the same group structure. The same behavior can be observed between the Symmetry Molecular Groups and the Non-Rigid Molecular Groups for the double rotation in acetone and the double rotation plus another degree of freedom. It may be concluded that the Non-Rigid Molecular Groups are more descriptive and contain more information than the Longuet-Higgins ones.

In addition, a NRG group of a sytem of higher dimension is seen to possess the same structure as its homologue with lower dimension. Only the operations are different. As a result, the Character Tables can be easily deduced, and from the Character Tables the Symmetry Eigenvectors. In this paper, we deduced the Symmetry Eigenvectors, on the basis of the triple free rotor solutions as irreducible representations, for the Non-Rigid Groups corresponding to the double rotation plus one of the four possible large amplitude vibrations of a frame of C_{2v} symmetry, i.e., the symmetric in-plane bending, the out-of-plane wagging, the asymmetric in-plane bending, and the torsion.

This extension of the application of the NRG groups of two-dimensional systems to three-dimensional ones can be generalized to problems of still higher dimensions.

References

[1] M. Baba, I. Hanazaki and V. Nagashima: *J. Chem. Phys.* **82** (1985) 3938.
[2] Y.G. Smeyers, A. Niño and D.C. Moule: *J. Chem. Phys.* **93** (1990) 5786.
[3] Y.G. Smeyers, M.L. Senent and D.C. Moule: *Int. J. Quantum Chem.* **S-24** (1990) 835.
[4] D.C. Moule, Y.G. Smeyers, M.L. Senent, D.J. Clouthier, J. Karolczak and R.H. Judge: *J. Chem. Phys.* **95** (1991) 3137.
[5] Y.G. Smeyers, M.L. Senent, V. Botella and D.C. Moule: *J. Chem. Phys.* **98** (1993) 2754.
[6] Y.G. Smeyers: "Teoría de grupos para moléculas no-rígidas. El grupo local. Aplicaciones", in *Memorias*, Vol. 13, Real Academia de Ciencias, Madrid (1989).
[7] Y.G. Smeyers: *Fol. Chim. Theoret. Lat.* **17** (1989) 15.
[8] Y.G. Smeyers: *Introduction to Non-Rigid Molecule Theory*, Adv. Quantum Chem., Vol. 23, Academic Press, New York (1992), pp. 1–77.
[9] H.C. Longuet-Higgins: *Mol. Phys.* **6** (1963) 445.
[10] M. Born and J.R. Oppenheimer: *Ann. Physik* **84** (1927) 457.
[11] C.M. Woodman: *Mol. Phys.* **19** (1970) 753.
[12] J.T. Hougen: *J. Chem. Phys.* **37** (1962) 1433.
[13] S.L. Altmann: *Proc. Roy. Soc.* **A298** (1967) 184.
[14] H. Frei, A. Bauder and Hs.H. Günthard: *Large Amplitude Motion in Molecules*, Topics in Current Chemistry, Vol. 81, Springer, Berlin (1979).
[15] Y.G. Smeyers and A. Hernandez-Laguna: *Int. J. Quantum Chem.* **22** (1982) 681.
[16] J. Maruani, A. Hernandez-Laguna and Y.G. Smeyers: *J. Chem. Phys.* **63** (1975) 4515.
[17] J. Maruani, Y.G. Smeyers and A. Hernandez-Laguna: *J. Chem. Phys.* **76** (1982) 3123.
[18] Y.G. Smeyers: *J. Mol. Struct. (Theochem.)* **107** (1984) 3.
[19] P.R. Bunker: *Molecular Symmetry and Spectroscopy*, Academic Press, New York (1979).
[20] Y.G. Smeyers and M.N. Bellido: *Int. J. Quantum Chem.* **19** (1981) 553.
[21] Y.G. Smeyers, M.L. Senent, J. Peñalver and D.C. Moule: *J. Mol. Struct. (Theochem.)* **287** (1993) 117.

NON-RIGIDITY IN HEPTACOORDINATE COMPLEXES

JEAN BROCAS
Université Libre de Bruxelles,
Department of Organic Chemistry,
Faculty of Sciences,
Avenue F.D. Roosevelt, 50,
1050 Bruxelles,
Belgium

Abstract. The dsd model is applied to four static geometries of heptacoordinate complexes. These dsd interconversions are described by permutational analysis and compared to the most significant dynamic NMR results for these complexes.

1. Introduction

Twenty five years ago, it was proposed that heptacoordinate stereochemistry should be discussed in terms of three static geometrical structures: the D_{5h} pentagonal bipyramid, the C_{2v} 4-capped trigonal prism and the C_{3v} capped octahedron. The less common C_s tetragonal base-trigonal base 4:3 geometry has also been proposed [1]. Interconversion paths connecting the D_{5h}, C_{2v} and C_{3v} polyhedra, shown in Figure 1, have been discussed later [2].

More recent theoretical calculations based on the minimization of the total repulsion of the ML bonds in the ML_7 framework confirm these early predictions since the above C_{2v}, C_{3v} and D_{5h} polyhedra appear to be the most stable geometries and are of comparable stability [3]. *Static* NMR observations are also in agreement with these considerations: as we will see below the only static structures observed are the C_{2v}, C_{3v}, D_{5h} and the 4:3 geometries. We will also see that the interpretation of the *dynamic* NMR observations can be related to some of the interconversions of Figure 1 but other pathways have been invoked.

The aim of the present contribution is to compare these dynamic NMR results to the so-called *dsd model*. In Section 2, we recall some results of the permutational analysis in connection with dynamic NMR results. In Section 3, the dsd model

Y.G. Smeyers (ed.), Structure and Dynamics of Non-Rigid Molecular Systems, 153–180.

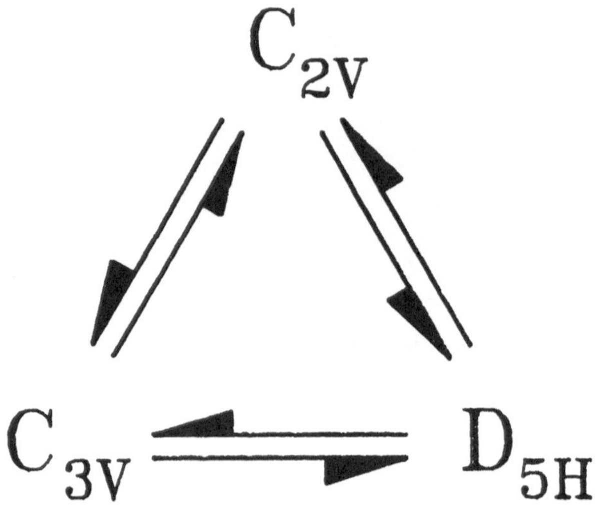

Figure 1. Hypothetical heptacoordinate interconversions.

for heptacoordinate interconversions is described. The most significant dynamic NMR observations are discussed in Section 4. We conclude in Section 5 by comparing these results to the dsd predictions.

2. Permutational Analysis

We recall briefly some of the results of permutational analysis. Details and further references may be found for instance in Reference [4]. It is convenient to label the sites of the molecular skeleton and the ligands disposed on it and supposed to be identical, at least for the moment. The skeleton numbering of three of the four observed skeleta is shown on Figure 2. The 4:3 geometry will be discussed later since it corresponds to several polyhedra [5]. We will use permutation groups to describe the molecular symmetry: a permutation (123) means that ligand on site 1 replaces ligand on site 2, that ligand on site 2 replaces ligand on site 3 and that ligand on site 3 replaces ligand on site 1. Let G be the permutation group expressing the molecular symmetry and A its subgroup containing only proper symmetry operations i.e. rotations.

For chiral molecules

$$G = A \tag{1}$$

whereas for achiral ones

$$G = A \cup \sigma A \tag{2}$$

where $\cup$ means union and where σ is any improper operation. It is easy to write down the symmetry group of each of the polyhedra of Figure 1.

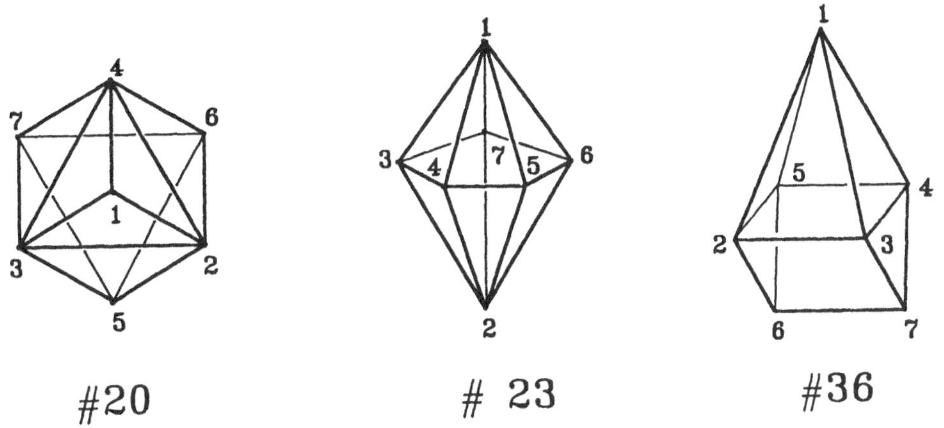

Figure 2. Skeleton site labelling for three polyhedra.

For the C_{2v} 4-capped trigonal prism, if I means identity

$$\begin{aligned} A &= \{I, (24)(35)(67)\} \\ A\sigma &= \{(23)(45)(67), (25)(34)\} \end{aligned} \tag{3}$$

for the C_{3v} capped octahedron

$$\begin{aligned} A &= \{I, (234)(576), (243)(567)\} \\ A\sigma &= \{(23)(67), (24)(57), (34)(56)\} \end{aligned} \tag{4}$$

and for the D_{5h} pentagonal bipyramid

$$\begin{aligned} A = \; & \{I, (34567), (35746), (36475), (37654), (12)(36)(45), \\ & (12)(34)(57), (12)(37)(46), (12)(35)(67), (12)(47)(56)\} \end{aligned} \tag{5}$$

and $A\sigma = A(12)$ is easy to write down.

Assuming that complete ligand scrambling is possible, there exist $n!$ ligand distributions on the molecular skeleton sites (here $n = 7$). A configuration is the set of distributions related by a proper symmetry operation. The number p of configurations is

$$p = \frac{n!}{|A|} \tag{6}$$

where $|A|$ is the number of elements in the set A.

Heptacoordinate complexes are non-rigid chemical species: interconversion of the configurations may be described by using permutations. Let x be a permutation among the set of 7! permutations of S_7, the symmetric group of degree 7. Let $M(x)$ be the set of permutations which are indistinguishable from x because (a) they generate the same final configuration as x, starting from a given fixed one, and/or (b) they must occur with the same probability as x.

Musher [6] has proposed to call such a set a "mode of rearrangements". The above conditions (a) and (b) have been expressed mathematically by Hässelbarth, Ruch [7] and Klemperer [8] in their pioneer papers. There are numerous examples of applications of these concepts in dynamic stereochemistry. The reader might find details and recent examples in a textbook [4] and a review [9] on this subject. The set $M(x)$ may be expressed in terms of the operations of the group G:

$$M(x) = (AxA) \cup (A\sigma x\sigma^{-1}A) \tag{7}$$

where $\cup$ means "union".

Muetterties [10] has defined the connectivity of a mechanism as the number of configurations reached from a given configuration in one step of this mechanism. The connectivity δ_x of the mode $M(x)$ is the connectivity of the mechanism whose result is expressed by the permutation x. It has been shown that [7]:

$$\delta_x = \frac{|A|}{|(x^{-1}Ax \cap A|} \tag{8}$$

when AxA and $A\sigma x\sigma^{-1}A$ are identical sets and that

$$\delta_x = 2\frac{|A|}{|(x^{-1}Ax) \cap A|} \tag{9}$$

when these sets are distinct ($\cap$ means "intersection").

The same authors have also shown that the group of $n!$ permutations may be partitioned into z sets $M(x)$. In other words, z is the number of a priori possible modes of rearrangements given by [7]:

$$z = \frac{|n!|}{|A|\,|G|} \sum_y \frac{|A \cap D(y)|^2 + |A\sigma \cap D(y)|^2}{|D(y)|} . \tag{10}$$

In the last equation, $D(y)$ is the conjugacy class of the element y in S_n.

Until now we did not discuss the relationship between the above definition of $M(x)$ and detailed balance. Let us consider x^{-1}, the inverse of x. If x transforms a starting configuration into a final one, x^{-1} clearly describes the reverse rearrangement. Since starting and final configurations have identical energies, x and x^{-1} describe equiprobable rearrangements. Accordingly, x and x^{-1} should belong to the same mode of rearrangement (see condition (b) above). This is however not always the case, as we shall see below.

Indeed, according to an argument of Nourse [11], one should distinguish between self-inverse (SI) and non-self-inverse (NSI) rearrangements: a rearrangement described by a permutation x is SI if and only if the set Rx contains at least either a self-inverse permutation (inversion) or a pair of mutually inverse permutations (inversion). The symbol R represents the Hougen group [12]

$$R = A \cup A\sigma J \tag{11}$$

where J is the inversion about the centre of mass. Hence, Rx is the set of permutations and permutations-inversion representing the configuration generated by the permutation x. It is easy to show [13] that, for SI rearrangements only,

$M(x) = M(x^{-1})$, or equivalently that x and x^{-1} belong to the same mode, as required. For NSI rearrangements however, x and x^{-1} do not belong to the same mode. Hence the definition of the mode has to be modified. Let $M_{\text{ext}}(x)$ be

$$M_{\text{ext}}(x) = M(x) \cup M(x^{-1}). \tag{12}$$

This last definition of the mode has to be used instead of Equation (7) when the mode is NSI [14, 15, 16].

Until now we did not discuss rearrangement mechanisms. Indeed, permutations or permutations-inversion connect an initial configuration to the final one without connotations about the reaction path. It has however been shown that within the PSD/TS description there exists a rather strict relation between permutations and the symmetry of the reaction path. Indeed the so-called "selection rules for the transition states of chemical reactions" [17] state that the symmetry of the TS (transition state) can be obtained from the symmetries of the reactants and products and from the permutation describing the reaction. In the paper of Pechukas [18] it was assumed that the TS is a simple saddle point of the potential surface, i.e. a stationary point with one principal axis of negative curvature, according to an argument of Murrell, Laidler and Pratt [19]. The path of steepest descent (PSD) plays a special role in this description: it is a curve that follows the potential gradient. From the TS along the principal axis of negative curvature and in two opposite directions, there are two PSD leading to two minima (one for the "reactant" and one for the "product" of the reaction). It has been show that nuclear symmetry is conserved along the PSD and may increase at its endpoints [18]. As a consequence, the PSD can only have symmetries shared by reactants and products.

This last condition is now made more explicit. Therefore, we need another notation for permutations: let [123] mean ligand 1 is replaced by ligand 2, ligand 2 by 3, and 3 by 1. In this convention, the permutational expression of the symmetry operations depends on the positions of the ligands on the skeleton. (This was not the case with the convention used until now; *vide supra*). For instance, in Figure 3, we consider the D_{5h} structure and we assume that it undergoes a rearrangement described by the permutation $x = [1326]$. If s and f denote the starting and final configurations, it is easy to verify that

$$\begin{aligned} A_s = \{ & I, [34567], [35746], [36475], [37654], [12][36][45], \\ & [12][34][57], [12][37][46], [12][35][67], [12][47][56]\} \end{aligned} \tag{13}$$

i.e. the same expression as Equation (5) except that ( ) has been replaced by [ ]. Similarly

$$\begin{aligned} A_f = \{ & I, [17245], [15427], [12574], [14752], [36][24][57], \\ & [36][17][25], [36][15][47], [36][14][27], [36][12][45]\} \end{aligned} \tag{14}$$

We are now able to express that the group $\tilde{A}$ of proper symmetry operations of the PSD is the set of proper operations shared by the reactant (s) and the

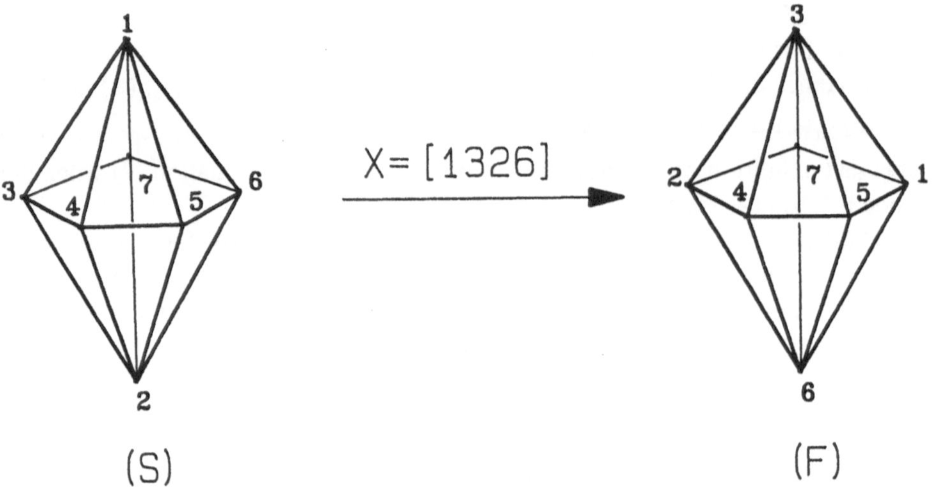

Figure 3. A rearrangement of the pentagonal bipyramid.

product (f):

$$\tilde{A} = A_s \cap A_f = \{I, [12][36][45]\}. \qquad (15)$$

It is easy to show that the group $\tilde{R}$ (see Equation (11)) for the PSD can be defined analogously and is given by

$$\tilde{R} = R_s \cap R_f = \{I, [12][36][45]\}. \qquad (16)$$

These two last equations mean that the only operation conserved along the PSD is the C_2 axis about ligand 7. No improper operations are conserved: the PSD is chiral.

What about the TS ? As discussed by Pechukas [18], the TS displays all the symmetries of the PSD plus extra symmetries which consist of the operations (if any) exchanging reactants and products. Such operations only exist if the rearrangement is SI. In this case indeed xR – which has to be used instead of Rx in the present convention – contains at least either a self-inverse permutation (inversion) or a pair of mutually inverse permutations (inversion). Let y and y^{-1} be this pair. Clearly y and y^{-1} transform s into f since they belong to xR. Their inverses, i.e. y^{-1} and y respectively, transform f into s. Hence, y and y^{-1} are operations exchanging reactants and products. Any pair of this type consists of extra symmetries of the TS. This argument holds of course for self-inverse permutations (inversion) i.e. when $y = y^{-1}$.

These considerations are easily applied to the rearrangement of Figure 3. We first calculate $xR = xA \cup xA\sigma J$, as show by Equation (11). Using Equation (5) for A and the fact that $A\sigma = A$ [12], we obtain successively

$$xA = \{[1326], [1345][267], [13574][26], [13][26475], [137][2654],$$
$$[1623][45], [16][234][57], [164][237], [167][235], [165][23][47]\} \qquad (17)$$

and

$$xA\sigma J = J\{\underline{[16][23]}, [1672345], [1623574], [1647523], [1654237], \underline{[13][26][45]}, [13426][57], [137264], [135267], [13265][47]\}. \quad (18)$$

In Equations (17) and (18), the only self-inverse or mutually inverse elements are underlined. Since they appear only in $xA\sigma J$ they correspond to improper operations of the TS. The TS has no proper extra symmetries, hence the group A_T of proper operations of the TS coincides with $\tilde{A}$ in the present case and is given by Equation (15). The Hougen group R_T for the TS is given by:

$$R_T = \{I, [12][36][45]\} \cup J\{[16][23], [13][26][45]\} \quad (19)$$

and includes the self-inverse elements of Equation (18). The geometric meaning of this result will be discussed in the next Section, but it is already clear that the TS is achiral.

It could also happen that the self-inverse or mutually inverse elements appear in xA or in both xA and $xA\sigma J$. This corresponds to proper extra symmetries of the TS or to both proper and improper extra symmetries, respectively.

Application of these theoretical considerations to heptacoordinate stereochemistry will be described in the next section.

It is convenient to rewrite Equations (8) and (9) for the connectivity in a different form. We first notice that $x^{-1}Ax \cap A = \tilde{A}$. Moreover, $|\tilde{R}| = |\tilde{A}|$ when AxA and $A\sigma x\sigma^{-1}A$ are distinct and $|\tilde{R}| = 2|\tilde{A}|$ when they are identical. It follows that

$$\delta_x = \frac{|R|}{|\tilde{R}|} \quad (20)$$

since $|R| = 2|A|$ for achiral skeleta.

3. Deltahedra and DSD Mechanisms

The static properties of ML_n molecules can be discussed in terms of the shape of the n-vertex polyhedron formed by the n ligands. When n increases, the number of topologically distinct polyhedra increases very rapidly. Table I shows this evolution for n = 4 to 8.

The list of the 34 and 257 polyhedra for $n = 7$ and 8 respectively has been established twenty years ago [20]. The 34 polyhedra for n = 7 have been obtained independently by King [21] from the polyhedra with 7 faces obtained by Federico [22]. In Table I, we listed some typical polyhedra whose chemical interest is discussed elsewhere [3]. In Table II, some topological properties of the 34 polyhedra for n = 7 are given [23]. The columns BD and K refer to the numbering of Britton-Dunitz and King respectively. In the present work #11 will mean polyhedron 11 in King's numbering. The column SYM gives the maximum possible symmetry of the polyhedron, i.e. its graph symmetry; v_i, f_j and e_{kl} (i, j, k, l = 3 to 6) are the number of vertices of degree i, the number of j-gonal

TABLE I. Topologically distinct polyhedra for $n = 4$ to 8.

n	number of polyhedra	some typical polyhedra
4	1	tetrahedron
5	2	trigonal bipyramid, tetragonal pyramid
6	7	octahedron, trigonal prism, pentagonal pyramid, etc.
7	34	pentagonal bipyramid, hexagonal pyramid, capped octahedron, 4-capped trigonal prism, etc.
8	257	hexagonal bipyramid, heptagonal pyramid, cube, square antiprism, D_{2d} dodecahedron, etc.

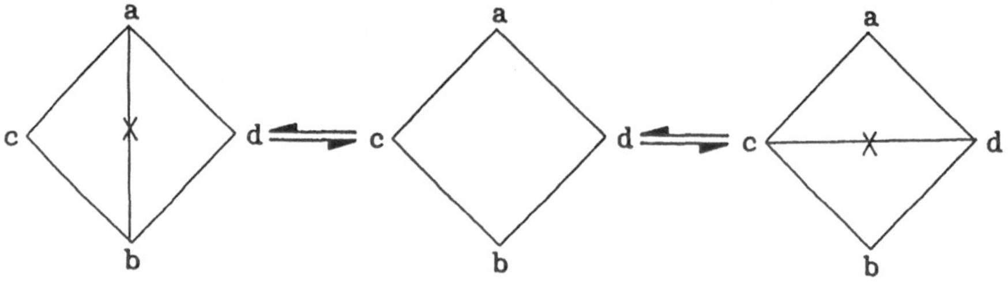

Figure 4. The single dsd.

faces and the number of edges connecting a vertex of degree k to one of degree l. These numbers are sufficient to characterize polyhedra up to $n = 6$. For $n = 7$, this is no longer true since #26 and #27 have identical v_i, f_j and e_{kl}. They have however different graph symmetry (C_1 and C_2 respectively) and their two edges connecting vertices of degree 3 are connected in #26 and disconnected in #27 [21]. Note also that the graph symmetries C_2 and C_s have been attributed to #30 [20, 21]. In fact, the only symmetry element of this polyhedron is a plane containing two edges: the first edge is shared by the two quadrilateral faces and the second one connects the vertex of degree four of the first edge to the vertex of degree five of the graph (see Figure 4, Reference [20]).

Permutational analysis has been applied successfully in penta- and hexacoordination to analyse the various mechanistic possibilities and to discriminate between them (see e.g. Reference [4] and work cited there). At first sight, one

TABLE II. The 34 polyhedra with $n = 7$ (from Reference [23] with permission).

	v_i				f_j				e_{kl}											
BD	6	5	4	3	6	5	4	3	33	34	35	36	44	45	46	55	56	66	K	SYM
1	1	3	0	3	0	0	0	10	0	0	6	3	0	0	0	3	3	0	# 12	C_{3v}
2	2	0	3	2	0	0	0	10	0	2	0	4	2	0	6	0	0	1	# 11	C_{2v}
3	1	2	2	2	0	0	0	10	0	2	2	2	0	4	2	1	2	0	# 13	C_2
4	0	3	3	1	0	0	0	10	0	0	3	0	3	6	0	3	0	0	# 20	C_{3v}
5	0	2	5	0	0	0	0	10	0	0	0	0	5	10	0	0	0	0	# 23	D_{5h}
6	1	1	2	3	0	0	1	8	0	4	2	3	0	2	2	0	1	0	# 14	C_s
7	1	1	2	3	0	0	1	8	1	2	2	3	1	2	2	0	1	0	# 15	C_1
8	0	3	1	3	0	0	1	8	1	1	6	0	0	3	0	3	0	0	# 21	C_s
9	0	3	1	3	0	0	1	8	0	2	7	0	0	2	0	3	0	0	# 22	C_s
10	1	0	4	2	0	0	1	8	0	4	0	2	4	0	4	0	0	0	# 16	C_{2v}
11	0	2	3	2	0	0	1	8	0	2	4	0	2	6	0	0	0	0	# 24	C_s
12	0	2	3	2	0	0	1	8	0	3	3	0	2	5	0	1	0	0	# 25	C_1
13	0	1	5	1	0	0	1	8	0	2	1	0	7	4	0	0	0	0	# 28	C_s
14	1	0	2	4	0	1	0	7	2	4	0	4	1	0	2	0	0	0	# 17	C_s
15	0	1	3	3	0	1	0	7	1	5	2	0	2	3	0	0	0	0	# 29	C_s
16	1	0	2	4	0	0	2	6	2	4	0	4	1	0	2	0	0	0	# 18	C_{2v}
17	0	2	1	4	0	0	2	6	2	2	6	0	0	2	0	1	0	0	# 26	C_1
18	0	2	1	4	0	0	2	6	2	2	6	0	0	2	0	1	0	0	# 27	C_2
19	0	1	3	3	0	0	2	6	1	4	3	0	3	2	0	0	0	0	# 33	C_1
20	0	1	3	3	0	0	2	6	1	5	2	0	2	3	0	0	0	0	# 32	C_1
21	0	1	3	3	0	0	2	6	0	7	2	0	1	3	0	0	0	0	# 30	C_s
22	0	1	3	3	0	0	2	6	0	6	3	0	2	2	0	0	0	0	# 31	C_s
23	0	0	5	2	0	0	2	6	1	4	0	0	8	0	0	0	0	0	# 36	C_{2v}
24	0	0	5	2	0	0	2	6	0	6	0	0	7	0	0	0	0	0	# 37	C_2
25	1	0	0	6	1	0	0	6	6	0	0	0	0	0	0	0	0	0	# 19	C_{2v}
26	0	1	1	5	0	1	1	5	4	3	4	0	0	1	0	0	0	0	# 34	C_1
27	0	0	3	4	0	1	1	5	3	6	0	0	3	0	0	0	0	0	# 38	C_s
28	0	1	1	5	0	0	3	4	4	3	4	0	0	1	0	0	0	0	# 35	C_s
29	0	0	3	4	0	0	3	4	3	6	0	0	3	0	0	0	0	0	# 40	C_{3v}
30	0	0	3	4	0	0	3	4	2	8	0	0	2	0	0	0	0	0	# 42	C_1
31	0	0	3	4	0	0	3	4	2	8	0	0	2	0	0	0	0	0	# 41	C_2
32	0	0	3	4	0	0	3	4	3	6	0	0	3	0	0	0	0	0	# 39	C_{3v}
33	0	0	1	6	0	1	2	3	7	4	0	0	0	0	0	0	0	0	# 43	C_s
34	0	0	1	6	0	0	4	2	7	4	0	0	0	0	0	0	0	0	# 44	C_{2v}

should like to apply these methods to heptacoordination, but, as shown in Table III, the number p of configurations (see Equation (6)) and the number z of rearrangement modes are no longer manageable when $n = 7$.

TABLE III. The number of configurations (p) and of rearrangement modes (z).

n	SYM	p	z
5	D_{3h}	20	6
6	O_h	30	5
7	D_{5h}	504	40
	C_{3v}	1680	308
	C_{2v}	2520	648

A crucial question arises: among the numerous rearrangements modes in heptacoordination, which one(s) should be tested against experimental results?

King [21, 24] has proposed a model which is inspired by the pioneer work of Lipscomb [25] on boranes and carboranes. The first aspect of this model is that deltahedra, i.e. polyhedra with triangular faces only, are the most favourable static geometries. The second aspect is that one assumes that the deformation of the polyhedra should not depart too much from the ideal situation, i.e. complete triangulation. The simplest process consistent with this requirement is the so-called single dsd shown in Figure 4, where it is seen that the edge ab shared by two of the triangular faces of the deltahedron is first suppressed and then replaced by an edge cd shared by two new triangular faces. The intermediate polyhedron has only one non-triangular face. From now on a cross will denote an edge present in the starting or final polyhedron but not in the intermediate one.

The assumption that the deltahedra are the most stable geometries is questionable since minimization of the total repulsion of the ML bonds leads to the conclusion [3] that the D_{5h} pentagonal bipyramid (#23), the C_{3v} capped octahedron (#20) and the C_{2v} 4-capped trigonal prism (#36) have all three maximal stability in spite of the fact that #36 has *two* quadrilateral faces whereas #20 and #23 are deltahedra! Three of the five deltahedra i.e. (#11, #12, #13) do not seem to be favourable structures. None of the polyhedra with *one* quadrilateral face (#14, #15, #16, #21, #22, #24, #25, #28) seems to be stable.

In fact, the only observed static structures are the capped octahedron (#20), the 4-capped trigonal prism (#36), the pentagonal bipyramid (#23). The so-called 4:3 geometry has also been detected but is consistent with several polyhedra (see below). A more detailed discussion and references about the static structure determinations may be found elsewhere [5].

Hence, the simplest and most reasonable attitude consistent with these theoretical and experimental facts was to investigate the deformations of the above four observed geometries by suppressing and adding edges. However an important restriction has to be taken into account: dynamic NMR observations for the

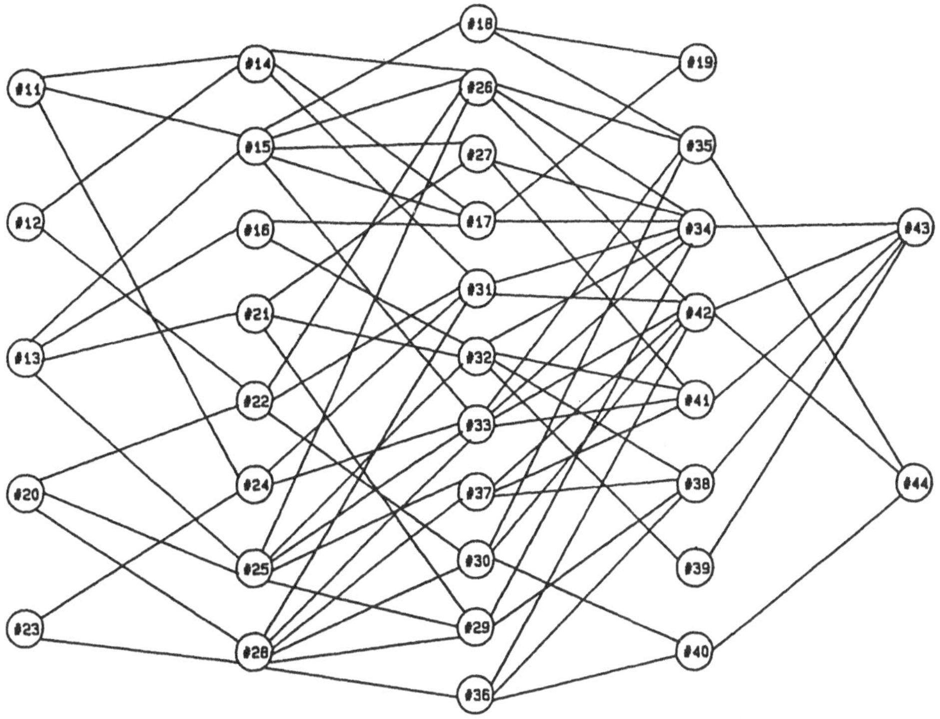

Figure 5. Successive edge suppression on the deltahedra (from [23] with permission).

heptacoordinate molecules (see Reference [5] and work cited there) show that ligand scrambling due to non-rigidity does not lead to isomerization. Therefore, we may limit ourselves to "degenerate" deformations i.e. where the starting and final configurations are identical, up to a permutation or a permutation-inversion of identical ligands. These degenerate deformations are precisely described by the formalism recalled in Section 2. Suppressing and adding edges is described on the graph of Figure 5. It has been constructed from a matrix given by Britton and Dunitz [20]. The vertices of this graph are the 34 polyhedra. In the left column are the deltahedra. They have 10 faces and 15 edges as imposed by the Euler relation $v + f = e + 2$ (v, f, e are the numbers of vertices, faces and edges) and by the fact that for deltahedra $2v = 3f$. In the next column, we find the polyhedra with 14 edges, 8 triangular faces and one quadrilateral one; then those with 13 edges with either 6 triangles and 2 squares or 7 triangles and one pentagon, etc. The edges of the graph represent an edge suppression or addition on the polyhedron.

3.1. DEFORMATIONS OF THE DELTAHEDRA

These have been considered previously for the five deltahedra (#11, #12, #13, #20, #23) [21, 23]. In fact, only the pentagonal bipyramid (#23) and the capped octahedron (#20) are of chemical interest. It appears that neither #20 nor #23 have single degenerate dsd. For instance, if we suppress the edge 23 the capped octahedron (see Figure 2) we obtain #28 (see Figure 5) and if we replace it by edge 15, we indeed perform a single dsd but this process in non-degenerate since the initial polyhedron had one vertex of degree 3 (vertex 1) and the final polyhedron has no such vertex. It is easy to realize that the final polyhedron is in fact #23 (see Figure 5).For the reasons invoked above, such a non-degenerate process is not interesting in the present context. Using the graph of Figure 5, it may be shown that the capped octahedron may be deformed by double-degenerate dsd processes, i.e. the combination of two processes shown in Figure 4. When the two suppressed edges share a vertex, the process generates an intermediate polyhedron with a pentagonal face instead of two square faces. This is a 4,5-pyramidal process rather than a double dsd [21]. In Figure 6, the possible degenerate double dsd and 4,5-pyramidal process of the capped octahedron are shown. A symbol such as #20/30 means that the starting and final polyhedron is #20 and the intermediate one, #30.

Using similar arguments, it is easy to show that the pentagonal bipyramid (#23) has no single degenerate dsd. The only possible degenerate double dsd is shown in Figure 7.

3.2. DEFORMATIONS OF THE NON-DELTAHEDRA

We first recall the arguments for the 4-capped trigonal prism (#36) (for details see Reference [5]). As seen on Figure 5, suppression of one edge in #36 leads to #38, #40 or #42. From these three polyhedra, one more edge suppression generates #43 or #44. These two polyhedra have the minimum number of edges for 7 vertex polyhedra, i.e. 11 edges.

Suppression of one edge in #36 may lead either to a polyhedron with 3 quadrilateral faces and 4 triangular ones such as #40 or #42 or to a polyhedron with 1 pentagonal face, 1 quadrilateral face and 5 triangular faces such as #38. The *new* faces may then be diagonalized. The diagonalization of the new quadrilateral in #40 or #42 are the familiar single dsd processes but #36 has no degenerate single dsd. The diagonalization of the new pentagonal face in #38 is a so-called 5-pyramidal [21] process but #36 has no degenerate five-pyramidal process. Since the starting polyhedron #36 has already 2 quadrilateral faces, the suppression of one edge of #36 may be followed by the diagonalization of one of the two *old* quadrilateral faces. We call such a process a dss′d′ process since the face s' which is diagonalized into d' is *different* from the face s arising from the edge removal in d. There exists one such degenerate dss′d′ for #36. It is shown on Figure 8 where the removed edge is 34 leading to the intermediate polyhedron #40. The original quadrilateral face 2376 of #36 is then diagonalized by adding the edge 36.

We now discuss the processes resulting from the removal of two edges. The

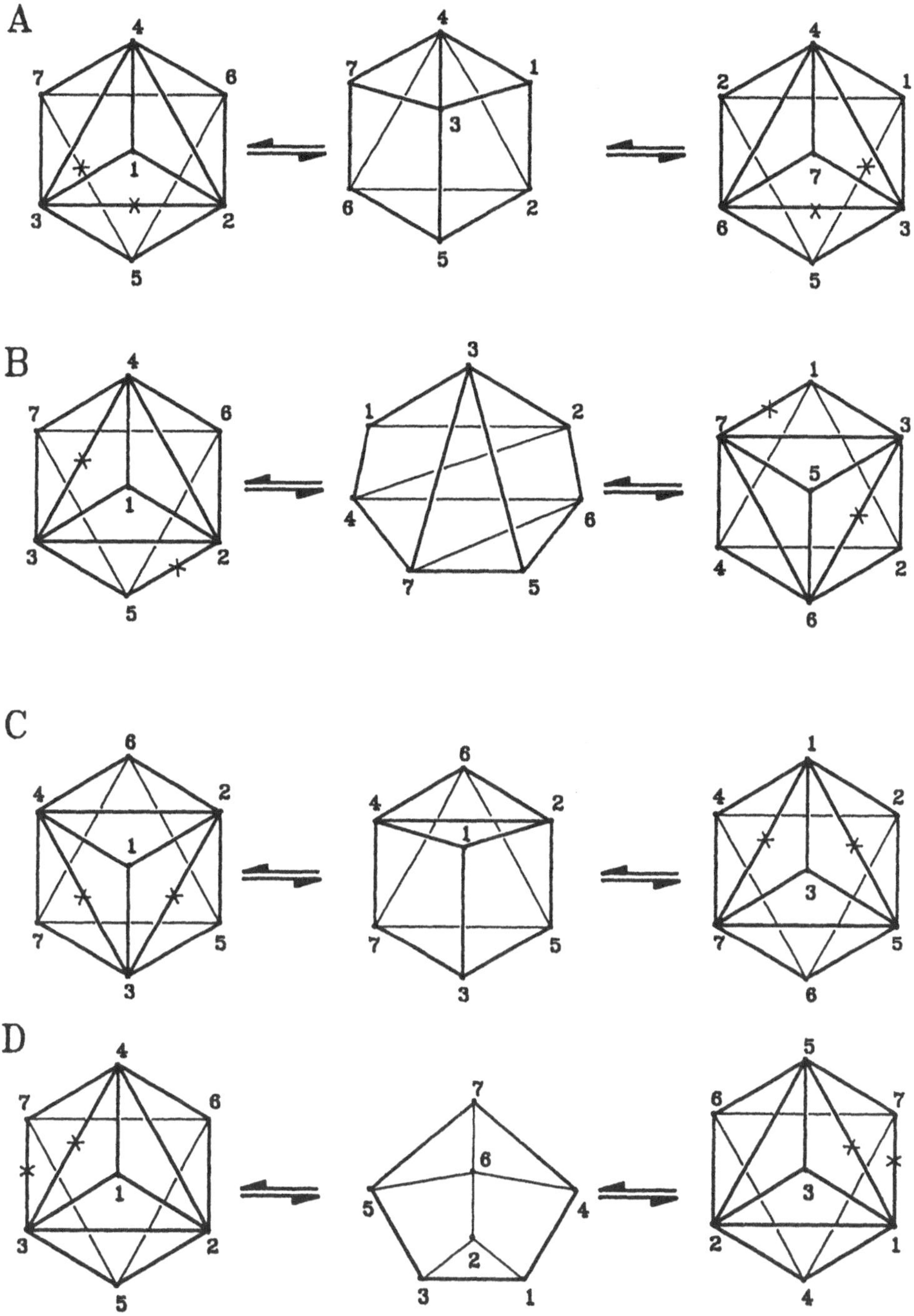

Figure 6. Degenerate double dsd (A, B, C) and 4,5-pyramidal process (D) for the capped octahedron (from [23] with permission).

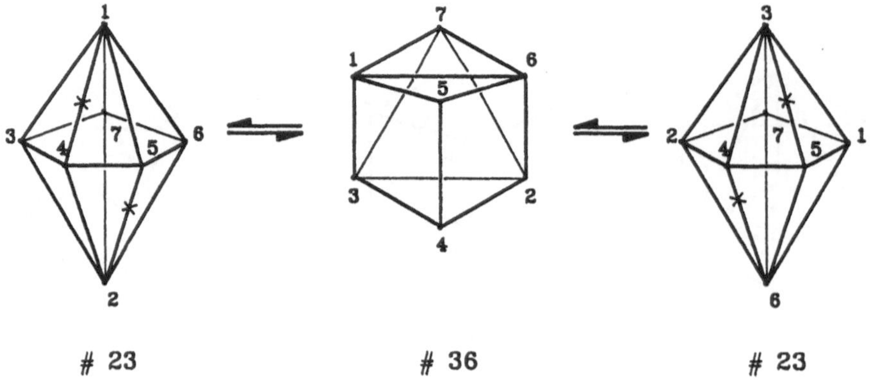

Figure 7. Degenerate double dsd for the pentagonal bipyramid (from [23] with permission).

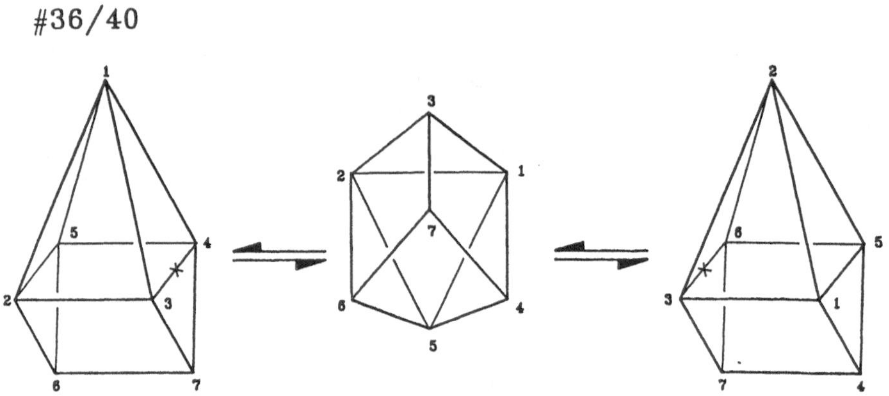

Figure 8. A single degenerate dss′d′ process for the 4-capped trigonal prism (from [5] with permission).

three processes described above – dsd, dss'd' and 5-pyramidal processes – may be combined pair-wise. The following combinations give rise to degenerate processes:

(a) a combination of dsd and dss′d′ is shown in Figure 9. The quadrilateral face 1374 is the intermediate of the dsd component whereas the dss′d′ component consists of the removal of the edge 12 in the "diamond" 1523 followed by the inclusion of an edge 46 in the old quadrilateral face 4576. The intermediate polyhedron is #44.

(b) a combination of two dss′d′ is shown in Figure 10. Indeed the two quadrilateral faces i.e. 2367 and 4567 of the starting polyhedron are diagonalized in the final polyhedron, by adding the edges 27 and 46. The removed edges are 12 and 34. The intermediate polyhedron is #44.

(c) a dss′d′ and a 5-pyramidal process are combined in Figure 11. The edges 23 and 14 are first removed. The intermediate polyhedron #43 as a pentagonal

#36/44(a)

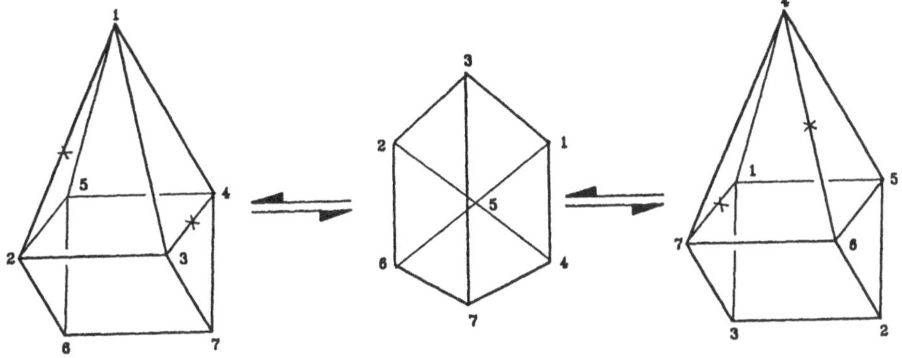

Figure 9. A degenerate dsd/dss′d′ process.

#36/44(b)

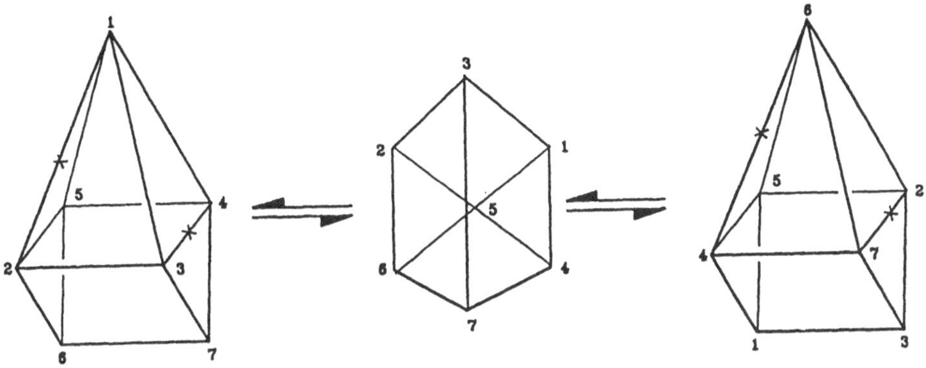

Figure 10. A degenerate dss′d′/dss′d′ process.

face 13762 which is diagonalized by adding the edge 27. The quadrilateral face 4567 of the starting polyhedron is diagonalized by adding the edge 46.

The relation between the 4:3 geometry and the dsd mechanisms will be discussed in the next section.

We now discuss briefly the permutation character and the symmetry properties of the PSD and TS of the above processes. From the Figures 6, 7, 8, 9, 10 and 11, it is easy to obtain a representative permutation for each of the considered processes. Using the definition given in Equations (7) and (12), it is possible to show that the permutations describing the interconversion processes of the capped octahedron belong to different modes. The same statement holds for the 4-capped trigonal prism. Each of the modes are moreover SI. The PSD and TS symmetries follow from the procedure leading to Equations (15), (16) and (19) and yield $\tilde{A}$,

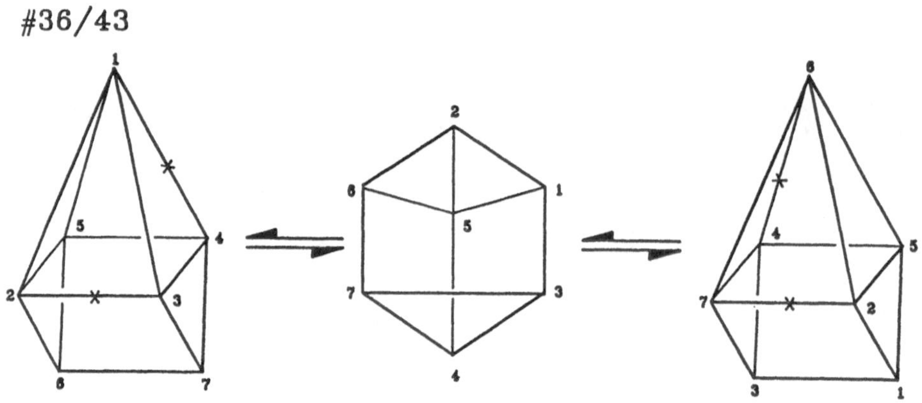

Figure 11. A degenerate dss′d′/5-pyramidal process.

$\tilde{R}$, A_T and R_T. Finally, the connectivity of each process is obtained via Equation (20). The results are summarized in Table IV, where the last column refers to the Schönfliess symbol [26] of R_T. Note also that there are two distinct #36/44 processes noted (a) and (b) (see Figures 9 and 10).

We also want to point out that the TS symmetry – as obtained by the PSD/TS procedure – does not always coincide with the graph symmetry of the intermediate polyhedron. For instance, in the #36/40 process the intermediate polyhedron has a C_{3v} graph symmetry (see Table II) as might be realized on Figure 8. The symmetry of the TS is only C_s. This means that the other elements of C_{3v} are incompatible with the requirement that #40 correspond to a single saddle point between the starting and final #36 polyhedra of Figure 8. Similar considerations hold for the two #36/44 processes of Table IV. For the other processes of Table IV, the graph symmetry of the intermediate polyhedron and the TS symmetry coincide. Other examples where they do not coincide may be found elsewhere [17, 27].

4. NMR Results

A systematic analysis of NMR results of heptacoordinate complexes has been performed in previous work [5]. In the complexes for which interesting NMR informations have been obtained, the central atom M is not surrounded by 7 identical ligands, but by different ligands, some of which being bridging bischelates. For this reason, these complexes could, *in principle* at least, exist as a mixture of permutational isomers having the same geometry, say pentagonal bipyramidal, but different dispositions of the ligands. *In practice*, however, the NMR results at hand show that each of these complexes exist as a single isomer. It will be useful to discuss the NMR results in terms of the concept of *dsd invariant isomer*: for a given geometry, an isomer will be called dsd invariant if there exist at least one degenerate dsd process of this geometry which can occur without isomerization.

TABLE IV. PSD and TS symmetries of the degenerate dsd processes.

Process	Permutation	δ_x	$\tilde{A}$	$\tilde{R}$	$\tilde{A}_T$	$\tilde{R}_T$	
#20/30	(16327)	6	I	$\tilde{A}$	$\tilde{A}$	$\tilde{A} \cup (17)(26)J$	C_s
#20/29	(123)(45)(67)	6	I	$\tilde{A}$	$\tilde{A}$	$\tilde{A} \cup (13)(45)J$	C_s
#20/37	(15)(27)(46)	6	I	$\tilde{A}$	$\tilde{A} \cup (15)(27)(46)$	A_T	C_2
#20/36	(13)(27)(45)	3	I	$\tilde{A} \cup (24)(57)J$	$\tilde{A} \cup (13)(27)(45)$	$A_T \cup \{(13)(25)(47), (24)(57)\}J$	C_{2v}
#23/36	(1623)	10	I, (12)(36)(45)	$\tilde{A}$	$\tilde{A}$	$\tilde{A} \cup \{(13)(26)(45), (16)(23)\}J$	C_{2v}
#36/40	(152)(346)	4	I	$\tilde{A}$	$\tilde{A}$	$\tilde{A} \cup (12)(46)J$	C_s
#36/44(a)	(154)(27)(36)	4	I	$\tilde{A}$	$\tilde{A}$	$\tilde{A} \cup (14)(26)(37)J$	C_s
#36/44(b)	(17536)	4	I	$\tilde{A}$	$\tilde{A} \cup (16)(24)(37)$	A_T	C_2
#36/43	(16)(25)(374)	4	I	$\tilde{A}$	$\tilde{A}$	$\tilde{A} \cup (16)(37)J$	C_s

If no such process exist, we will speak of *non-dsd invariant* isomer.

4.1. PENTAGONAL BIPYRAMID #23

Many complexes of this type have been observed. The only possible degenerate dsd process is #23/36 (see Table IV).

4.1.1. The following isomers are *non dsd invariant* for this process [5]:

4.1.1.1. *Trischelates.* In $Mo(NO)(S_2CNMe_2)_3$ [28], $Mo(N_2R)(S_2CNMe_2)_3$ (R = aryl), $TiCl(S_2CNMe_2)_3$ [29] the monodentate ligand is axial, two chelates are equatorial and the third one is axial-equatorial. A similar structure holds for $TiCl(SOCNR_2)_3$ (R = Me, Et) [30] where the three S atoms are on the same triangular face. Ligand scrambling has been observed for these isomers.

4.1.1.2. *Bischelates.* An approximate C_{2v} structure has been observed for TaH $[P(C_6H_5)_2]_2(dmpe)_2$: the PPh_2 ligands are axial, H and the two dmpe chelates are equatorial. Coalescence of dmpe methyl signals has been observed [31].

4.1.2. The following isomers are *dsd invariant* [5]:

4.1.2.1. *Bischelates.* The two chelates are axial-equatorial in $ReH_3(L–L)_2$, where L–L is dppe or dpae. Two of the three H atoms occupy adjacent equatorial positions, the third one is adjacent to the two equatorial chelate sites, a C_2 structure [32]. At -50°C, the four P atoms in $ReH_3(dppe)_2$ are equivalent. This is compatible with the degenerate double dsd of Figure 7.

In $TaX(\eta^4$-naphtalene)$(dmpe)_2$ the naphtalene chelating unit is ligated in such a way that its complexed 1,3-diene moiety plays the role of the neighbouring hydrides in $ReH_3(L–L)_2$. Therefore, these trischelate and bischelate have similar behaviour and are best discussed along the same lines: for the trischelate ^{31}P NMR line shape analysis is in agreement with the degenerate double dsd of Figure 7 [33].

4.1.2.2. *Monochelates.* The complex HMo $[P(OCH_3)_3]_4[O_2CCF_3]$ has an average structure with two axial phosphite ligands and the two other phosphites in the equatorial plane and separated by the hydride ligand. The trifluoroacetate chelate is equatorial. Axial and equatorial phosphorous signals coalesce above -20°C. This could be due to the degenerate double dsd of Figure 7 (also called pair-wise exchange) but detailed line shape analysis shows that the observed spectra are consistent with non-pair-wise exchange and not with the degenerate double dsd [34].

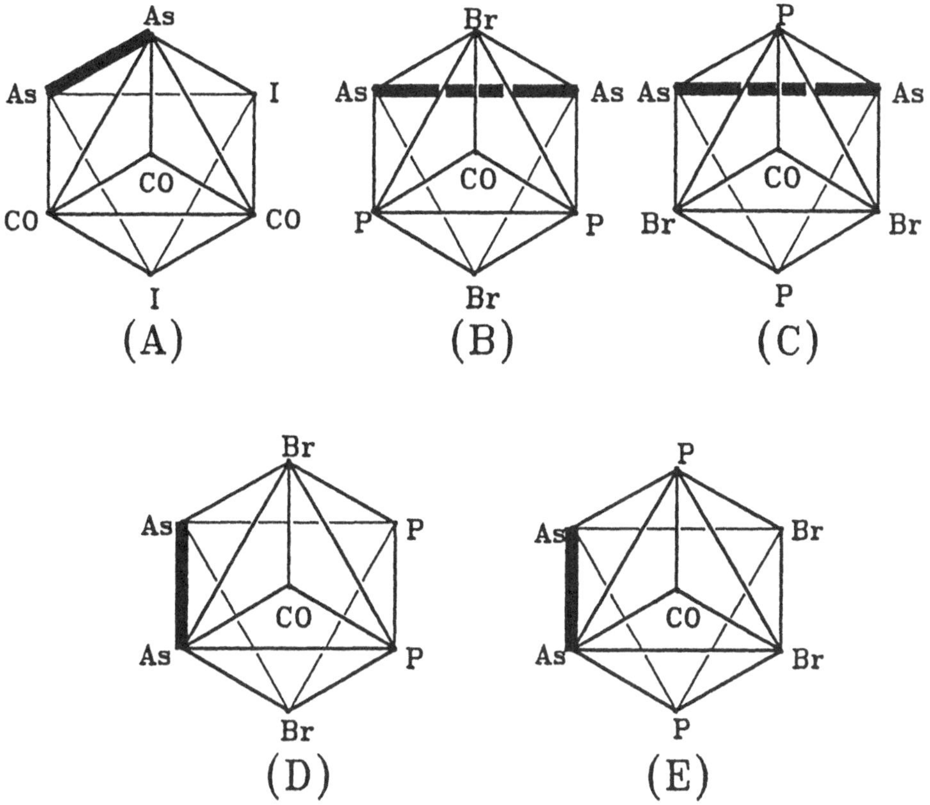

Figure 12. Monochelate capped octahedra (from [5] with permission).

4.1.2.3. *Monodentate ligands.* The $CrH_2[P(OCH_3)_3]_5$ complex has a C_{2v} structure: the hydride ligands are in distal equatorial positions [35]. According to the line shape analysis performed by Van Catledge, Ittel and Jesson, there are two basic permutational sets compatible with the experimental results, one of these sets describes the degenerate double dsd of the pentagonal bipyramid (see Figure 7).

4.2. CAPPED OCTAHEDRON #20

There are four degenerate dsd processes for this geometry i.e. #20/29, #20/30, #20/36 and #20/37 (see Table IV). The isomers whose non-rigidity has been observed by NMR are shown along the C_3 axis of the framework in Figures 12 and 13. They are all *non-dsd invariant* [5].

In Figure 12, the central metal is W in (A) and Mo in the other isomers. The heavy As–As lines represent the cis-$(CH_3)_2AsC(CF_3) = C(CF_3)As(CH_3)_2$ chelate,

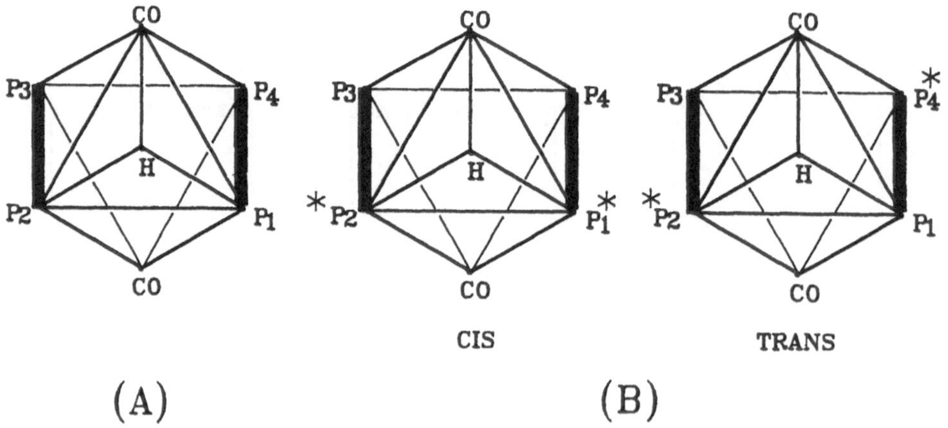

Figure 13. Bischelate capped octahedra (from [5] with permission).

P represents the phosphorus atom of $P(OCH_3)_3$ or $P(OCH_3)_2$ (C_6H_5) in (B), of $P(OC_2H_5)_3$ in (C) and $P(CH_3)_2$ (C_6H_5) in (D) and (E). These complexes have been studied by ^1H and ^{13}C NMR and are non-rigid structures [36].

In Figure 13, we represent $TaH(CO)_2(dmpe)_2$ (A) and $MoH(CO)_2$ $(P\text{–}P^*)_2^+$ (B) where P–P* is $R_2PCH_2CH_2ER'_2$ (R = R′ = Me, E = P or R = R′ = Ph, E = P or R = Me, R′ = Et, E = P or R = Me, R′ = Ph, E = P or R = R′ = Me, E = As or R = R′ = Ph, E = As or R = Ph, R′ = Me, E = As). The Ta complex [37] and the Mo complexes [38] are non rigid, as shown by ^1H and ^{31}P NMR line shape analysis. Cis-trans isomerization has been reported for unsymmetrically chelated complexes.

4.3. 4-CAPPED TRIGONAL PRISM #36

There are four degenerate dsd processes for this geometry: # 36/40, #36/43, #36/44(a) and #36/44(b) (see Table IV). Among the compounds having this geometry, $[(t\text{–}C_4H_9NC)_7Mo]^{2+}$ is of course *dsd invariant.* It shows only one ^{13}C resonance down to -135°C [3, 39]. This does not allow discrimination between the degenerate dsd processes and other mechanisms leading to scrambling of the three types of positions in #36.

The following isomers are *non-dsd invariant* [5]:

4.3.1. $[(t\text{–}C_4H_9NC)_6MoX]^+$ (X = Br, I) have a 4-capped trigonal prism geometry with capping X [40, 41] and are stereochemically non-rigid [39].

4.3.2. $[M(CO)_2(dmpe)_2I]^+$ (M = Mo, W) a C_{2v} structure with capping I; the CO ligands occupy the ends of the edge common to the two quadrilateral faces, each dmpe chelate spans an edge parallel to the previous one. This is also a non-rigid structure [42]. Cis-trans isomerization has been observed for the parent $MX(CO)_2(L\text{–}L')_2$ (M = Mo, Ta; X = Cl, I; L–L′: unsymmetric bidentate phosphine) of similar structure [43].

4.3.3. $MoI_2(CO)_3[o\text{-}(As(CH_3)(C_6H_5))_2C_6H_4]$ has a capping iodine ligand. Two of the COs and the chelate are disposed as in the previous complexes. The third CO and the second I occupy the remaining vertices of the capped face. The complex is non-rigid [44].

4.4. 4:3 GEOMETRY

This so-called piano stool geometry requires a more detailed discussion (see Reference [5]). It has been proposed in several occasions [1], [3], [45], [46], [47]. The NMR results to be discussed below indicate that the observed complexes show the following characteristics [46, 47]:

(a) they have only one tetragonal face and 8 triangular ones
(b) this tetragonal base is parallel to one of the triangular faces (the trigonal base).

As indicated previously [5], these requirements do not specify unequivocally the polyhedron: in the list established by Britton and Dunitz [20] there are 8 polyhedra with 7 vertices having one tetragonal face and 8 triangular ones. Among these polyhedra, 3 have a triangular face which does not touch the tetragonal face i.e. #15 (C_1), #22 (C_s) and #28 (C_s) (see Figure 14). A C_s structure where a trigonal base edge and a tetragonal base edge are parallel has been considered previously [1]. It implies the existence of a second tetragonal face and therefore it will not be discussed here.

We now use more specific information about some complexes of 4:3 geometry i.e. $W(CO)_3(S_2CNR_2)_2$ and $W(CO)_2L(S_2CNR_2)_2$ (R = Me, Et; L = PPh_3, PEt_3, $P(OEt)_3$). The structures of the di- and tricarbonyl frameworks are shown in Figure 15 where (C_3) denotes the third carbonyl carbon of the tricarbonyl compound. On this figure, it is impossible to decide to which of the #15, #22 or #28 geometries this structure corresponds, since the edges connecting the bases have not been drawn. Clearly the edges PS_4, S_2C_2 and S_1S_3 lead to non-convex polyhedra. The four remaining possible polyhedra are displayed in Figure 16 (dots and circles represent vertices of degree 5 and 3 respectively), the other vertices have degree four. Polyhedron #15 of Figure 14 is incompatible with the structural data about these complexes. The polyhedra of Figure 16 are compatible with the low temperature static NMR spectra of the di- and tricabonyl complexes [46, 47]. We now discuss the dynamic aspects of the NMR information.

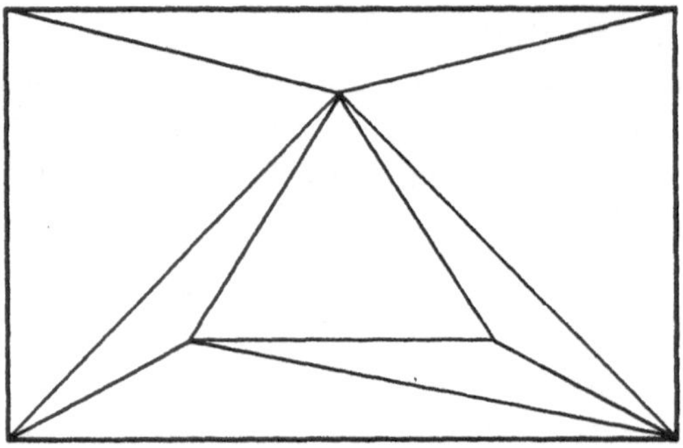

15

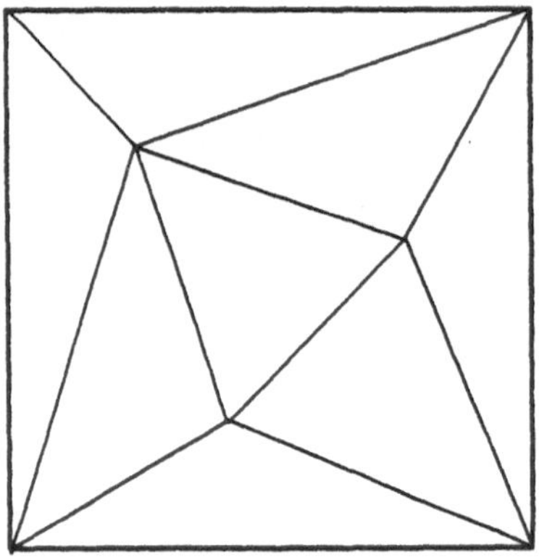

28

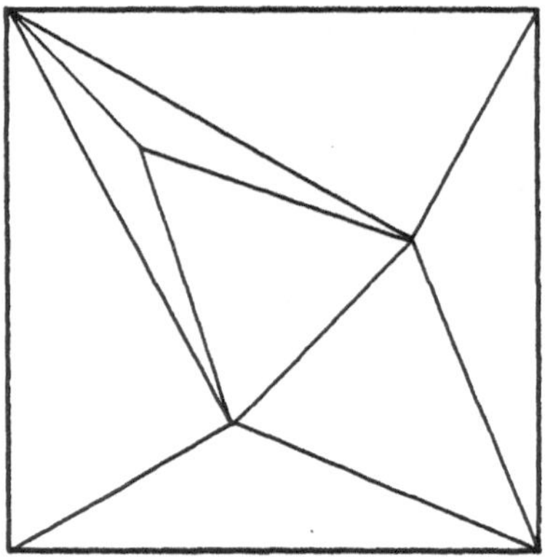

22

Figure 14. Possible 4:3 polyhedra(from [5] with permission).

For the dicarbonyl, a mechanism shown in Scheme II of Reference [47] has been proposed to account for its coalescence pattern: the phosphine ligand is assumed to jump from the upper SSC triangle to which it is coplanar to the lowest one (see Figure 15). We want to discuss the possibility to interpret this mechanism in terms of degenerate dsd processes and therefore we need to apply it to the four structures of Figure 16, where all the edges have been represented. On the structures #28 (b) and #28 (c), the jump mechanism may be visualized by suppressing the edge S_3C_2 and by adding an edge S_2C_1.

In doing this, the degree of the vertices C_1 and C_2 which are equal to 4 in #28 (b) become equal to 5 and 3 respectively after the edge modifications. In #28 (c), on the contrary, the final degree of C_1 and C_2 will be equal to 4 instead of 5 and 3 respectively. This amounts to an unobserved isomerization of the complex.

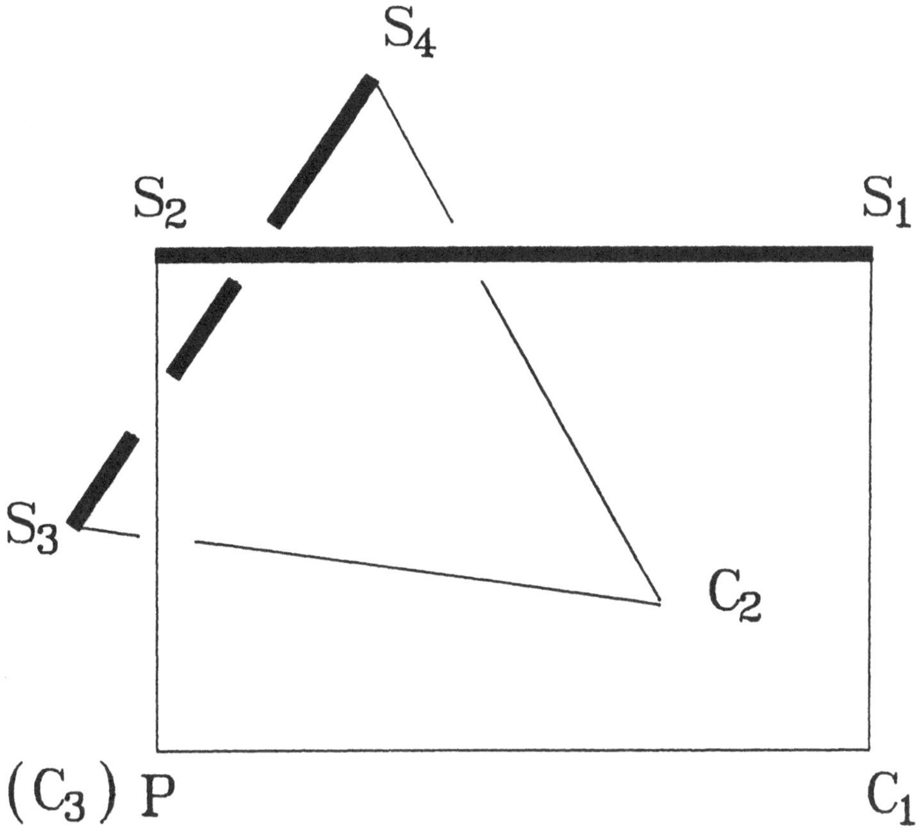

Figure 15. A 4:3 structure (from [5] with permission).

However, it has been shown [5] that the jump may be realized without isomerization if other edges are modified. The result is shown on Figure 17. It appears that the phosphine jump is equivalent to a degenerate combination of a double dsd and a dss′d'. Indeed, the double dsd consists of two edge switchings in the "diamonds" $S_1S_2S_3S_4$ and $S_1C_1C_2S_4$ of #28 (b) and (c). Moreover, again in #28 (b) and (c), the edge S_3C_2 is suppressed in the "diamond" $PS_3S_4C_2$ and the edge S_2C_1 is added to diagonalize the quadrilateral face $PS_2S_1C_1$, the dss′d′ component. The intermediate polyhedron is #44 when starting from both #28 (b) or 28 (c).

The polyhedra #22 and #28 (a) in Figure 16 could be analysed along the same lines, but the P vertex is now of degree 3 instead of 4 in #28 (b) and (c). For this reason, the phosphine jump implies at least 5 edge modifications. The intermediate figure is no longer a polyhedron since it has only 9 edges.

We now turn our interest towards the tricarbonyl compound. Scheme I in Reference [46] is a proposal to interpret the intermediate temperature coalescence

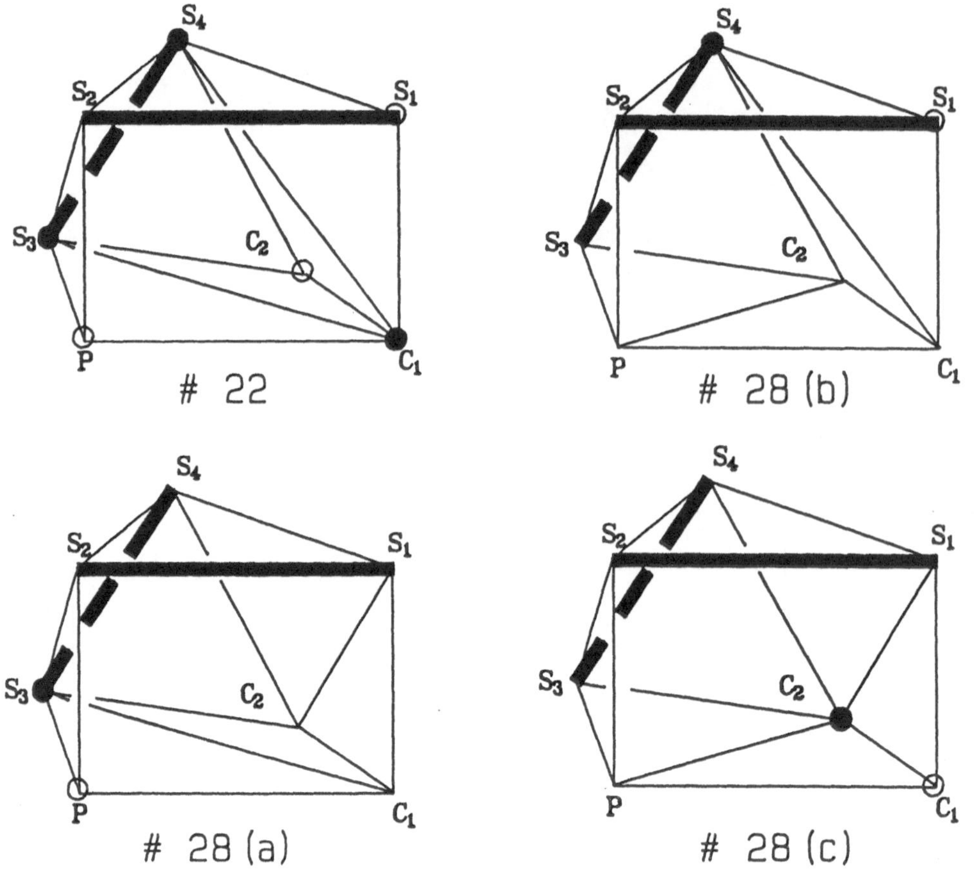

Figure 16. Possible 4:3 bischelate polyhedra (from [5] with permission).

of carbonyls C_1 and C_3 (see Figure 15) by a twisting of the trigonal base relative to the tetragonal base. We perform this twist on the tricarbonyl derivative #28 (b) (see Figure 16, where P is now replaced by C_3). As seen on Figure 18, this motion is in fact a degenerate single dsd, with #36 as intermediate polyhedron, a 4-capped trigonal prism with C_2 as capping ligand. The high temperature coalescence of the tricarbonyl derivative has been assumed to occur via a $2\pi/6$ rotation of the three carbonyl carbons (Scheme II, Reference [46]) in this intermediate polyhedron. In Figure 19, it appears that this rotation is equivalent to a degenerate dsd/dss′d′ process with a new intermediate polyhedron, namely #44.

5. Conclusions

We did mention that the NMR results show that, in general, isomerization is not observed. Two exceptions should be kept in mind: cis-trans isomerization of

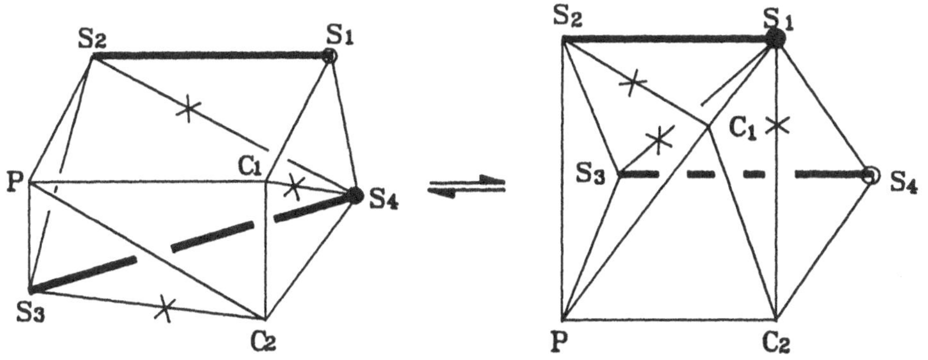

28 (b)

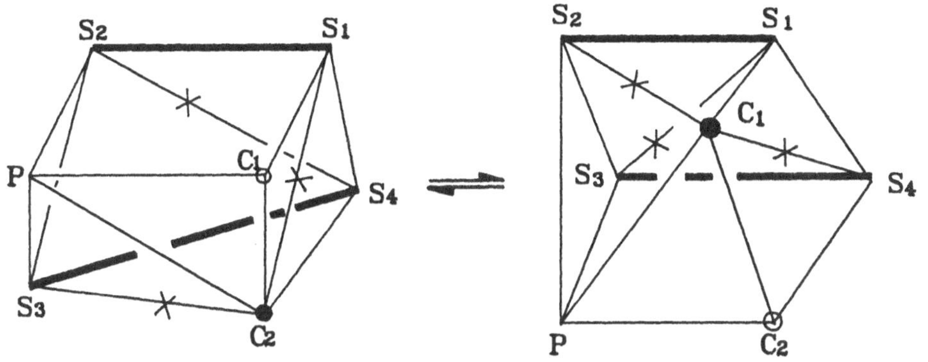

28 (c)

Figure 17. Interconversion of 4:3 polyhedra (from [5] with permission).

unsymmetric bischelate complexes of capped octahedral [38] or 4-capped trigonal prismatic geometries [43] (see Sections 4.2. and 4.3.). Hence the fact that *non-dsd invariant* isomers are non-rigid but do not isomerize shows that their non-rigidity does not occur via dsd processes. This is the case for the *non-dsd invariant* pentagonal bipyramids (see Sections 4.1.1.1. and 4.1.1.2.). This is also the case for all the observed capped octahedra and for all the observed 4-capped trigonal prisms, except the *dsd invariant* $[(t-C_4H_9NC)_7Mo]^{2+}$ complex [39]. The cis-trans isomerizations do not modify these considerations, since they are not compatible

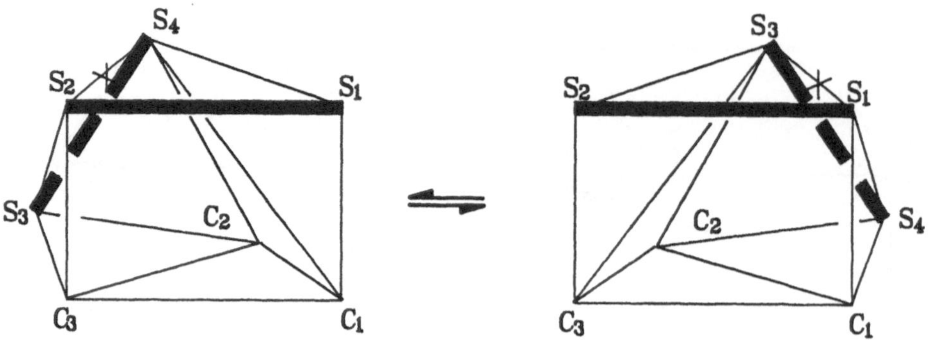

Figure 18. Twisting motion on #28 (b) (from [5] with permission).

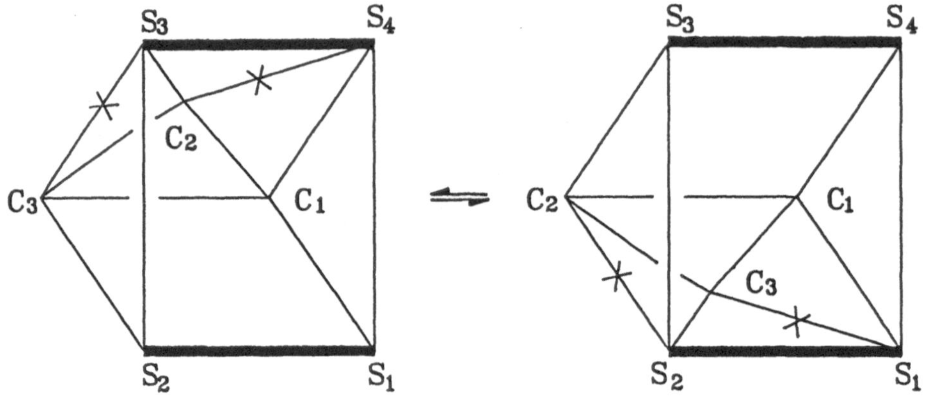

Figure 19. Edge rearrangement in the intermediate polyhedron of Figure 18 (from [5] with permission).

with dsd mechanisms.

Among the dsd invariant complexes, we do not discuss $[(t\text{–}C_4H_9NC)_7Mo]^{2+}$ any longer since it does not allow discrimination between dsd and other processes. The dsd invariant pentagonal bipyramidal bischelate $ReH_3(dppe)^2$, the similar $TaX(\eta^4$-naphtalene)$(dmpe)_2$ and $CrH_2[P(OCH_3)_3]_5$ also of pentagonal bipyramidal form show NMR line shape behaviour compatible with dsd processes. For the 4:3 geometry, the proposed [47] phosphine jump of $W(CO)_2\ L(S_2CNR_2)_2$ is equivalent to a degenerate combination of a double dsd and a dss′d′ whereas the assumed [46] trigonal base/tetragonal base twist is in fact a degenerate single dsd of $W(CO)_3(S_2CNR_2)_2$. The polyhedron #36 (4-capped trigonal prism) could undergo a $2\pi/6$ rotation of its tricarbonyl subunit [46] which is equivalent to a degenerate dsd/dss′d′ process.

The above results suggest that dsd processes are rather favourable *dynamic* pathways and that dsd behaviour is only prevented for *static* reasons i.e. when these processes lead to isomers which are high energy lying for steric and/or electronic reasons [48, 49]. The only counterexample is the pentagonal bipyramidal $HMo[P(OCH_3)_3]_4[O_2CCF_3]$: it has a behaviour which is incompatible with the degenerate double dsd in spite of the fact that it is dsd invariant and that there are no static arguments against dsd in this case.

We think that NMR line shape analysis of new *dsd invariant* complexes should be useful in order to decide if dsd behaviour is a general rule for these complexes.

References

[1] E.L. Muertterties and C.M. Wright: *Quart. Rev.* **21** (1967) 109.
[2] E.L. Muertterties and L.G. Guggenberger: *J. Amer. Chem. Soc.* **96** (1974) 1748.
[3] D.L. Kepert: *Inorganic Stereochemistry*, Springer, New York (1982).
[4] J. Brocas, M. Gielen and R. Willem: *The Permutational Approach to Dynamic Stereochemistry*, McGraw-Hill, New York (1983).
[5] J. Brocas: *J. Math. Chem.* **14** (1993) 153.
[6] J.I. Musher: *J. Amer. Chem. Soc.* **94** (1972) 5662.
[7] W. Hässelbarth and E. Ruch: *Theoret. Chim. Acta* **29** (1973) 259.
[8] W.G. Klemperer: *J. Chem. Phys.* **56** (1972) 5478; *J. Amer. Chem. Soc.* **94** (1972) 8360.
[9] R. Willem: *Progress in NMR Spectroscopy*, Pergamon, London (1987), Vol. 20, p. 1.
[10] E.L. Muetterties: *J. Amer. Chem. Soc.* **91** (1969) 1636, 4115.
[11] J.G. Nourse: *J. Amer. Chem. Soc.* **102** (1980) 4883.
[12] J.T. Hougen: *J. Phys. Chem.* **37** (1962) 1433.
[13] J. Brocas and R. Willem: *J. Amer. Chem. Soc.* **105** (1983) 2217.
[14] W.G. Klemperer: *J. Amer. Chem. Soc.* **94** (1972) 6940.
[15] J. Brocas, R. Willem, D. Fastenakel and J. Buschen: *Bull. Soc. Chim. Belges* **84** (1975) 483.
[16] D.J. Klein and A.H. Cowley: *J. Amer. Chem. Soc.* **97** (1975) 1633.
[17] R.E. Stanton and J.W. McIver Jr.: *J. Amer. Chem. Soc.* **97** (1975) 3632.
[18] P. Pechukas: *J. Chem. Phys.* **64** (1976) 1516.
[19] J.N. Murrel and K.J. Laidler: *Trans. Faraday Soc.* **64** (1968) 371; J.N. Murrel and G.L. Pratt: *Trans. Faraday Soc.* **66** (1970) 1680.
[20] D. Britton and J.C. Dunitz: *Acta Cryst.* **29A** (1973) 362.
[21] R.B. King: *Inorg. Chem.* **24** (1985) 1716.
[22] P.J. Federico: *Geometriae Dedicata* **3** (1975) 469.
[23] J. Brocas and M. Bauwin: *J. Math. Chem.* **6** (1991) 281.
[24] R.B. King: *Inorg. Chim. Acta* **49** (1981) 237.
[25] W.N. Lipscomb: *Science* **153** (1966) 373.
[26] F.A. Cotton: *Chemical Applications of Group Theory*, Wiley, New York (1990).
[27] J. Brocas: in *Advances in Dynamic Stereochemistry*, M. Gielen (Ed.), Freund Publishing House, London (1985), Vol. 1, Chapter 2, p. 43.
[28] R. Davis, M.N.S. Hill, C.E. Holloway, B.F.G. Johnson and K.H. Al-Obaidi: *J. Chem. Soc.* (A) (1971) 994.
[29] E.O. Bishop, G. Butler, J. Chatt, J.R. Dilworth, G.J. Leigh, D. Orchard and M.W. Bischop: *J. Chem. Soc. Dalton Trans.* (1978) 1654.
[30] S.L. Hawthorne and R.C. Fay: *J. Amer. Chem. Soc.* **101** (1979) 5268.
[31] P.J. Domaille, B.M. Foxman, T.J. Mc Neese and S.S. Wreford: *J. Chem. Soc.* **102** (1980) 4114.
[32] A.P. Ginsberg and M.E. Tully: *J. Amer. Chem. Soc.* **95** (1973) 4749.
[33] J.O. Albright, S. Datta, B. Dezube, J.K. Kouba, D.S. Marynick, S.S. Wreford and B.M. Foxman: *J. Amer. Chem. Soc.* **101** (1979) 611.
[34] S.S. Wreford, J.K. Kouba, J.F. Kirner, E.L. Muetterties, I. Tavanaiepour and V.W. Day: *J. Amer. Chem. Soc.* **102** (1980) 1558.
[35] (a) F.A. Van-Catledge, S.D. Ittel, C.A. Tolman and J.P. Jesson: *J. C. S. Chem. Comm.* (1980)

254; (b) F.A. Van-Catledge, S.D. Ittel, J.P. Jesson: *Organometallics* **4** (1985) 18.
[36] W.R. Cullen and L.M. Mihichuk: *Canad. J. Chem.* **54** (1976) 2548.
[37] P. Meakin, L.J. Guggenberger, F.N. Tebbe and J.P. Jesson: *Inorg. Chem.* **13** (1974) 1025.
[38] S. Datta, B. Dezube, J.K. Kouba and S.S. Wreford: *J. Amer. Chem. Soc.* **100** (1978) 4404.
[39] S.J. Lippard: *Progr. Inorg. Chem.* **21** (1978) 91.
[40] D.F. Lewis and S.J. Lippard: *Inorg. Chem.* **11** (1972) 621.
[41] C.T. Lam, M. Novotny, D.L. Lewis and S.J. Lippard: *Inorg. Chem.* **17** (1978) 2127.
[42] J.A. Connor, G.K. McEwen and C.J. Rix: *J. Chem. Soc. Dalton Trans.* (1974) 589.
[43] L.D. Brown, S. Datta, J.K. Kouba, L.K. Smith and S.S. Wreford: *Inorg. Chem.* **17** (1978) 729.
[44] K. Henrick and S.B. Wild: *J. Chem. Soc. Dalton Trans.* (1974) 2500.
[45] E.B. Dreyer, C.T. Lam and S.J. Lippard: *Inorg. Chem.* **18** (1979) 1904.
[46] J.L. Templeton and B.C. Ward: *Inorg. Chem.* **19** (1980) 1753.
[47] J.L. Templeton and B.C. Ward: *J. Amer. Chem. Soc.* **103** (1981) 3743.
[48] R. Hoffman, B.F. Beier, E.L. Muetterties and A.R. Rossi: *Inorg. Chem.* **16** (1977) 511.
[49] R.J. Gillespie: *Chem. Soc. Rev.* (1992) 59.

JET-COOLED FLUORESCENCE EXCITATION SPECTRA AND CARBONYL WAGGING POTENTIAL ENERGY FUNCTIONS OF CYCLIC KETONES IN THEIR ELECTRONIC EXCITED STATES

J. LAANE, J. ZHANG, W.-Y. CHIANG, P. SAGEAR

AND

C. M. CHEATHAM
Department of Chemistry,
Texas A&M University,
College Station, TX 77843,
U.S.A.

Abstract. The jet-cooled fluorescence excitation spectra of 2-cyclopenten–1-one and its 5,5-d_2 isotopomer have been recorded in the 370 to 340 nm region. The electronic origin for the undeuterated species occurs at 27210 cm^{-1} for the $S_1(n,\pi^*)$ electronic excited state. The vibrational frequencies for the three carbonyl motions and the nine ring modes were observed for the excited state. Bands at 67, 158, and 256 cm^{-1} for the d_0 species, at 63, 147, and 240 cm^{-1} for the 5-d_1 isotopomer, and at 59, 138, and 227 cm^{-1} for the d_2 species were assigned to the ring-puckering motion in the S_1 state. A single one-dimensional potential energy function accurately fits the data for all three isotopomers. This function is nearly purely quartic in character and shows the ring to be planar in the electronic excited state. However, it has become less rigid, and this is ascribed to a decrease in initial angle strain within the ring. The C=O and C=C stretching frequencies occur at 1418 and 1357 cm^{-1} for the d_0 molecule. The ring-twisting frequency for the S_1 state occurs at 274 cm^{-1}. Previous electronic absorption measurements had resulted in a misassignment for this motion.

The jet-cooled fluorescence excitation spectra of the $n \rightarrow \pi^*$ transitions of cyclopentanone, 3-cyclopenten–1-one, and cyclobutanone have been analysed to determine the vibrational energy spacings in the $S_1(n,\pi^*)$ electronic excited states for the out-of-plane carbonyl wagging motions. A double-minimum potential energy function was determined for each and the barriers were found to be 680, 926, and 1940 cm^{-1}, respectively. The carbonyl wagging angles were determined

Y.G. Smeyers (ed.), Structure and Dynamics of Non-Rigid Molecular Systems, 181–201.

to be 22°, 26°, and 41°, respectively. The out-of-plane ring modes were also assigned for each molecule.

1. Introduction

In recent years the combination of tunable high power lasers and pulsed supersonic jet techniques have provided the means of obtaining high resolution electronic spectra that are greatly simplified due to the sample cooling provided by the supersonic jet. The cooling results in the removal of hot bands and the narrowing of the bandwidths for the remaining bands. The lack of spectral congestion allows a much more accurate and thorough analysis of the low-frequency vibrations in electronic excited states than previously possible using fluorescence or absorption techniques.

A number of carbonyl compounds have been analysed in their $S_1(n,\pi^*)$ states. In this electronic state a carbonyl compound typically distorts from a planar to a pyramidal configuration about the carbonyl carbon atom. This was predicted for formaldehyde in 1953 by Walsh [1] and verified spectroscopically by Brand in 1956 [2]. Formaldehyde's six vibrational degrees of freedom and its large rotational constants made the analysis of much of its vibronic spectra feasible even before high-resolution or jet spectroscopy techniques were available. The $S_1(n,\pi^*)$ state of acetaldehyde has been studied by fluorescence and absorption methods since 1954 as a prototype of larger carbonyl compounds [3–5]. However, its electronic absorption spectra are complex and ill-resolved, and little agreement existed on the interpretation. Only recently, with the vibrational and rotational cooling obtained in a supersonic jet, did it become possible to make accurate assignments [6–8]. The $S_1(n,\pi^*)$ states of oxalyl fluoride [9], acetone [7, 10], cyclopentanone [11], cyclobutanone [11], 2-indanone [12] and benzophenone [13] have also been studied using jet cooling and fluorescence excitation spectroscopy (FES). Recently Laane and co-workers [14–16] have recorded and analysed the FES data for 2-cyclopenten–1-one (2CP), 3-cyclopenten–1-one (3CP), cyclopentanone (CP), and cyclobutanone (CB), and these results will be presented here.

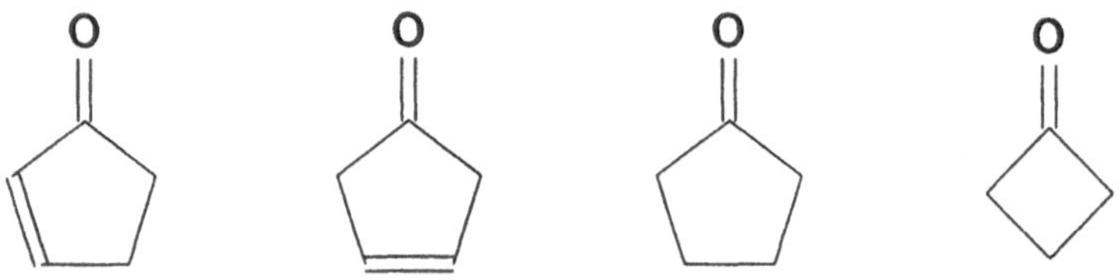

2CP 3CP CP CB

The electronic absorption spectra of 3CP and 2CP at room temperature have previously been reported by Gordon and Orr [17]. These workers concluded that in the excited state of 3CP the carbonyl group becomes pyramidal with an out-of-plane angle of 32 degrees and an inversion barrier of 780 cm^{-1}. The 2CP absorption spectrum was assigned mainly in terms of the carbonyl in-plane and out-of-plane wags and the ring-puckering and ring-twisting vibrations. However, the spectra were complicated by the presence of many hot bands from vibrationally excited states, and these were not readily distinguishable from the transitions originating from the ground state. The assignment of the ring-puckering spectrum incorrectly utilized hot-band transitions, and the authors predicted a small barrier to planarity in the electronic excited state. Moreover, they incorrectly assigned the ring-twisting frequency to 147 cm^{-1} (instead of 287 cm^{-1}) [18] in the ground state and to 160 cm^{-1} (instead of 274 cm^{-1}) in the electronic excited state.

Cheatham and Laane [18] have recently completed a reanalysis of the ring-puckering and ring-twisting far-infrared spectra of 2CP and two of its isotopomers. The puckering and twisting states, along with many combination states, were well defined, and a two-dimensional vibrational potential energy surface was determined in terms of these two large-amplitude motions. This detailed knowledge of the ground state energy levels was essential for the analysis of the electronic excited state of 2CP to be discussed here.

Cheatham and Laane [14] also recorded the $S_1(n, \pi^*)$ laser-induced FES spectra of jet-cooled 2CP-d_0 and 2CP-d_2 and reassigned many of the low-frequency vibrational bands in the excited state. The determination of the ring-puckering levels made it possible to establish the vibrational potential energy function for this motion (and thus the molecular conformation) in the electronic excited state. The ring-twisting and carbonyl-bending motions were also correctly identified.

In addition to our findings for 2CP, we will also present our recent results [15, 16] on the three other cyclic ketones, 3CP, CP, and CB. Each molecule has C_{2v} symmetry for a planar structure for which the purely electronic transition is symmetry forbidden ($^1A_2 \leftarrow {}^1A_1$). There is no conjugation in these molecules and thus the carbonyl group is not expected to remain co-planar with the ring system in any of these molecules. Sufficient data for the out-of-plane C=O wagging motion for each molecule was obtained so that we could determine the double-minimum potential energy functions which govern these vibrations. Data were also obtained for the out-of-plane ring motions.

2. Experimental

The experimental procedures utilized were similar to those previously described [14]. Frequency calibration of the system was accomplished by recording the optogalvanic spectrum of neon in a Cr–Ne hollow cathode lamp (see references 19 and 20 for details). Our frequency accuracy is ± 1 cm^{-1}. Further details on the experimental apparatus and experimental conditions can be found elsewhere [14,

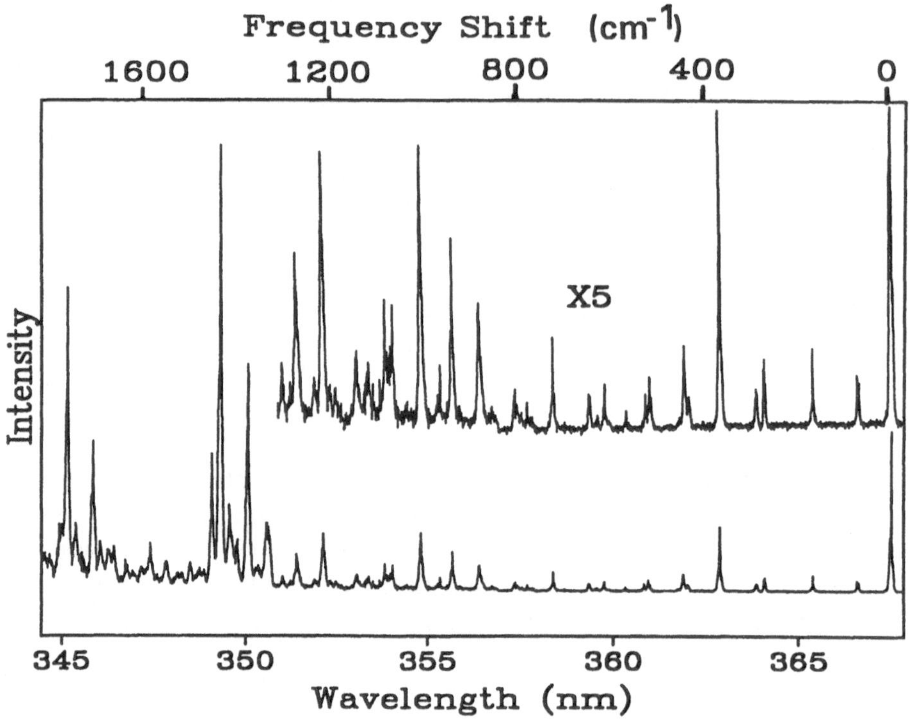

Figure 1. Fluorescence excitation spectrum of jet-cooled 2CP-d_0; $T_R \simeq 17$ K; $T_V \simeq 50$ K.

21]. Comparisons with previously published spectra by others show that we are able to get improved signal/noise ratios and sensitivity with our system.

3. Results and Discussion

3.1. 2-CYCLOPENTEN-1-ONE (2CP)

3.1.1. *Vibrational Assignments*

Figure 1 shows that low-resolution FES of 2CP-d_0, and that of 2CP-d_2 can be found elsewhere [14]. The intensities of the bands in the upper spectrum have been expanded by a factor of five. The spectra were recorded in a region from below the electronic origin to about 1800 cm^{-1} beyond it. Figures 2 and 3 show the expanded low energy region of the FES; the $S_1(n, \pi^*)$ excited state vibrational frequencies are indicated. The upper curve in Figure 3 shows a weaker series assigned to the 2CP-d_1 isotopic impurity ($\sim$ 10%) in the 2CP-d_2 sample. The frequency shifts are indicated from the 2CP-d_1 origin at 27206 cm^{-1}. The 2CP-d_0 and 2CP-d_2 electronic origins were observed at 27210 and 27203 cm^{-1}, respectively, in agreement with the values of 27211 and 27206 cm^{-1} from the lower resolution absorption spectra [17]. In comparison to 3CP and CP, which will be considered

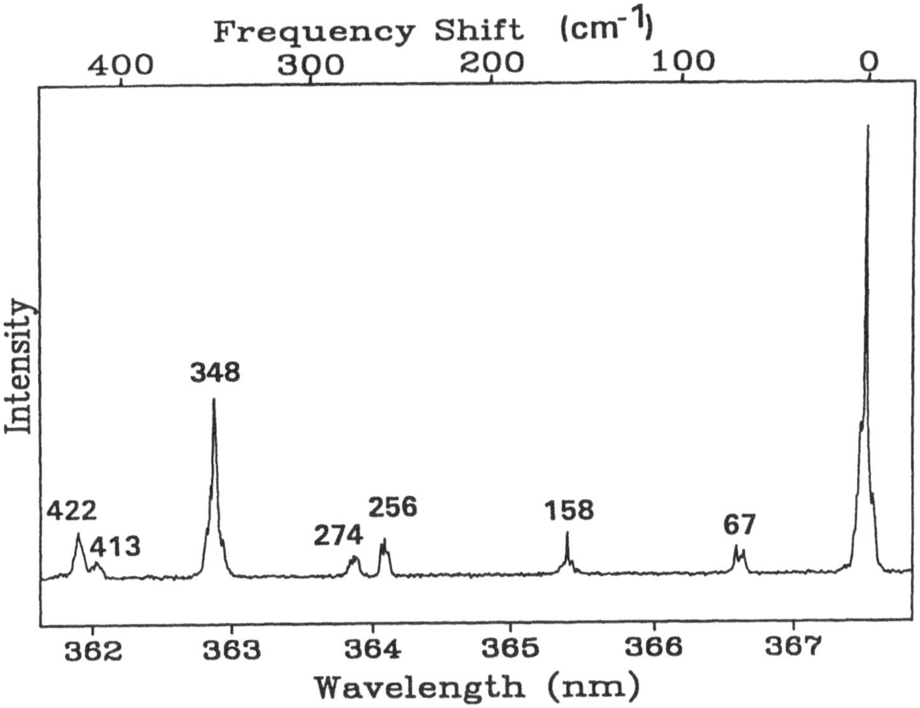

Figure 2. Low-energy region of the FES spectrum of 2CP-d_0 showing the ring-puckering and ring-twisting bands.

later, the 2CP frequency is considerably lower. This is the expected result from the conjugation between the C=O and C=C groups which results in a lower energy π^* orbital. The electronic origin for 2CP is extremely intense in contrast to the $S_1(n,\pi^*)$ origins of similar molecules [8, 22]. This is the result of a planar excited state structure and a high Franck-Condon factor.

The frequencies of the relevant fundamental vibrations determined for the electronic ground state [23] and S_1 excited state [14] are given in Table I. The FES assignments can be found elsewhere [14]. Thirteen of the thirty S_1 excited state vibrational frequencies have been determined from the fluorescence excitation spectrum. In order to obtain additional spectral information and in order to identify the hot bands (those originating from vibrational excited states in the S_0 ground state) in the FES and absorption spectra, we also recorded the fluorescence spectra at several different temperatures. The data, which are available elsewhere [14], clearly show that the bands at -27, 139 and 180 cm^{-1} are hot bands.

For the higher frequency vibrations our FES assignments agree well (± 10 cm^{-1}) with the absorption study of Gordon and Orr [17]. However, we disagree on the assignments involving the ring-puckering (ν_{30}), ring-twisting (ν_{29}), and carbonyl out-of-plane wag (ν_{28}). Gordon and Orr accepted a value of 170 cm^{-1}

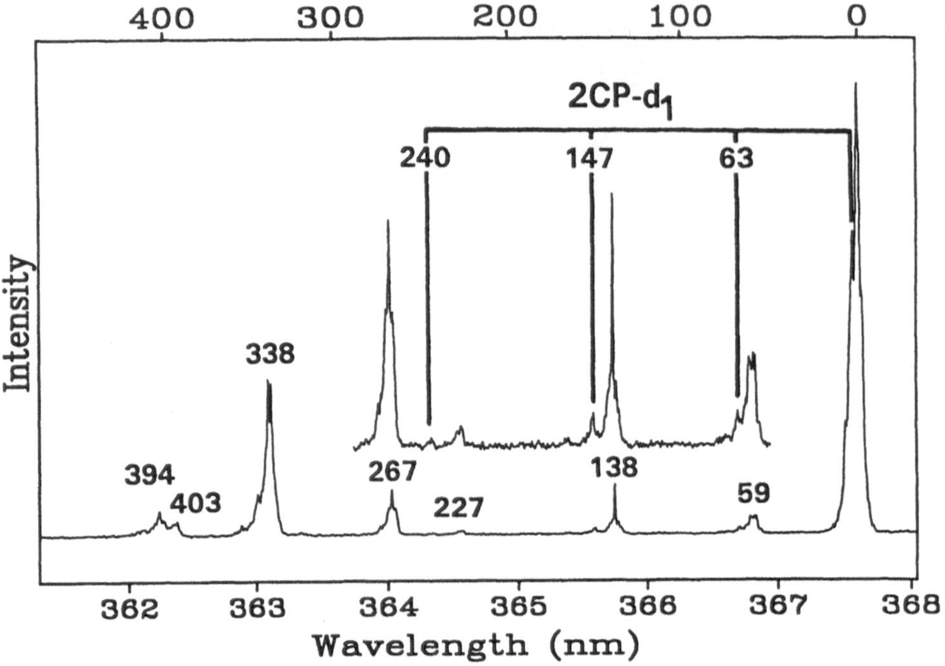

Figure 3. Low-energy region of the FES spectrum of 2CP-d_2 showing the ring-puckering and ring-twisting bands. The expanded region shows the ring-puckering bands from a small amount of 2CP-d_1 present.

for ν_{29} in the ground electronic state from an estimate based on a tricky microwave intensity measurement [23]. This estimate was in conflict with other assignments which placed the frequency at 300 cm^{-1} in the Raman spectrum of the liquid [23, 25]. A very recent and comprehensive analysis of the far-infrared spectra of 2CP-d_0, 2CP-d_1, and 2CP-d_2 showed the twisting bands of the vapor to lie in the 280 to 290 cm^{-1} range [18]. Their misassignment caused Gordon and Orr not only to use the wrong value for ν_{29} in the ground and excited electronic states, but also to misassign ν_{28} and the excited vibrational states of ν_{30} in both the S_0 and S_1 electronic states. Their use of a hot band at 179 cm^{-1} as their 30^2_0 assignment also resulted in an incorrectly determined ring-puckering potential energy function for the S_1 state. The 179 cm^{-1} band is actually a $29^1_0 30^0_1$ band, whereas 158 cm^{-1} corresponds to the 30^2_0 transition. We have also observed the 30^3_0 transition at 256 cm^{-1}. Our far-infrared work [18] clearly demonstrated that ν_{29} occurs at 287 cm^{-1} for the electronic ground state, and the fluorescence data reported here show that this is shifted to 274 cm^{-1} in the electronic excited state.

Figure 4 shows the ring-puckering data for the ground and excited electronic states for the d_0, d_1, and d_2 isotopomers. As will be seen, the ring-puckering series of bands at 67, 158, and 258 cm^{-1} can be fit nicely with a one-dimensional potential energy function. A fourth band observed at 360 cm^{-1} in the absorption

TABLE I. Vibrational frequencies (cm^{-1}) for the ground and excited $S_1(n, \pi^*)$ states of 2CP-d_0 and 2CP-d_2.

Approx. Description	2CP-d_0 Ground[a]	2CP-d_0 Excited	2CP-d_2 Ground[a]	2CP-d_2 Excited
ν_5 C=O stretch	1748	1357	1743	1360
ν_6 C=C stretch	1599	1418	1602	1421
ν_{13} Ring mode	1094	1037	1114	1037
ν_{14} Ring mode	999	974	957	960
ν_{15} Ring mode	912	906	851	843
ν_{16} Ring mode	822	849	810	814
ν_{17} Ring mode	753	746	746	727
ν_{18} Ring mode	630	587	630	580
ν_{19} C=O def (‖)	464	348	449	338
ν_{26} α-CH bend	750	768	737	762
ν_{28} C=O def (⊥)	537	422	?	403
ν_{29} C=C twist	287[b]	274	281[b]	267
ν_{30} Ring-puckering	94[b]	67	85[b]	59

[a]Ground state frequencies are from ref. 23 unless otherwise noted.
[b]Ref. 18.

spectrum may correspond to the 0 → 4 transition, and a FES hot band at 265 cm^{-1}, tentatively assigned to the 1 → 4 transition, appears to confirm the position of the fourth puckering level.

For the ring twisting, in addition to the 29^1_0 transitions at 274 and 267 cm^{-1}, respectively, for the d_0 and d_2 isotopomers, we have observed the overtones at 543 and 530 cm^{-1} demonstrating that the ring-twisting vibration is nearly harmonic, as it is in the ground state [18]. Of the more than fifty fluorescence excitation bands observed for 2CP-d_0, thirty-four involve ν_{30} and twenty involve ν_{29} (several are associated with both). At higher energies, combinations with the C=O in-plane (ν_{19}) and out-of-plane (ν_{28}) wags, which occur at 348 and 422 cm^{-1}, are common. These modes are both considerably lower in frequency than in the electronic ground state reflecting the decrease in π character of the C=O bond.

The intense bands at 1357 and 1418 cm^{-1} for 2CP-d_0 are clearly due to the C=O and C=C stretches, respectively. These two vibrations are Franck–Condon active due to the increased bond lengths for both bonds. Intense combination bands for each of these stretches were observed with the carbonyl in-plane wag, 19^1_0. All of the other ring mode transitions have also been observed for both the d_0 and d_2 species and many of them were also found to be associated with combination bands. The only other excited state frequency determined from the FES data was for ν_{26}, a CH out-of-plane bending motion, at 768 cm^{-1}.

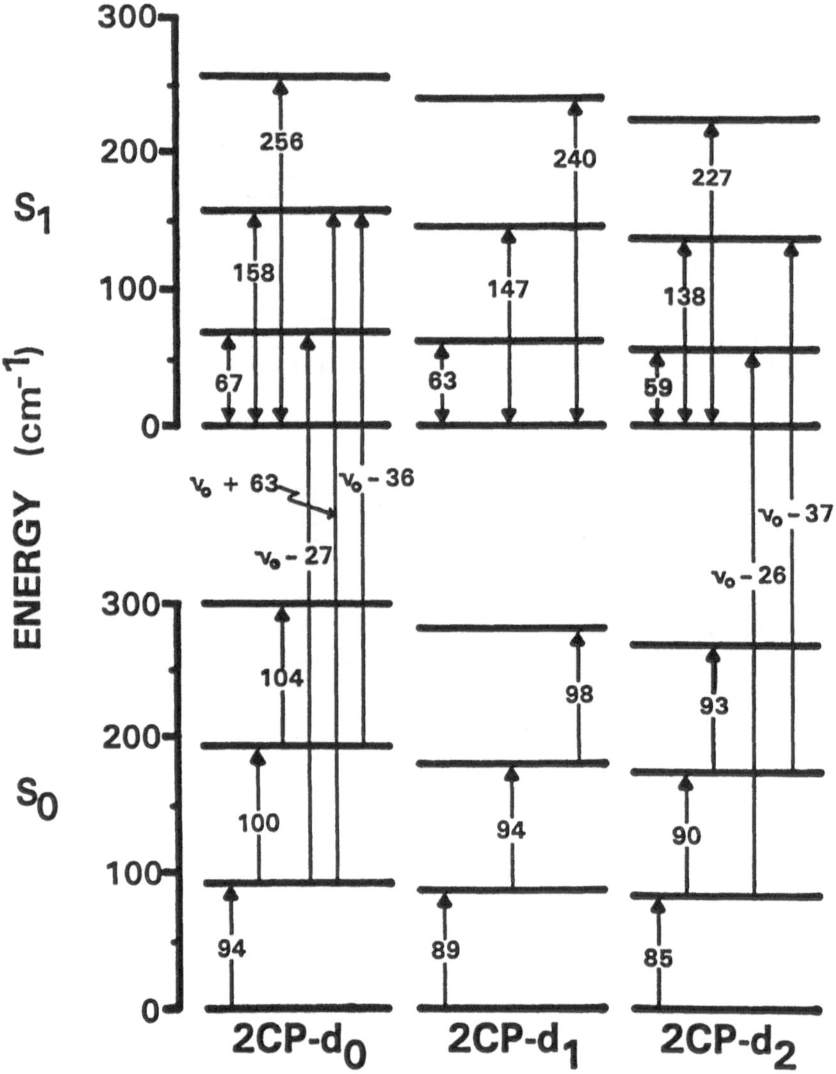

Figure 4. Energy level diagram for the ring-puckering vibrations of 2CP-d_0, 2CP-d_1, and 2CP-d_2 in the ground and $S_1(n, \pi^*)$ excited electronic states.

3.1.2. *Vibrational Hamiltonian*

The one-dimensional vibrational Hamiltonian for the ring-puckering has the form

$$H(x) = (-\hbar^2/2)d/dx\; g_{44}(x)d/dx + V(x), \tag{1}$$

where x is the puckering coordinate, $g_{44}(x)$ is the reciprocal reduced mass expansion [26–28], and the potential energy is given by

$$V(x) = ax^4 + bx^2. \tag{2}$$

The utilization of these types of functions to represent ring-bending motions have been considered in detail elsewhere [29, 30].

3.1.3. *Kinetic Energy Functions*

The kinetic energy (reciprocal reduced mass) expansions $g_{44}(x)$ for the ring-puckering vibration of the three 2CP isotopomers are expressed in the form

$$g_{44}(x) = g_{44}^{(0)} + g_{44}^{(2)}x^2 + g_{44}^{(4)}x^4 + g_{44}^{(6)}x^6. \quad (3)$$

The values of the coefficients can be found elsewhere [14]. It was most gratifying that the same model (which included a small amount of CH_2 rocking) precisely predicted the correct isotopic shifts observed for the electronic excited state as well as the ground state [14].

3.1.4. *Ring-Puckering Potential Energy Functions*

The ring-puckering energy levels and observed transitions for the ground and excited electronic states are shown in Figure 4. These were used to determine the coefficients a and b in Equation (2). The potential function, which fit the data for all three isotopic species, was determined to be

$$V(\text{cm}^{-1}) = 2.5 \times 10^6 x^4 + 1.8 \times 10^3 x^2. \quad (4)$$

In the electronic ground state the coefficients a and b are 0.6×10^6 and 2.6×10^4, respectively. The electronic excitation has caused the potential energy function to change from one with a substantial quadratic contribution in the S_0 state to one that is almost purely quartic in the S_1 state. The potential energy functions for both the ground and excited electronic states are shown together in Figure 5. The result of the increased quartic nature of the potential is to flatten the curve near the minimum and to increase the slope of the walls of the potential at higher energies. Thus, 2CP has less resistance to ring-puckering in the $S_1(n, \pi^*)$ state than in the ground state.

3.1.5. *Rotational Contours*

We have used an asymmetric-top band contour computer program by van der Veken [31] to simulate the low-temperature contours observed in the FES spectra. The rotational constants were calculated using molecular structures predicted by the molecular mechanics program (MMP2) of Allinger and Burkett [32]. The ground state constants calculated with this computer program ($A' = 0.2455$, $B' = 0.1212$, $C' = 0.0838$ cm^{-1}) are essentially the same as those from the microwave study [24]. The excited state constants were calculated by inputing new equilibrium bond lengths into MMP2 according to the changes observed for acrolein [33, 34]. Both the C=O and C=C bond lengths were increased by 0.1 Å, and the length of the C–C bond between them was decreased by 0.1 Å. The excited state rotational constants A, B, and C were calculated to be 0.2565, 0.1196, and 0.0831 cm^{-1}, respectively. This procedure resulted in good agreement between

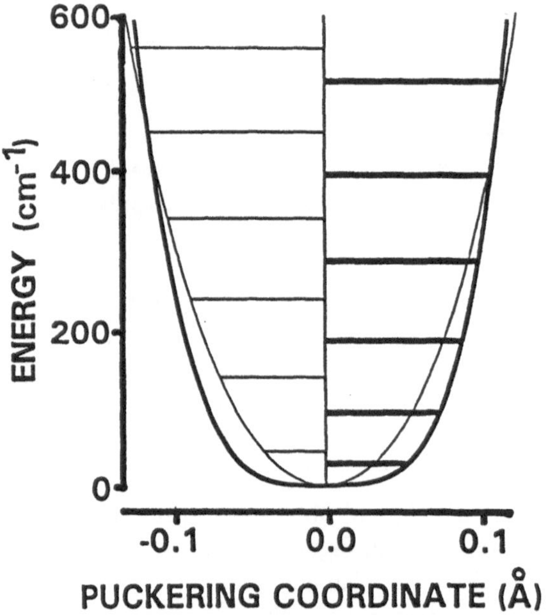

Figure 5. Potential energy function and energy levels for the ring-puckering vibration of 2CP-d_0 in the ground (thin lines) and $S_1(n,\pi^*)$ excited (thick lines) electronic states.

observed and calculated band contours [14] and allowed the rotational temperature of the spectra to be calculated (17° K).

3.1.6. *Bonding in the Electronic Excited State*

Figure 6 shows a simplified molecular orbital energy diagram for the n and π orbitals of 2CP. Excitation to the $S_1(n,\pi^*)$ excited electronic state removes a non-bonded electron from the carbonyl oxygen and transfers it to a π^* orbital, which is anti-bonding between carbon atoms 2 and 3 and also between carbon 1 and the oxygen atom. However, an increase in the bond order between carbons 1 and 2 is expected. A simple Hückel calculation [14] for the bond orders for the ground and excited states gave the following result:

$$\underset{\text{Ground}}{\overset{1.8\quad\;\; 1.5\quad\;\; 1.9}{O=C \;\;-\;\; C=C}} \qquad \underset{\text{Excited}}{\overset{1.5\quad\;\; 1.6\quad\;\; 1.7}{O=C \;\;-\;\; C=C.}}$$

While the calculation is only approximate, it provides a semi-quantitative picture of the bonding changes expected following the excitation to the S_1 electronic state. As can be seen in Table I, the decreases in the C=O and C=C bond orders manifest themselves in the ν_5 and ν_6 stretching frequencies which are decreased from 1748 to 1357 cm^{-1} and from 1599 to 1418 cm^{-1}, respectively. The expected increase in the C(1)–C(2) bond stretching force constant has less effect on the other ring mode frequencies since these vibrations also depend on the force constants of the other

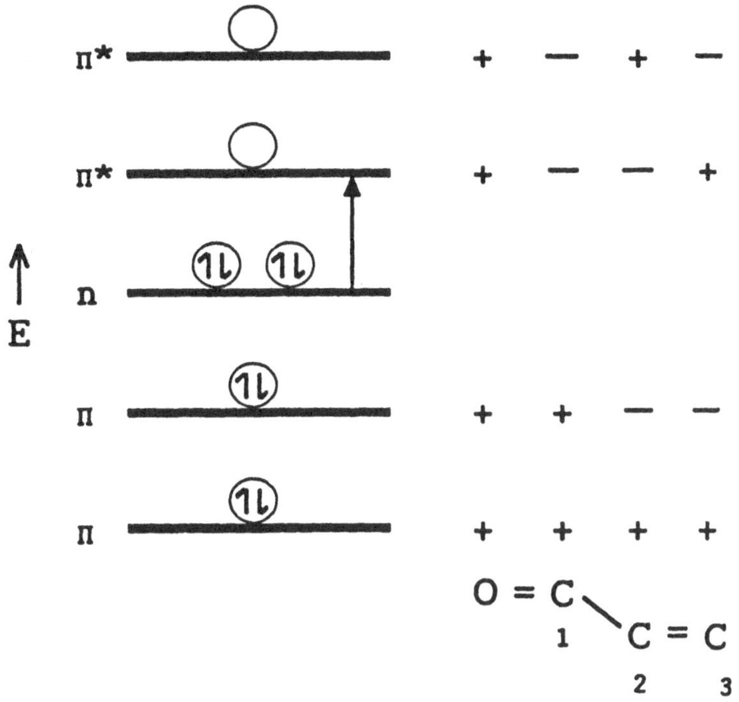

Figure 6. Molecular orbital diagram for 2CP.

C–C bonds and C–C–C angles. Nonetheless, the observed frequency increase for ν_{16} from 822 to 849 cm^{-1} does appear to reflect this effect. Although the value of 1357 cm^{-1} for ν_5 in the electronic excited state has dropped considerably relative to the ground state, this decrease is smaller than that for 3CP (1239 cm^{-1}) and 2CP (1227 cm^{-1}). For these non-conjugated ring molecules the π^* orbital is localized on the carbonyl group and the excited state bond order is decreased to a greater extent.

The ring-puckering potential energy functions for 2CP in both the ground and excited states have the form given in Equation (2) and are compared to each other in Figure 5. We have previously shown [35, 36] that the potential energy coefficients a and b both have contributions from angle strain (a_s and b_s) and torsional (a_t and b_t) forces:

$$a = a_s + a_t \qquad a_s >> a_t \tag{5}$$

$$b = b_s + b_t. \tag{6}$$

Furthermore, we have shown that b_s depends both on the initial strain in the ring and on the angle-bending force constants. For 2CP in the electronic ground state b_s is positive and greater in magnitude than b_t, which is negative. The latter contribution arises primarily from the CH_2–CH_2 torsion between carbon atoms 4 and 5, and this effect is at a maximum for a planar ring. In the electronic excited

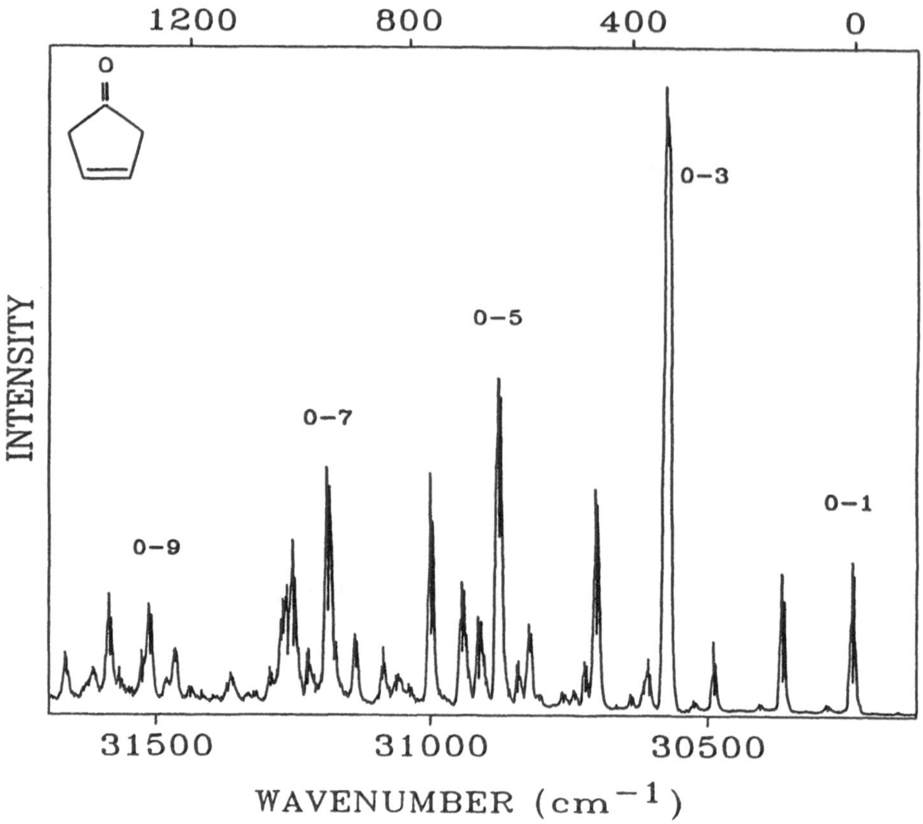

Figure 7. Fluorescence excitation spectrum of 3-cyclopenten–1-one.

state, b_t is expected to be little changed. However, the initial strain at the carbonyl carbon atom should be significantly reduced, and the magnitude of b_s should be lowered. This is confirmed by our experimental results here which show that the sum of b_s and b_t has been reduced nearly to zero, and the quadratic constant has barely maintained a slight positive value. (The value of b has dropped from 2.6×10^4 to 0.2×10^4 cm^{-1} Å^{-2}.) Thus, the ring remains planar in the electronic excited state, but its rigidity has been substantially reduced.

3.2. 3-CYCLOPENTEN–1-ONE (3CP)

Figure 7 shows the jet-cooled fluorescence excitation spectrum [15] of 3-cyclopenten–1-one. The band origin is observed at 30238 cm^{-1}. Each $v = 0$ (in the electronic ground state) to $v = n$ (in the A_2 electronic excited state) transition of the C=O wag has B_2 vibrational symmetry (assuming the molecule to lie in the xz plane) for n = odd, but has A_1 vibrational symmetry for n = even. Only transitions to the n = odd states can be observed. These show up as intense Type

B bands arising from $A_2 \times B_2 = B_1$ symmetry. The first five of these transitions are labelled in Figure 7. It should be noted that since the $v = 0$ and $v = 1$ levels in the $S_1(n, \pi^*)$ state are near-degenerate, the band origin lies very close to the $0 \rightarrow 1$ frequency. The other bands in the spectrum include many combinations of the C=O wag with the ring-puckering vibration and combinations of these with other fundamentals including the C=O stretch. Of particular note is the $3^1_0 29^4_0 30^1_0$ band at 32211 cm^{-1}, shifted 746 cm^{-1} from the band origin and 746 cm^{-1} from the C=O stretching (v_3) value at 1227 cm^{-1}. The 1973 cm^{-1} value corresponds to the sum of the $0 \rightarrow 4$ wagging transition (v_{29}) and the ring-puckering (v_{30}) frequency of 127 cm^{-1}. This shows the $0 \rightarrow 4$ spacing to be 619 cm^{-1}, which is 21 cm^{-1} less than the $0 \rightarrow 5$ spacing.

In order to analyse the C=O wagging vibration in the electronic excited state, we have utilized our computer programs, described previously, for calculating the reduced masses [26, 27] and energy levels [29, 30] for the Hamiltonian given in Equation (1) and the potential energy function given by Equation (2). Here x is used for the C=O wagging coordinate given in terms of the wagging angle ϕ and the C=O bond distance R by [26]

$$x = R\phi. \tag{7}$$

The reciprocal reduced mass expansion g_{44} for this coordinate has the form given in Equation (3).

For 3-cyclopenten–1-one the reduced mass and the carbonyl wagging potential energy parameters which best fit the observed frequency separations are given in Table II. The experimentally determined potential energy function is shown in Figure 8 along with both the observed and calculated frequency separations. The minimum energy corresponds to a wagging angle of ±26° and the barrier to inversion is 926 cm^{-1} (2.65 kcal/mole). Using their absorption spectra, Gordon and Orr [17] also assigned transitions for the C=O out-of-plane wag. Their assignments are partially in agreement with our data but led to a calculated inversion barrier of 780 cm^{-1}. These workers estimated the wagging angle to be ±33° at the energy minimum.

For 3CP the ring-puckering frequency of 127 cm^{-1} in the S_1 state is considerably higher than the value of 83 cm^{-1} in the ground state.

3.3. CYCLOPENTANONE (CP)

The survey spectrum of cyclopentanone [15, 16] is shown in Figure 9. The band origin is at 30276 cm^{-1}. In the ground state this molecule is twisted [37] and, in the C_{2v} approximation, the vibrational ground state is nearly doubly degenerate with symmetry species A_1 and A_2. The twisting conformation (and degeneracy) carries through to the electronic excited state as demonstrated by the similarity in the ring-bending and twisting frequencies in the S_1 state. The purely electronic transition is again $^1A_2 \leftarrow {}^1A_1$ which is forbidden in the C_{2v} approximation. However,

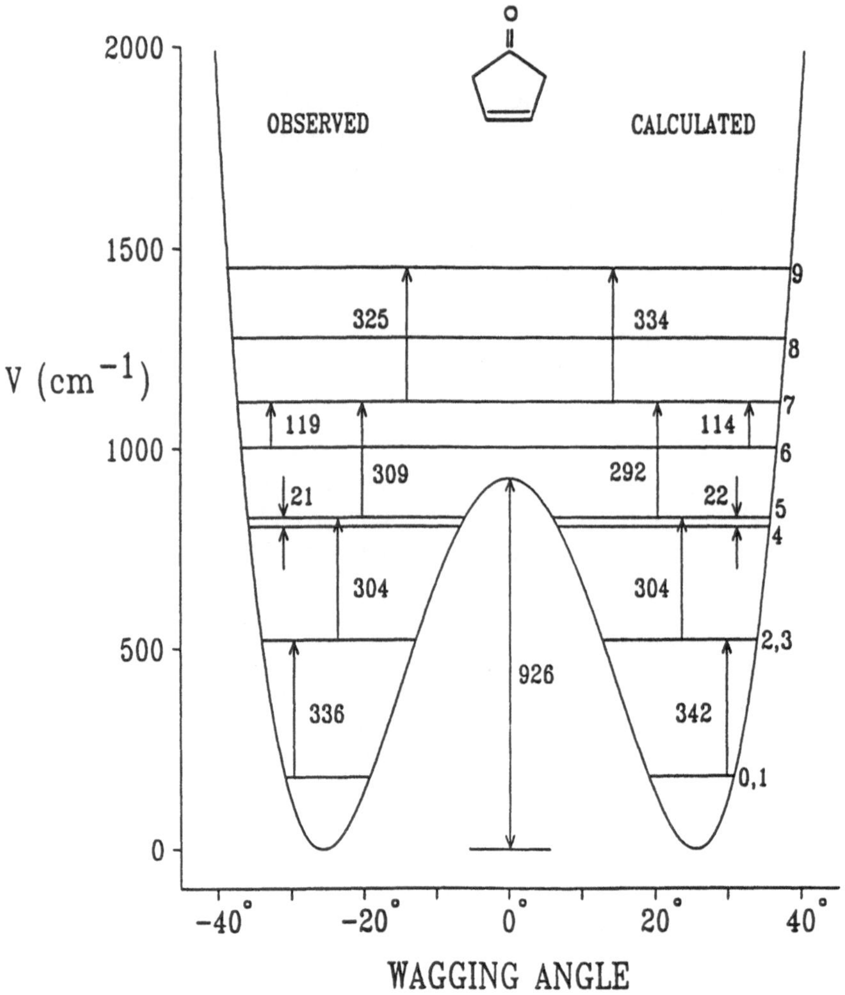

Figure 8. Vibrational potential energy function for the C=O out-of-plane wagging vibration of 3-cyclopenten–1-one.

combination with odd quantum transitions of the B_2 C=O wagging results in Type B bands from B_1 symmetry. If either the ground or excited electronic state is also in combination with the near-degenerate A_2 twisting state, the even quanta C=O wagging transitions can also be observed as Type A (A_1) bands $[A_2 \times A_2 \times (B_2)^n = A_1$ for n = even]. Figure 9 shows that transitions for both even and odd quantum states of the C=O wag in the S_1 state are readily observed. The band contours for the Type A and B bands are as expected (Figure 10), and even overlapped bands can be fit with the contour calculation.

Figure 11 shows the C=O wagging potential energy function for cyclopentanone with a barrier of 680 cm^{-1} and the energy minima at ±22°. The kinetic

TABLE II. Potential energy parameters and reduced masses for C=O wagging vibrations in the $S_1(n, \pi^*)$ electronic state.

Molecule	μ (au)	$V = ax^4 + bx^2$ a (cm^{-1}/Å^4)	b(cm^{-1}/Å^2)	Barrier (cm^{-1})	ϕ_{min}
CP	5.569[a]	10.49×10^3	-5.34×10^3	680	22°
3CP	5.260[b]	8.11×10^3	-5.48×10^3	926	26°
CB	4.244[c]	2.47×10^3	-4.38×10^3	1940	41°

[a] $g_{44} = 0.17957 - 0.049144x^2 + 0.014227x^4 - 0.002181x^6$
[b] $g_{44} = 0.19012 - 0.054853x^2 + 0.016335x^4 - 0.002554x^6$
[c] $g_{44} = 0.23565 - 0.076454x^2 + 0.024096x^4 - 0.003922x^6$

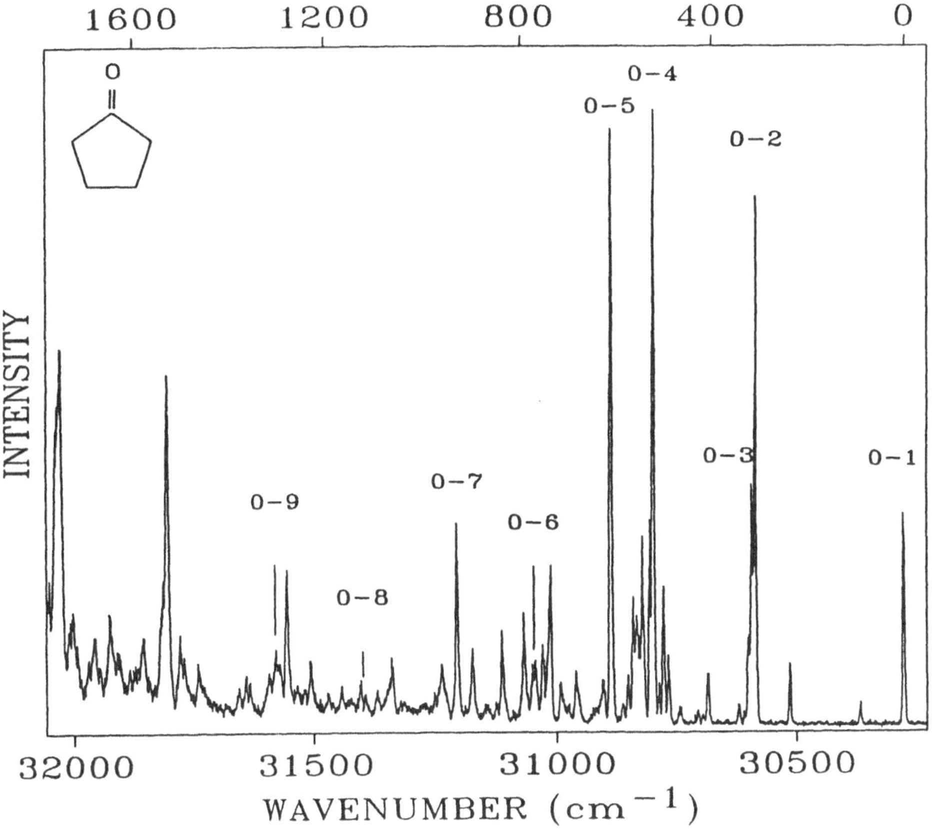

Figure 9. Fluorescence excitation spectrum of cyclopentanone.

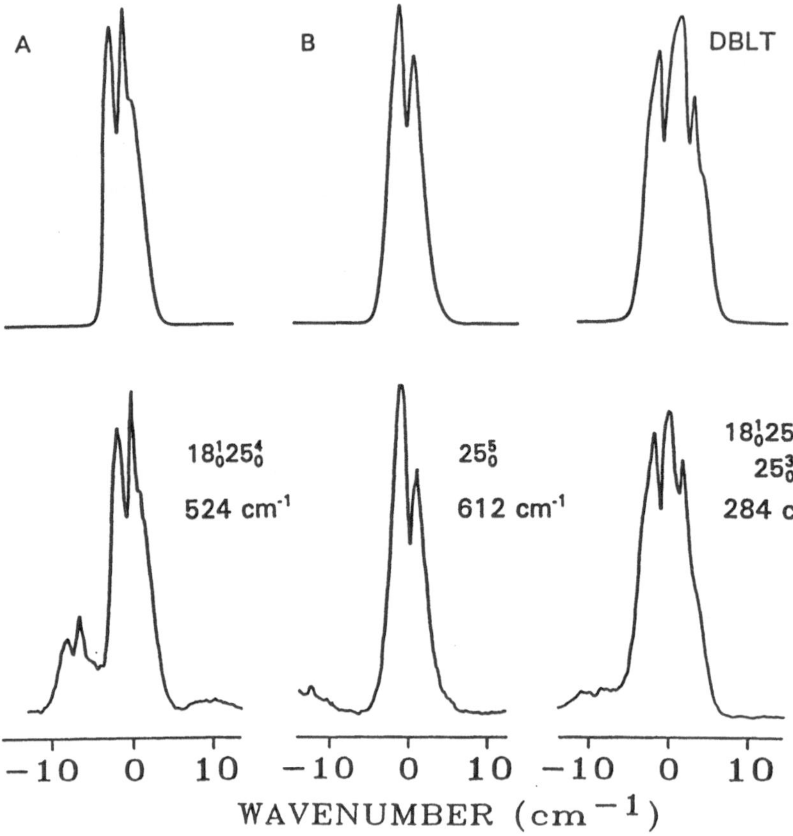

Figure 10. Observed (bottom) and calculated band contours for Type A and B bands of cyclopentanone and for an A/B doublet of the d_4 isotopomer.

and potential energy terms are given in Table II. The comparison between the S_0 and S_1 states for the fundamental vibrational frequencies of several other modes is given in Table III.

Table IV compares the observed data for the ring-twisting and ring-bending motions of CP and CP-d_4 in the S_0 and S_1 states. The fundamental frequencies for these two modes are changed little in the two states since the ring has a similar twisted conformation for each state. However, as shown in Table IV, the data for the ring modes indicates that in the S_1 state there is a barrier to pseudorotation of about 550 cm^{-1}. This value is a measure of the energy difference between the lower energy twisted conformation and the bent structure. The twisting data suggest that the barrier to planarity (the energy difference between the planar and twisted forms) may exceed 1000 cm^{-1}. Thus, the two-dimensional potential energy surface for the out-of-plane ring modes is substantially different from that proposed for the ground state [37].

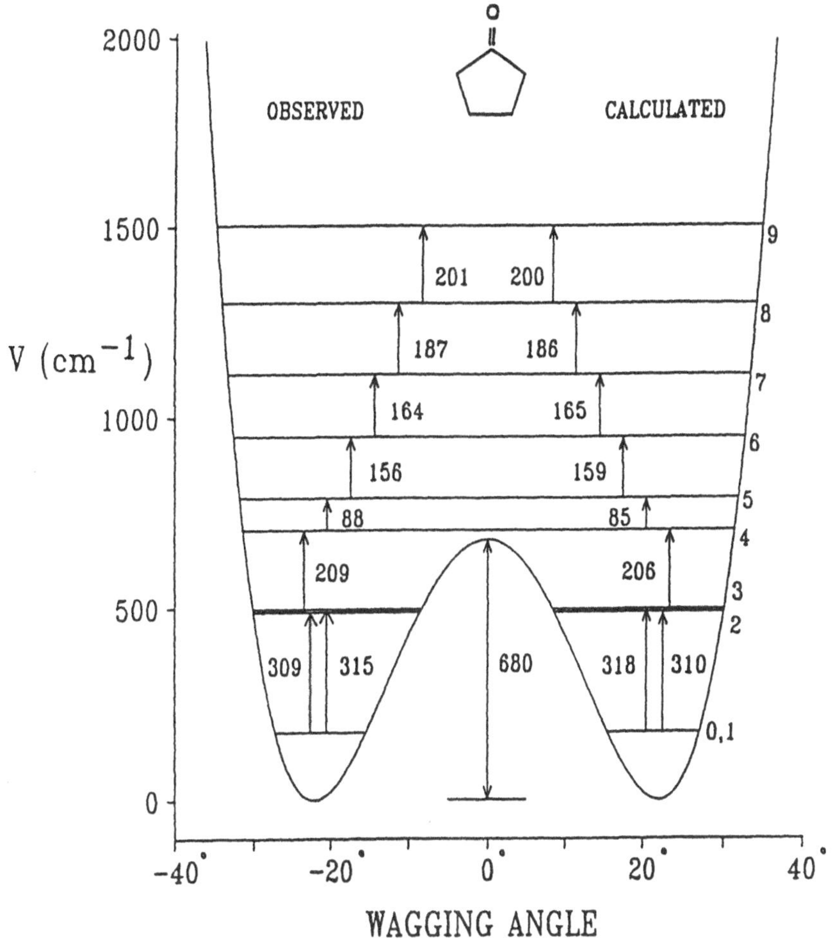

Figure 11. Vibrational potential energy function for the C=O out-of-plane wagging vibration of cyclopentanone.

3.4. CYCLOBUTANONE (CB)

The fluorescence excitation spectrum of cyclobutanone [15] is shown in Figure 12, with the band origin at 30292 cm^{-1}. Only transitions involving the odd quantum levels of the wag are allowed as Type B bands ($A_2 \times B_2 = B_1$ which gives the Type B band type). As it turns out, the barrier to inversion of cyclobutanone is sufficiently high that the lowest six pairs of levels are each nearly doubly degenerate with vibrational symmetry species A_1 and B_2. Even though transitions involving only the vibrational states of B_2 symmetry can be observed, these have essentially identical frequencies with those involving the A_1 states. Thus no significant information is lost.

TABLE III. Vibrational frequencies (cm^{-1}) for the ground S_0 and excited $S_1(n, \pi^*)$ electronic states of cyclopentanone and cyclopentanone-d_4.

	Approx. Description	CP-d_0 Ground	CP-d_0 Excited	CP-d_4 Ground	CP-d_4 Excited
A_1 ν_3	C=0 stretch	1770	1230	1769	1231
ν_{11}	Ring angle bending	705	532	624	461
A_2 ν_{18}	Ring twisting	238	238	220	223
B_1 ν_{25}	C=0 out-of-plane wagging	446	309	375	282
ν_{26}	Ring bending	95	91	89	83
B_2 ν_{36}	C=0 in-plane wagging	467	342	438	316

TABLE IV. Vibrational frequencies for ring-twisting (ν_{18}) and ring-bending (ν_{26}) in the S_0 and S_1 states for cyclopentanone-d_0 and -d_4.

	CP-d_0 Ground	CP-d_0 Excited		CP-d_4 Ground	CP-d_4 Excited	
	Obs	Obs	Calc[a]	Obs	Obs	Calc[b]
Ring twisting						
0-2	236[c]	238	238	220	223	223
2-4	---	228	229	---	214	213
4-6	---	218	218	---	202	202
Ring bending						
0-1	95.0	91	92[d]	84.5	83	83[d]
1-2	93.0	88	86	82.9	79	79
2-3	91.1	81	80	81.4	71	73
3-4	89.3	71	72	79.9	--	67

[a] Using $V = 21.6(z^4-16.3z^2)$ in reduced coordinates. Barrier=1433 cm^{-1}.
[b] Using $V = 20.9(z^4-15.4z^2)$ in reduced coordinates. Barrier=1240 cm^{-1}.
[c] Ref. 37.
[d] Using $V = \frac{1}{2}V_2(1-\cos 2\phi)+\frac{1}{2}V_4(1-\cos 4\phi)$ with V_2=550 cm^{-1}=barrier and V_4=-50 cm^{-1}. $B(d_0)$=3.20 cm^{-1}; $B(d_4)$=2.55 cm^{-1}

Figure 13 shows the C=O wagging potential energy function with a barrier of 1940 cm^{-1} and energy minima at ±41°. Table II presents the reduced mass and potential energy parameters for cyclobutanone.

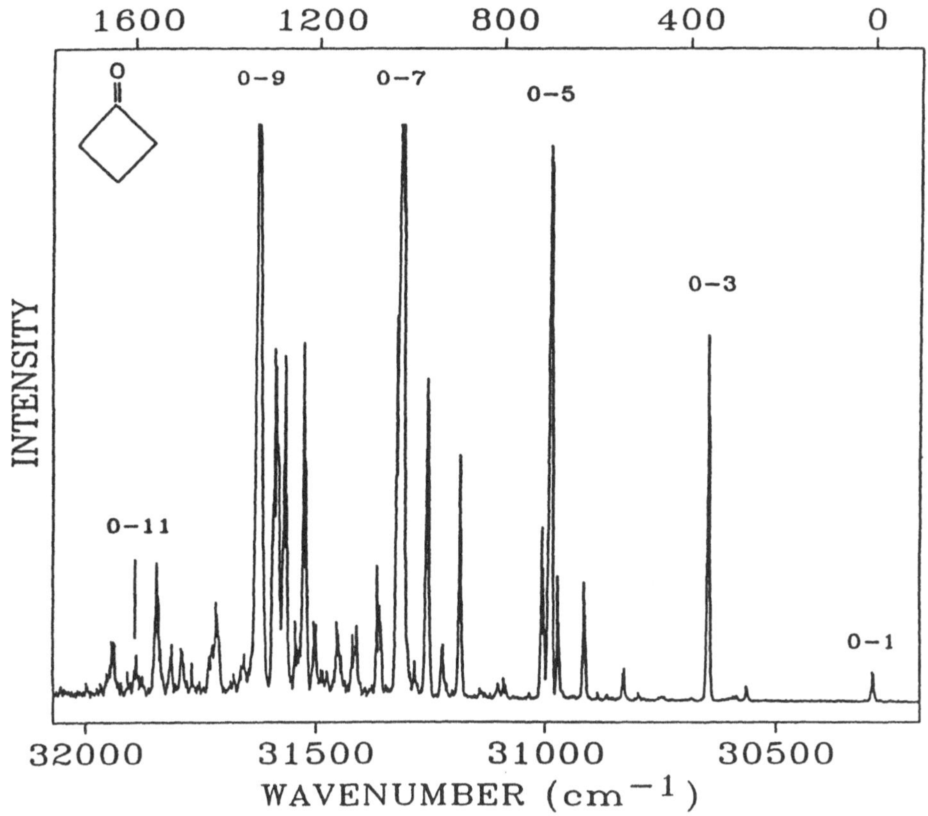

Figure 12. Fluorescence excitation spectrum of cyclobutanone.

The ring-puckering frequency shows a dramatic increase in going from 37 cm^{-1} in the S_0 state to 106 cm^{-1} in the excited state. The high ring angle strain appears to be a factor here.

4. Summary

We have determined the vibrational frequencies for all of the carbonyl and ring modes of 2-cyclopenten–1-one and its 5,5-d_2 isotopomer in the $S_1(n, \pi^*)$ electronic excited state. The expected decreases in the C=O and the C=C bond strengths were observed. The ring-puckering potential energy function for the excited state was determined, and this shows that the molecule remains planar but becomes less rigid in the excited state. We have also used our data [14] to reassign previously reported electronic absorption spectra [17] of 2CP and its isotopomers. In particular, the ring-twisting and the C=O bending modes had previously been incorrectly identified.

Table II compares the barriers to inversion and the equilibrium carbonyl wag-

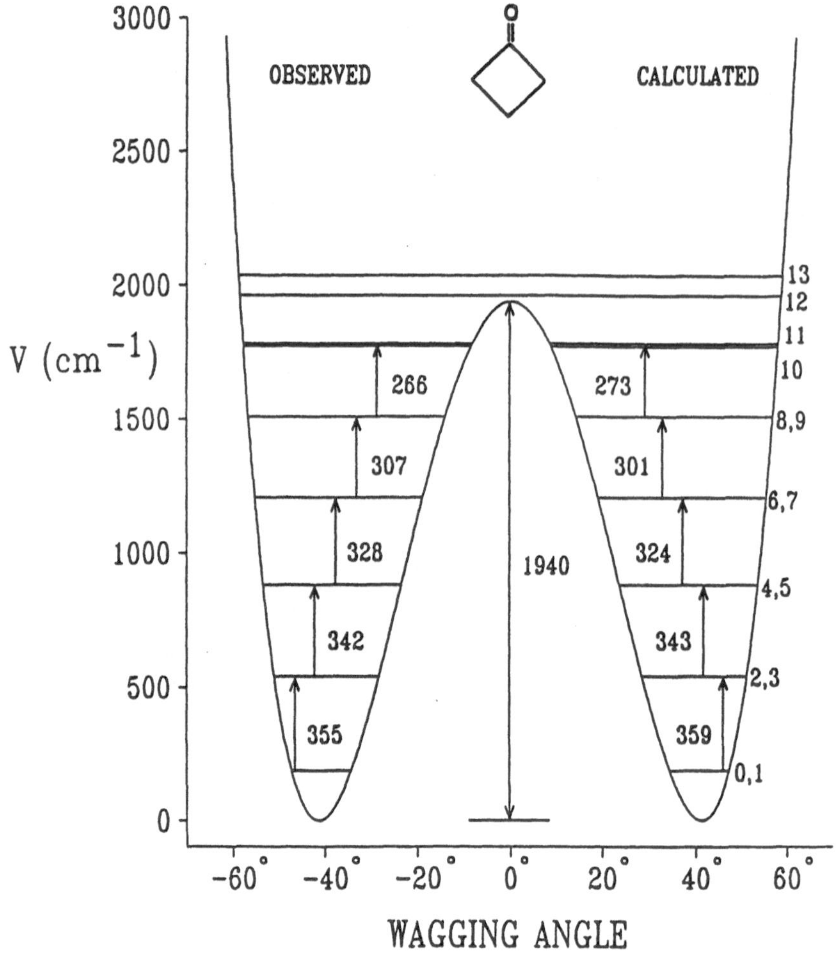

Figure 13. Vibrational potential energy function for the C=O out-of-plane wagging vibration of cyclobutanone.

ging angles for the cyclic ketones (3CP, CP, and CB) in their $S_1(n,\pi^*)$ electronic excited states. 2-Cyclopenten–1-one (2CP) retains sufficient conjugation to remain planar in the S_1 state. For each of the other three molecules studied there is a barrier to C=O inversion in the S_1 state and its magnitude increases as the ring-angle strain increases. The ring-bending vibrational frequencies in the S_1 states, reflecting the angle strain in these cyclic molecules, show the largest increases for the more strained molecules. The ring-bending frequencies change as follows in going from the S_0 to S_1 states: 94 to 67 cm^{-1} for 2CP, 83 to 127 cm^{-1} for 3CP, 95 to 92 cm^{-1} for CP, and 37 to 106 cm^{-1} for CB.

Acknowledgments

The authors are grateful to the National Science Foundation, the Robert A. Welch Foundation, and the Texas ARP program for financial support.

References

[1] A.D. Walsh: *J. Chem. Soc.* (1953) 2306.
[2] J.C.D. Brand: *J. Chem. Soc.* (1956) 858.
[3] V.R. Rao and I.A. Rao: *Indian J. Phys.* **28** (1954) 491.
[4] K.K. Innes and L.E. Giddings, *J. Mol. Spectrosc.* **7** (1961) 435.
[5] L.M. Hubbard, D.F. Bocian and R.R. Birge: *J. Am. Chem. Soc.* **103** (1981) 3313.
[6] M. Noble and E.K.C. Lee: *J. Chem. Phys.* **81** (1984) 1632.
[7] M. Baba, I. Hanazaki and U. Nagashima: *J. Chem. Phys.* **82** (1985) 3938.
[8] M. Noble, E.C. Apel and E.K.C. Lee: *J. Chem. Phys.* **78** (1983) 2219.
[9] M.G. Liverman, S.M. Beck, D.L. Monts and R.E. Smalley: *J. Chem. Phys.* **70** (1979) 192.
[10] M. Baba and I. Hanazaki: *Chem. Phys. Lett.* **103** (1983) 93.
[11] M. Baba and I. Hanazaki: *J. Chem. Phys.* **81** (1984) 5426.
[12] M. Baba: *J. Chem. Phys.* **83** (1985) 3318.
[13] J.H. Frederick, E.J. Heller, J.L. Ozment and D.W. Pratt: *J. Chem. Phys.* **88** (1988) 2169.
[14] C.M. Cheatham and J. Laane: *J. Chem. Phys.* **94** (1991) 7743.
[15] J. Zhang, W.Y. Chiang, P. Sagear, and J. Laane: *Chem. Phys. Lett.* **196** (1992) 573.
[16] J. Zhang, W.Y. Chiang, and J. Laane: *J. Chem. Phys.* **98** (1993) 6129; **100** (1994) 3455.
[17] R.D. Gordon and D.R. Orr: *J. Mol. Spectrosc.* **129** (1988) 24.
[18] C.M. Cheatham and J. Laane: *J. Chem. Phys.* **94** (1991) 5394.
[19] G.A. Bickel and K.K. Innes: *Appl. Opt.* **24** (1985) 3620.
[20] K. Narayama, G. Ullas, and S.B. Rai: *Chem. Phys. Lett.* **156** (1989) 55.
[21] C.M. Cheatham: Ph.D. Dissertation, Texas A&M University, College Station, Texas (1990).
[22] N. Mikami, A. Hiraya, I. Fujiwara, and M. Ito: *Chem. Phys. Lett.* **74** (1980) 531.
[23] T.H. Chao and J. Laane: *J. Mol. Spectrosc.* **48** (1973) 266.
[24] D. Chadwick, A.C. Legon, and D.J. Millen: *J. Chem. Soc., Faraday Trans. 2* **75** (1979) 302.
[25] R. Cataliotti, G. Paliani, and S. Santini: *J. Mol. Spectrosc.* **103** (1984) 56.
[26] J. Laane, M.A. Harthcock, P.M. Killough, L.E. Bauman, and J.M. Cooke: *J. Mol. Spectrosc.* **91** (1982) 286.
[27] M.A. Harthcock and J. Laane: *J. Mol. Spectrosc.* **91** (1982) 300.
[28] R.W. Schmude, M.A. Harthcock, M.B. Kelly, and J. Laane: *J. Mol. Spectrosc.* **124** (1987) 369.
[29] J. Laane: *Pure and Appl. Chem.* **59** (1987) 1307.
[30] J. Laane: *Appl. Spectrosc.* **24** (1970) 73.
[31] B.J. van der Veken: in *Vibrational Spectra and Structure*, J.R. Durig (Ed.), Marcel Dekker, New York (1986), Vol. 15, pp. 313–400.
[32] U. Burkett and N.L. Allinger: *Molecular Mechanics*, American Chemical Society Monograph, Vol. 177, Washington DC, (1982).
[33] J.M. Hollas: *Spectrochim. Acta.* **19** (1963) 1425.
[34] J.C. Brand and D.G. Williamson: *Diss. Far. Soc.* **35** (1963) 184.
[35] J.D. Lewis and J. Laane: *J. Mol. Spectrosc.* **53** (1974) 417.
[36] J. Laane: *J. Chem. Phys.* **50** (1969) 776.
[37] T. Ikeda and R.C. Lord: *J. Chem. Phys.* **56** (1972) 4450.

Acknowledgements

The authors are grateful to the [illegible] Foundation and the [illegible] ARP [illegible]

References

[illegible]

PHOTOFRAGMENTATION DYNAMICS OF VAN DER WAALS COMPLEXES

G. DELGADO-BARRIO
Instituto de Matemáticas y Física Fundamental,
C.S.I.C.,
C/ Serrano, 123,
28006 Madrid,
Spain

AND

J. ALBERTO BESWICK
LURE,
Université de Paris-Sud
F-91405 Orsay,
France

Abstract. The quantum theory of intramolecular dynamics in van der Waals molecules excited with a single infrared or visible photon, is presented. Vibrational, rotational, and electronic predissociation processes are described. Several quantum approximations, as well a quasiclassical trajectory technique, are discussed.

1. Introduction

The experimental study of van der Waals (vdW) molecules and small clusters has developed considerably in the last decade with the combined use of supersonic expansions to synthesize the complexes and laser techniques to investigate their energetics and dynamics [1–9].

VdW molecules are weakly bound complexes formed by atoms and/or molecules held together not by ordinary chemical bonds but by electrostatic and/or dispersion forces. The primary characteristics of this species are:

– Low dissociation energies, from 10 to 500 cm^{-1}
– Large (intermolecular) bondlength, from 2 to 5 Å
– Retention of the individual properties of the constituents within the aggregate.

There are a variety of reasons for the interest in the study of vdW com-

Y.G. Smeyers (ed.), Structure and Dynamics of Non-Rigid Molecular Systems, 203–247.

plexes. Detailed spectroscopic studies provide information on the intermolecular interactions which in turn are very important in determining a great variety of gas and bulk properties. In addition, of considerable interest is the study of the intramolecular dynamics of vdW molecules in electronically, vibrationally or rotationally excited states. A central and most interesting feature of excited-state intramolecular relaxation processes in such systems involves the breaking of the vdW bond. This new class of photochemical photofragmentation via electronic, vibrational, or rotational excitation of one of the constituents of the vdW complex, provides ideal model systems for the study of energy redistribution in polyatomic molecules.

Consider for instance a triatomic vdW molecule consisting of a rare-gas atom X bound to a normal diatomic molecule BC. The vibrational excitation of BC within the complex provides more energy than necessary to break the van der Waals $X \cdots BC$ bond. At first approximation the intramolecular BC vibration and the intermolecular motion are decoupled and the BC molecule will vibrate many thousands (or millions) of vibrational periods before enough energy is transferred to the vdW bond to dissociate the complex. This provides an unambiguous example of vibrational predissociation (VP), a process which is intimately related to a broad class of phenomena involving intramolecular and intermolecular vibrational energy exchange in molecular systems and in condensed phases. In particular, studies of intramolecular vibrational energy flow in vdW clusters will be helpful in assessing the range of applicability of statistical theories.

It is the purpose of this article to summarize the theoretical treatments in the area of intramolecular dynamics in vdW molecules, exploring the different predissociation processes occurring in these weakly bound systems when excited with a single photon in the infrared or visible spectral region. We shall first discuss vibrational predissociation (VP) for which there is the largest body of theoretical and experimental results. We then present the theory for rotational predissociation (RP) and electronic predissociation (EP).

2. Vibrational Predissociation

VP of the vdW molecules provides central information on energy acquisition within a normal chemical bond followed by intramolecular vibrational energy redistribution which results in the fragmentation of the vdW bond. Experimental techniques have made possible the preparation and spectroscopic determination of VP rates of vdW molecules. In the vdW molecules the excitation process is well defined and can be controlled. The resulting intramolecular relaxation is then an almost unique example for VP on a single electronic potential energy surface. The VP of vdW molecules is a comparatively simple process, being a theoretically tractable model problem for state-to-state investigations of unimolecular reaction dynamics and intramolecular energy transfer. The understanding of VP of vdW molecules is relevant for the elucidation of the general features of bond-breaking processes in chemical systems. The simplest system to study VP is the triatomic $X \cdots BC$ where the diatomic molecule BC is attached to an atom X by a van der Waals bond. A general photo-predissociation process can be written as

$$X - BC + h\nu \longrightarrow X - BC^* \xrightarrow{\tau} BC + X + \Delta E \tag{1}$$

with $h\nu$ being the photon energy, and τ the lifetime of the excited complex. After predissociation the two fragments X and BC fly apart with a relative kinetic energy ΔE. Thus, there are two steps in this process: (1) excitation to a quasi-bound (metastable) state, and (2) decay. For the particular case of VP we have:

1. Photon excitation of BC within the complex to a well defined $BC^*(v)$ vibrational level v of a given electronic configuration;
2. Predissociation: $X - BC^*(v) \xrightarrow{\tau} BC(v') + X$ resulting in X and BC fragments with BC left in a vibrational level $v' < v$ of the same electronic state.

The first detailed experimental results on VP in vdW molecules were conducted by Levy and coworkers [10] on the He-I_2 complex. They determined the VP lifetimes of He-$I_2(B^3\pi_{0_u^+})$ from the linewidths Γ_v of the R branch heads in the fluorescence excitation spectra. Resolved fluorescence spectra also provide the final state vibrational populations of the I_2 fragments.

The following experimental data for VP of He-$I_2(B^3\pi_{0_u^+})$ were reported:

1. The dissociation energy D_0 in the ground vdW state is
$$13.6 \leq D_0 \leq 14.8 \text{ cm}^{-1}.$$
2. The energy difference between the first excited level and the zero-point vdW level is in the range $[5.47, 5.85]$ cm^{-1}.
3. For vibrational excitation of the I_2 vibration in the region $10 \leq v \leq 45$ the dependence of the linewidth, Γ_v, on the vibrational quantum number of $I_2(B^3\pi)$ follows the empirical relationship
$$\Gamma_v(\text{cm}^{-1}) = 0.555 \times 10^{-4} v^2 + 0.174 \times 10^{-5} v^3. \tag{2}$$
The corresponding lifetimes $\tau_v = \hbar/2\Gamma_v$ varied from 221 psec at $v = 12$ to 38 psec at $v = 26$.
4. Up to $v = 60$ the predissociation produces $I_2(B)$ fragments mainly in the vibrational state corresponding to the loss of one quanta:
$$v' = v - 1.$$
5. For $v > 60$ the most prominent channel is that corresponding to the loss of two quanta of vibration
$$v' = v - 2.$$
6. The experiments of W. Sharfin *et al.* [11] show deviations with respect to Equation 2 in the region $50 \leq v \leq 64$. In particular Γ_v presents a maximum at $v = 57$.

These results have been rationalized in terms of quantum mechanical calculations performed on a semiempirical potential energy surface [4, 12]. A T-shaped equilibrium configuration was assumed and the potential energy surfaces were represented by a sum of pairwise atom-atom interactions. Two different analytical forms for describing the He-I interaction have been used [12]. One was a Morse function

$$V_{AB} = D\{\exp[-2\alpha(R_{AB} - \bar{R}_{AB})] - 2\exp[-\alpha(R_{AB} - \bar{R}_{AB})]\} \tag{3}$$

which is a reasonable description at small and intermediate distances but certainly poor for long-range interactions, where the usual R^{-6} dependence due to dispersion forces is expected. The other analytical form used for describing the He-I interaction was a modified Buckingham potential, i.e.,

$$V_{AB} = H\exp(-\beta R_{AB}) - KR_{AB}^{-6} \tag{4}$$

the three parameters H, K and β were chosen such that the potential function and its second derivative at the equilibrium position be equal to the corresponding values for the Morse potential. The VP rates calculated with these potentials for different vibrational excitations of I_2 are insensitive to the form of the potential at large distances. The difference was in the order of 5%. The reason is that the VP rates are essentially determined by bound-continuum coupling with the bound state strongly localized around the minimum of the potential energy surface. Hence for the determination of VP rates the Morse potential provides a reasonable description of the vdW interaction. Recently, similar treatments have been used to interpret the experimental data for NeI_2 predissociation [13]. Other systems which have also been studied theoretically and experimentally are Cl_2, Br_2, ICl complexed with rare gas atoms [14, 15].

3. Quantum Treatments

3.1. GOLDEN RULE TREATMENT

In the framework of the first-order perturbation theory for electric dipole transitions, the cross section for the excitation from an initial bound state $|\Psi_i\rangle$, with energy E_i, to a final dissociative state $|\Psi_{fE}\rangle$ of energy E, is given by:

$$\sigma_{fE\leftarrow i} \quad \propto |\langle\Psi_{fE}|\mathbf{M}\cdot\mathbf{e}|\Psi_i\rangle|^2 \qquad \text{with} \qquad E = E_i + \hbar\omega \tag{5}$$

where $\mathbf{e}$ is the polarization vector of the incident photon, with energy $\hbar\omega$, $\mathbf{M}$ is the transition dipole moment and f is a collection of quantum numbers which specifies the final quantum state of the fragments. For a triatomic molecule the Hamiltonian for nuclear motion, after separation of the center of mass of the whole system, can be written as [4]

$$H = \frac{\hbar^2}{2m}\left(-\frac{\partial^2}{\partial R^2} + \frac{\ell^2}{R^2}\right) + \frac{\hbar^2}{2\mu}\left(-\frac{\partial^2}{\partial r^2} + \frac{\mathrm{j}^2}{r^2}\right) + U_{BC}(r) + V(R, r, \theta) \tag{6}$$

where R is the distance between X and the center of mass of BC, r is the internuclear distance for BC, and θ is the angle between the two vectors **r** and **R**. The reduced masses μ and m factors in Equation (6) are

$$\mu = \frac{m_B m_C}{m_B + m_C}; \qquad m = \frac{m_X(m_B + m_C)}{m_X + m_B + m_C} \tag{7}$$

while $\boldsymbol{\ell}$ and **j** are the angular momentum operators (in units of $\hbar$) associated with **R** and **r**, respectively.

In Equation (6) the potential energy surface for the nuclear motion has been written as the sum of $U_{BC}(r)$, the intramolecular potential interaction for the "free" diatomic molecule BC, and $V(R, r, \theta)$, the intermolecular interaction between the rare gas atom X and the diatomic molecule BC. We thus have $V \to 0$ as $R \to \infty$.

In the framework of the Golden rule treatment, a zero-order Hamiltonian H_0 is defined which has discrete, $|\Phi_\ell\rangle$, as well as continuum, $|f', E'\rangle$, eigenfunctions. The following diagonal and off-diagonal matrix elements of the total Hamiltonian H are then defined:

$$\begin{aligned} \langle \Phi_\ell | H | f, E \rangle &= V_{f,E} \\ \langle \Phi_\ell | H | \Phi_{\ell'} \rangle &= \langle \Phi_\ell | H_0 | \Phi_{\ell'} \rangle = E_\ell \delta_{\ell\ell'} \\ \langle f, E | H | f', E' \rangle &= \langle f, E | H_0 | f', E' \rangle = E \delta_{ff'} \delta(E - E'). \end{aligned} \tag{8}$$

In the neighbourhood of an isolated resonance $|\Phi_\ell\rangle$, the exact dissociative wave function $|\Psi_{fE}\rangle$ can be expressed as [16]

$$|\Psi_{fE}\rangle = a_f(E)|\Phi_\ell\rangle + \sum_{f'} \int dE' b_{f'E'}(f, E)|f', E'\rangle \tag{9}$$

Defining the "energy-shift" $\Delta(E)$ and the halfwidth $\Gamma(E)$ of the resonance

$$\begin{aligned} \Delta(E) &= \sum_{f'} \mathcal{P} \int dE' \frac{|V_{f'E'}|^2}{E - E'} \\ \Gamma(E) &= \sum_{f'} \pi |V_{f'E}|^2 \end{aligned} \tag{10}$$

where $\mathcal{P}$ indicates "principal part", the coefficient $a_f(E)$ of the bound component in Equation (9) takes the form

$$a_f(E) = \frac{V_{fE}}{(E - E_\ell) - \Delta(E) - i\Gamma(E)}. \tag{11}$$

For slow predissociation characterized by a lifetime of the complex much longer than its vibrational period, the quantities $\Gamma(E)$, $\Delta(E)$ and V_{fE} can be considered as slowly varying with energy and calculated at the resonance energy.

For van der Waals systems the equilibrium distances and frequencies associated to the van der Waals bond are usually very similar in the initial and final states. Under these conditions, the Franck–Condon factors strongly favour the transitions to the discrete states, whereas direct excitation of the continuum states will be extremely weak. Hence, for van der Waals predissociation, the total photoabsorption cross section can be approximated by:

$$\sum_f \sigma_{fE\leftarrow i} \propto |\langle\Phi_\ell|\mathbf{M}\cdot\mathbf{e}|\Psi_i\rangle|^2 \sum_f |a_f(E)|^2 = |\langle\Phi_\ell|\mathbf{M}\cdot\mathbf{e}|\Psi_i\rangle|^2 A(E) \quad (12)$$

where

$$A(E) = \frac{\Gamma/\pi}{(E-E_r)^2+\Gamma^2} = \sum_f |\langle\Psi_{fE}|\Phi_\ell\rangle|^2 \quad (13)$$

provides the characteristic Lorentzian lineshape with half-width at half-maximum given by Γ, centered at the resonance energy $E_r = E_\ell + \Delta$.

The final state distribution of the fragments is obtained as

$$P_f = \frac{|\langle\Psi_{fE}|\Phi_\ell\rangle|^2}{\sum_{f'} |\langle\Psi_{f'E}|\Phi_\ell\rangle|^2}. \quad (14)$$

The final states of the diatomic fragments are usually described in the basis set $\{\varphi_v(r)\Theta_{j\Omega}^{JMp}(\hat{r},\hat{R})\}$,where the radial functions $\varphi_v(r)$ are solutions of

$$\left\{-\frac{\hbar^2}{2\mu}\frac{\partial^2}{\partial r^2} + U_{BC}(r) - E_v\right\}\varphi_v(r) = 0 \quad (15)$$

while the $\Theta_{j\Omega}^{JMp}(\hat{r},\hat{R})$ functions are the body-fixed “free rotor” wave functions defined by [17]

$$\Theta_{j\Omega}^{JMp}(\hat{r},\hat{R}) = \left[\frac{2J+1}{8\pi(1+\delta_{\Omega 0})}\right]^{1/2} \left\{D_{M\Omega}^{J*}(\phi_R,\theta_R,0)Y_{j\Omega}(\theta,\phi) + p(-1)^J D_{M-\Omega}^{J*}(\phi_R,\theta_R,0)Y_{j-\Omega}(\theta,\phi)\right\} \quad (16)$$

$\mathbf{J} = \boldsymbol{\ell} + \mathbf{j}$ being the total angular momentum, M and $\Omega \geq 0$ its projections on the space-fixed and body-fixed z-axis, respectively, and $p = \pm 1$ the parity under total nuclear coordinates inversion. $D_{M\Omega}^{J*}(\phi_R,\theta_R,0)$ are the Wigner rotational functions and $Y_{j\Omega}(\theta,\phi)$ the spherical harmonics, θ_R and ϕ_R being the polar angles specifying the $\mathbf{R}$ vector with respect to the space fixed frame, while θ and ϕ are the polar angles of $\mathbf{r}$ in the body fixed frame. The $\mathbf{j}^2$ operator is diagonal in this basis set with eigenvalues $j(j+1)$ while $\boldsymbol{\ell}^2$ has diagonal and nondiagonal matrix elements in Ω:

$$\begin{aligned}\langle\Theta_{j\Omega}^{JMp}|\boldsymbol{\ell}^2|\Theta_{j'\Omega'}^{JMp}\rangle &= \delta_{jj'}\delta_{\Omega\Omega'}[J(J+1)+j(j+1)-2\Omega^2] \\ &- \delta_{jj'}\delta_{\Omega\Omega'\pm 1}[(J+1)-\Omega'(\Omega'\pm 1)]^{1/2} \\ &\times [(j+1)-\Omega'(\Omega'\pm 1)]^{1/2}.\end{aligned} \quad (17)$$

The “discrete” function $|\Phi_\ell\rangle$ is then expanded as

$$\Phi_\ell^{JMp}(\mathbf{r},\mathbf{R}) = \sum_v \sum_{j,\Omega} \sum_n a_{vj\Omega n}^\ell \chi_n(R)\varphi_v(r)\Theta_{j\Omega}^{JMp}(\hat{r},\hat{R}) \quad (18)$$

where $\chi_n(R)$ is a harmonic oscillator basis set with frequency and equilibrium distance fitted to describe the ground state of the electronic energy surface. The coefficients $a^{\ell}_{vj\Omega n}$ and the energy E_ℓ of $|\Phi_\ell\rangle$ are obtained by diagonalizing the Hamiltonian matrix in this basis set.

For the dissociative state, the wave function $|\Psi^{JMp}_{vj\Omega E}\rangle$ is expanded as

$$\Psi^{JMp}_{vj\Omega E}(\mathbf{r},\mathbf{R}) = \sum_{v'}\sum_{j'\Omega'} \Phi^{JMpvj\Omega}_{Ev'j'\Omega'}(R)\varphi_{v'}(r)\Theta^{JMp}_{j'\Omega'}(\hat{r},\hat{R}). \tag{19}$$

Introduction of Equation (19) in the time independent Schrödinger equation with the Hamiltonian given in Equation (16) leads to the close coupled equations:

$$\begin{aligned}\Big\{-\frac{\hbar^2}{2m}\frac{\partial^2}{\partial R^2} + E_v + B_v j(j+1) - E\Big\}\Phi^{JMpvj\Omega}_{Evj\Omega}(R) \\ = -\sum_{v',j',\Omega'}\Big\{\delta_{\Omega\Omega'}\langle\Theta^{JMp}_{j\Omega}|V_{vv'}|\Theta^{JMp}_{j'\Omega'}\rangle\Phi^{JMpvj\Omega}_{Ev'j'\Omega'}(R) \\ + \frac{\delta_{jj'}\delta_{vv'}}{2mR^2}\langle\Theta^{JMp}_{j\Omega}|\boldsymbol{\ell}^2|\Theta^{JMp}_{j'\Omega'}\rangle\Phi^{JMpvj\Omega}_{Ev'j'\Omega'}(R)\Big\}\end{aligned} \tag{20}$$

with

$$\begin{aligned} B_v &= \langle\varphi_v|\hbar^2/2\mu r^2|\varphi_v\rangle \\ V_{vv'}(R,\theta) &= \langle\varphi_v|V(r,R,\theta)|\varphi_{v'}\rangle \end{aligned} \tag{21}$$

where the usually small vibrational off-diagonal matrix elements of the $\mathbf{j}^2$ have not been included. After integration of the coupled equations (20), the $\Phi^{JMpvj\Omega}_{Ev'j'\Omega'}(R)$ functions are obtained by imposing the usual asymptotic conditions [17]:

$$\begin{aligned}\Psi^{JMp}_{vj\Omega E}(\mathbf{r},\mathbf{R}) \underset{R\to\infty}{\sim} \varphi_v(r)\Theta^{JMp}_{j\Omega}(\hat{r},\hat{R})e^{ik_{vj}R}/\sqrt{k_{vj}} \\ - \sum_{v'j'\Omega'} S^*_{vj\Omega,v'j'\Omega'}(E)\varphi_{v'}(r)\Theta^{JMp}_{j'\Omega'}(\hat{r},\hat{R})e^{-ik_{v'j'}R}/\sqrt{k_{v'j'}}\end{aligned} \tag{22}$$

where S is the scattering matrix and $k_{vj} = \sqrt{2m(E - E_v - B_v j(j+1)}$.

The potential matrix elements $V_{vv'}(R,\theta)$ are usually expanded in terms of Legendre polynomials as

$$V_{vv'} = \sum_\lambda V^\lambda_{vv'} P_\lambda(\cos\theta) \tag{23}$$

so that the angular matrix elements become [17]

$$\begin{aligned}\langle\Theta^{JMp}_{j\Omega}|V_{vv'}|\Theta^{JMp}_{j'\Omega'}\rangle \\ = \delta_{\Omega\Omega'}\sum_\lambda V^\lambda_{vv'}\left(\frac{2j'+1}{2j+1}\right)\langle j',0,\lambda,0;j,0\rangle\langle j',\Omega,\lambda,0;j,\Omega\rangle\end{aligned} \tag{24}$$

where the $\langle j_1,m_1,j_2,m_2;j_3,m_3\rangle$ are Clebsh–Gordan coefficients. Usually the resolution of the close coupled equations (20) involves too many channels [18] and in what follows some decoupling schemes among the different degrees of freedom will be discussed.

3.2. VIBRATIONAL DIABATIC APPROXIMATION (VDA)

For low vibrational diatomic states, the "fast" vibrational motion of BC can be diabatically decoupled from the "slow" intermolecular motions of $X \cdots BC$. Therefore a zero-order Hamiltonian for a given vibrational state can be defined as

$$H_v = \frac{\hbar^2}{2m}\left(-\frac{\partial^2}{\partial R^2} + \frac{\ell^2}{R^2}\right) + B_v \mathbf{j}^2 + E_v + V_{vv}(R,\theta) \tag{25}$$

which has discrete as well as continuum solutions. The bound levels of H_v can be expanded as

$$\Psi_{v,q}^{JMp}(\mathbf{r},\mathbf{R}) = \varphi_v(r)\sum_{j,\Omega}\sum_{n} a_{j\Omega n}^{v,q}\chi_n(R)\Theta_{j\Omega}^{JMp}(\hat{r},\hat{R}) \tag{26}$$

where the $a_{j\Omega n}^{v,q}$ coefficients and the corresponding energies, E_{vq}, are obtained by diagonalization techniques.

The continuum eigenfunctions of H_v can be similarly expanded as

$$\Psi_{j\Omega E}^{JMpv}(\mathbf{r},\mathbf{R}) = \varphi_v(r)\sum_{j'\Omega'} \Phi_{Ej'\Omega'}^{JMpvj\Omega}(R)\Theta_{j'\Omega'}^{JMp}(\hat{r},\hat{R}) \tag{27}$$

which, after substitution in the time-independent Schrödinger equation, provides the close coupling equations (20) with the sum in the right-hand side being restricted to just one v.

Using the approximations above, the rates for VP are expressed as

$$\Gamma_{vq} = \pi \sum_{v'\leq v, j', \Omega'} |\langle\Psi_{j'\Omega' E_{vq}}^{JMpv'}|V(r,R,\theta)|\Psi_{v,q}^{JMp}\rangle|^2 \tag{28}$$

where $|\Psi_{j'\Omega' E_{vq}}^{JMpv'}\rangle$ must be calculated on the energy-shell $E = E_{vq}$, and the final state distribution of the fragments are then given by

$$P_{v',j'}^{vq} = \sum_{\Omega'} \frac{\pi|\langle\Psi_{j'\Omega' E_{vq}}^{JMpv'}|V(r,R,\theta)|\Psi_{v,q}^{JMp}\rangle|^2}{\Gamma_{vq}} . \tag{29}$$

3.3. ROTATIONAL INFINITE ORDER SUDDEN APPROXIMATION (RIOSA)

Due to the anharmonicity of the BC diatomic potential, the higher the diatomic vibration, the closer the different v channels. Then, strong interactions appear between them and the diabatic separation of the intramolecular vibration becomes poor. In particular, there is a certain v for which the $v-1$ exit channel is closed and the $v-2$ becomes predominant. In these cases there appears a strong interaction among quasibound levels with different v quantum number. For He...I_2 that happens for $v \geq 60$. Since in the dissociation the motion along the stretching coordinate, R, is faster than the rotation of the BC fragment, the Rotational Infinite Order Sudden Approximation is expected to hold [19]. In this approximation, the total wave function is written as

$$\Psi_{vj\Omega E}^{JMp} = \Phi_E^v(r,R;\theta)F(\hat{r},\hat{R}) \tag{30}$$

where $\Phi_E^v(r,R;\theta)$ are eigenfunctions of the effective Hamiltonian:

$$H_\theta = -\frac{\hbar^2}{2m}\frac{\partial^2}{\partial R^2} - \frac{\hbar^2}{2\mu}\frac{\partial^2}{\partial r^2} + U_{BC} + V(R,r;\theta). \tag{31}$$

Equation (31) can be solved by the use of the following expansion:

$$\Phi_E^v(r,R;\theta) = \sum_{v'} \phi_{v'E}^v(R;\theta)\varphi_{v'}(r) \tag{32}$$

which leads to the coupled equations

$$\left\{-\frac{\hbar^2}{2m}\frac{\partial^2}{\partial R^2} + E_v - E\right\}\phi_{vE}^v(R;\theta) = -\sum_{v'} V_{vv'}(R;\theta)\phi_{v'E}^v(R;\theta). \tag{33}$$

By solving the close coupled equations (33) for a given fixed value of θ, the positions $E_r(\theta)$ of the quasibound states, their halfwidths $\Gamma(\theta)$ and the final state vibrational distributions $P_v(\theta)$ of the fragments, can be obtained by analysis of the energy dependence of the scattering matrix [19]. This procedure is repeated for a number of θ values and the function $E_{v,s}(\theta)$ is taken to be the effective potential for the bending motion. The angular function $F(\hat{r},\hat{R})$ in Equation (30), is then given by the eigenfunctions of

$$\left\{\frac{\hbar^2\boldsymbol{\ell}^2}{2m\bar{R}^2} + \frac{\hbar^2\mathbf{j}^2}{2\mu\bar{r}^2} + E_{v,s}(\theta) - E_{v,s,b}\right\}F_{v,s,b}^{JMp}(\hat{r},\hat{R}) = 0 \tag{34}$$

where b represents the quantum number associated to the bending motion. Since the bound wave functions are mainly sensitive to the details of the potential surface in the region of the well, in Equation (34) R and r have been replaced by their equilibrium values, $\bar{R}$ and $\bar{r}$, respectively. In order to solve Equation (34) the angular function $F_{v,s,b}^{JMp}(\hat{r},\hat{R})$ is expanded as

$$F_{v,s,b}^{JMp}(\hat{r},\hat{R}) = \sum_{j'\Omega'} a_{j'\Omega'}^{vsb}\Theta_{j'\Omega'}^{JMp}(\hat{r},\hat{R}), \tag{35}$$

the coefficients $a_{j'\Omega'}^{vsb}$ and the energies $E_{v,s,b}$ being obtained by diagonalization techniques. The rates for VP are then given by the matrix elements

$$\Gamma_{v,s,b} = \langle F_{v,s,b}^{JMp}(\hat{r},\hat{R})|\Gamma_{v,s}(\theta)|F_{v,s,b}^{JMp}(\hat{r},\hat{R})\rangle \tag{36}$$

and the final state ro-vibrational distribution of the diatomic fragment, are provided by

$$P_{v'j'}^{vsb} = \sum_{\Omega'} a_{j'\Omega'}^{vsb} \sum_{j''\Omega''} a_{j''\Omega''}^{vsb}\langle\Theta_{j'\Omega'}^{JMp}(\hat{r},\hat{R})|P_v'(\theta)|\Theta_{j''\Omega''}^{JMp}(\hat{r},\hat{R})\rangle. \tag{37}$$

3.4. VIBRATIONAL DIABATIC AND ROTATIONAL INFINITE ORDER SUDDEN APPROXIMATION (VD-RIOSA)

The two approaches described above can be combined to produce a stronger decoupled scheme [20, 21]. The bound states are written as

$$\Psi_{v,s,b}^{JMp}(\mathbf{r},\mathbf{R}) = \varphi_v(r)\phi_{v,s}(R;\theta)F_{v,s,b}^{JMp}(\hat{r},\hat{R}) \tag{38}$$

where, as defined above, v, s and b are the quantum numbers associated with the diatomic vibrational, van der Waals stretching and van der Waals bending motions, respectively. The functions $\phi_{v,s}(R;\theta)$ are discrete solutions, for a fixed value of θ, of the Schrödinger equation:

$$\left\{-\frac{\hbar^2}{2m}\frac{\partial^2}{\partial R^2}+V_{vv}(R;\theta)-W_{v,s}(\theta)\right\}\phi_{v,s}(R;\theta)=0 \tag{39}$$

with the eigenvalues $W_{v,s}(\theta)$ being an effective potential for the bending motion. Neglecting the effect of the angular momentum operator on the $\phi_{v,s}(R;\theta)$ functions, i.e., writing

$$\begin{aligned}\boldsymbol{\ell}^2\phi_{v,s}(R;\theta)F^{JMp}_{v,s,b}(\hat{r},\hat{R}) &= \phi_{v,s}(R;\theta)\boldsymbol{\ell}^2F^{JMp}_{v,s,b}(\hat{r},\hat{R})\\ \mathbf{j}^2\phi_{v,s}(R;\theta)F^{JMp}_{v,s,b}(\hat{r},\hat{R}) &= \phi_{v,s}(R;\theta)\mathbf{j}^2F^{JMp}_{v,s,b}(\hat{r},\hat{R})\end{aligned} \tag{40}$$

the $F^{JMp}_{v,s,b}(\hat{r},\hat{R})$ functions will be the solutions of:

$$\left\{\frac{\hbar^2}{4m}[\boldsymbol{\ell}^2,\bar{R}^{-2}_{v,s}(\theta)]+B_v\mathbf{j}^2+W_{v,s}(\theta)-E_{v,s,b}\right\}F^{JMp}_{v,s,b}(\hat{r},\hat{R})=0 \tag{41}$$

where $[A,B]=AB+BA$ is the anticommutator introduced to symmetrize the Hamiltonian [21] and

$$\bar{R}^{-2}_{v,s}(\theta)=\langle\phi_{v,s}(R;\theta)|\frac{1}{R^2}|\phi_{v,s}(R;\theta)\rangle. \tag{42}$$

In this VD-RIOS approximation the zero order continuum functions are written as [22]

$$\Psi^{JMp}_{v,j,\Omega,E}(\mathbf{r},\mathbf{R})=\varphi_v(r)\phi_{v,E}(R;\theta)\Theta^{JMp}_{j\Omega}(\hat{r},\hat{R}) \tag{43}$$

where $\phi_{v,E}(R;\theta)$ are the continuum eigenfunctions of Equation (39). The rates for VP from an initially excited quasibound level, labeled by the quantum numbers (v, s, and b), are given by

$$\Gamma_{v,s,b}=\pi\sum_{v',j',\Omega'}|\langle\Theta^{JMp}_{j'\Omega'}|V_{v'v}(\theta)|F^{JMp}_{v,s,b}\rangle|^2 \tag{44}$$

where

$$V_{v'v}(\theta)=\langle\phi_{v',E}(R;\theta)|V_{v'v}(R,\theta)|\phi_{v,s}(R;\theta)\rangle \tag{45}$$

are the relevant potential coupling terms. Using Equation (34), the final state ro-vibrational distribution of the BC fragments will then be given by

$$P_{v',j'}=\pi\sum_{\Omega'}\frac{|\langle\Theta^{JMp}_{j'\Omega'}|\langle V_{v'v}(\theta)\rangle|F^{JMp}_{v,s,b}\rangle|^2}{\Gamma_{v,s,b}}. \tag{46}$$

4. Quasiclassical Models

The classical trajectory method has been extensively applied in reactive and inelastic collisions [23]. It consists in solving the relevant classical Hamilton equations once the initial conditions (coordinates and momenta of every particle) have been established. For a $X \cdots BC$ three-atomic system the elimination of the coordinates of one of the atoms (B in this case), in the center of mass system, leads to the Hamiltonian

$$H = \frac{m_B + m_C}{2m_X M} P_X^2 + \frac{m_B + m_X}{2M \cdot m_C} P_C^2 - \frac{P_X \cdot P_C}{M} + V \tag{47}$$

where the P's are the conjugate momenta to the coordinates, M is the total mass, and V is the potential energy. The first initial coordinate selected is r according to the distribution of a nonrotating Morse oscillator [24, 25, 26]

$$r = \bar{r} - \frac{1}{\alpha} \ln \left\{ \frac{-2a}{b + (b^2 - 4ac)^{\frac{1}{2}} \sin(2\pi s)} \right\} \tag{48}$$

where $a = E - D$, $b = 2D$, $c = -D$ and s is a random number uniformly distributed in the range $[0, 1]$. Here E is the energy of the vibrational quantum level of BC relative to the minimum of the well and D, α, $\bar{r}$ are the Morse parameters of this diatomic molecule.

The angle θ is randomly chosen according to a uniform distribution in the range $[0, \frac{\pi}{2}]$ or $[0, \pi]$, and then a random length R is selected in the range $[3, 9]$ Å (also uniformly), where the van der Waals potential well is typically located. In order to calculate the atom-diatom relative momentum, the quantities r, R and θ must satisfy the following inequality

$$E_{v,s,b} - V(R, r, \theta) \geq 0 \tag{49}$$

where $E_{v,s,b}$ is the van der Waals energy. If this expression is not fulfilled, the random selection of θ and R is iterated while the value of r is kept fixed. After the initial coordinates are determined, the associated momenta can be calculated.

For each trajectory, the dissociation time is calculated as the elapsed time from the beginning till the last turning point R was reached. The distribution of trajectories are then fitted to an exponential law [27]

$$N_{ND} = N_T \exp(-t/\tau) \tag{50}$$

where N_T is the total number of trajectories, N_{ND} is the number of nondissociated trajectories at time t and τ is the lifetime of the complex. The half-width is calculated as

$$\Gamma = \frac{\hbar}{2\tau} \, . \tag{51}$$

A plot of $\ln(N_{ND}/N_T)$ versus time t can be useful in order to determine the reliability of the assumed exponential law of decay. A linear fitting of this function provides the lifetime and, by means of Equation (51), the linewidth. Finally, the probabilities P_j of producing a particular diatomic rotational level j are calculated using box quantization.

5. Vibrational Predissociation of Rare Gas-Halogen and Interhalogen van der Waals Complexes

In this section we present some results for the vibrational predissociation of $X \cdots I_2$ (X = He, Ne) and also Ne$\cdots$ICl. Several calculations for He$\cdots I_2(B)$ with different potential parameters have been performed in a wide range of vibrational quantum numbers up to $v = 63$.

For He$\cdots I_2$, Ne$\cdots I_2$ and Ne$\cdots$ICl, the VD approximation (see Section 3.2) gives very good results as compared to the exact calculations [18, 28, 29, 30]. The discrepancies become larger when increasing v due to the fact that the coupling among different v channels, neglected in the VD approximation, increases with the vibrational quantum number.

All the calculations predict an increasing of the rate for vibrational predissociation when the vibrational state of the BC fragment increases, in agreement with the experimental data [10]. This fact can be qualitatively rationalized in terms of the "energy gap law": due to the anharmonicity of the intramolecular potential, the kinetic energy of the fragments decreases when v is increased. As a consequence the associated continuum wave function presents less oscillations in the region of the well, giving rise to a better overlap with the quasibound state wave function.

The VD-RIOSA agree also very well with the VDA and with the "exact" calculations for all these systems. As an example we present in Figure 1 the experimental lifetimes vs. the vibrational excitation of I_2 obtained for Ne$\cdots I_2$ [13]: compared with the theoretical results obtained using the VD and the VD-RIOS approximations, that had given similar results, in both cases the agreement with the experimental data is very good.

The quasiclassical model has also been applied to the VP of He$\cdots I_2$. In Figure 2 the quasiclassical results together with the experimental data [24, 25] are represented. The overall agreement is good and the quasiclassical results reproduce the behaviour of the rates very well.

The experimental results of Sharfin *et al.* [11] show that the function $Bv^2 + Cv^3$ fits well with the linewidth dependence on the vibrational quantum number v of I_2 up to $v \simeq 45$, but deviates for higher values. In particular, they have determined a further enhancement of the rates peaking at $v \simeq 57$ followed by a pronounced decrease. They have interpreted this effect as the gradual closing of the $\Delta v = -1$ channel.

In the region $v > 45$ the quasiclassical calculations also reproduce satisfactorily the experimental behaviour. In order to understand the maximum in Γ the final vibrational distributions of I_2 fragments using a box-quantization procedure was also determined. The results are presented in Table I. We note that the distributions show very neatly the gradual closing of the $\Delta v = -1$ channel and therefore they support the interpretation of Sharfin *et al.* regarding the behaviour of the linewidths in this region.

The quasiclassical calculations show that for the Ne$\cdots I_2$ complex the line-

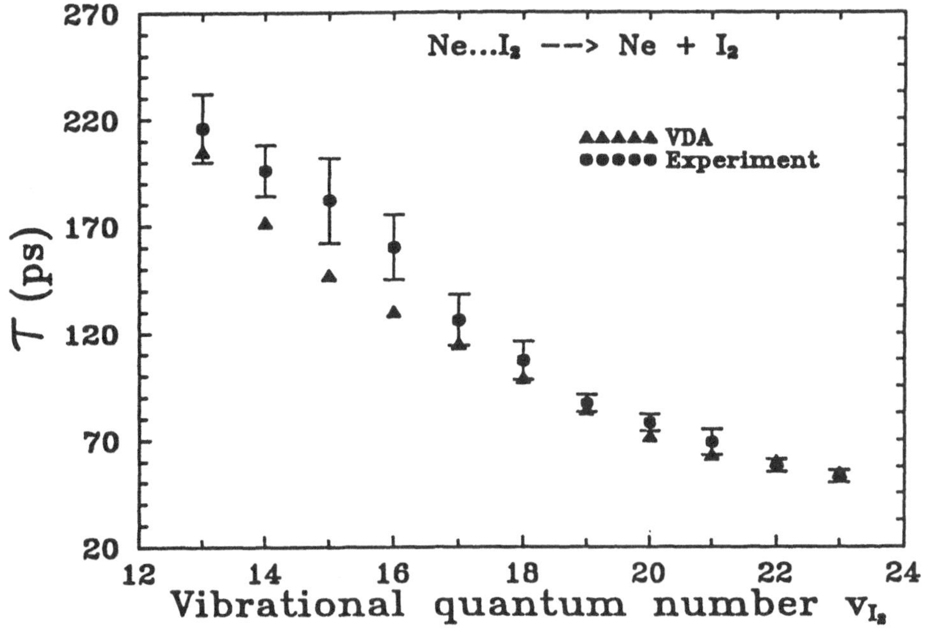

Figure 1. Experimental lifetimes vs. the vibrational excitation of I_2 obtained for Ne···I_2 [13], compared with the theoretical results obtained by VD approximation.

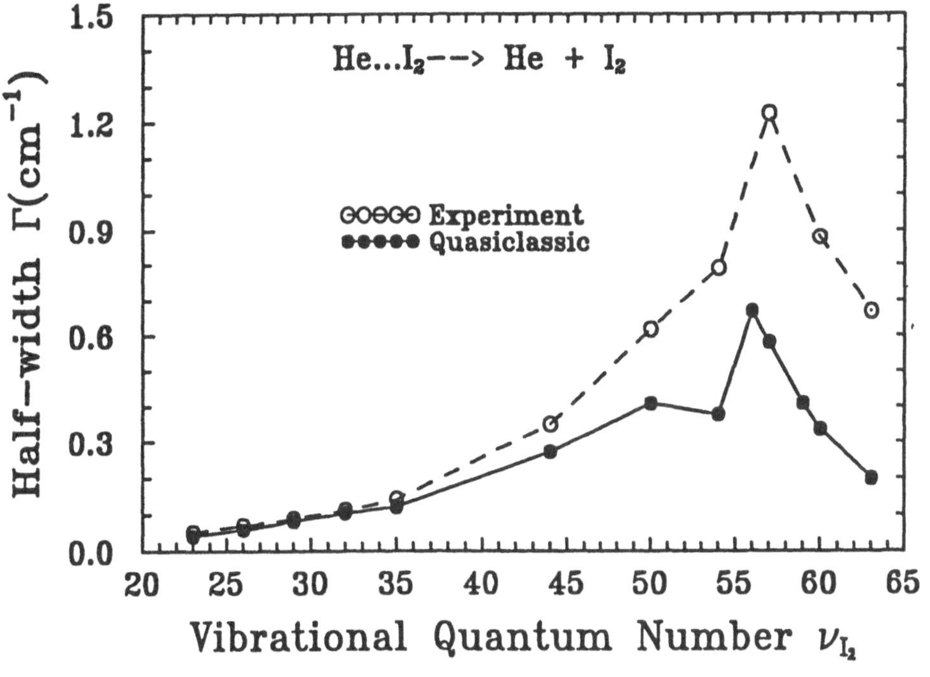

Figure 2. Vibrational predissociation linewidths (in cm^{-1}) of He···I_2. Quasiclassical results [24, 25] together with the experimental data [11].

TABLE I. Probabilities for the final vibrational channels ν'_{I_2} as a function of the vibration ν_{I_2} in the VP of He···I_2 complex.

v_{I_2} \ v'_{I_2}	v-1	$v-2$	$v-3$	$v-4$
50	74	24	2	
54	63	30	7	
56	52	36	9	3
59	33	44	16	7
60	20	52	21	7
63	-	47	33	20

widths also present maxima as a function of v. Using the IOS approximation, two maxima were found at $v = 30$ and $v = 33$ (see Figure 3). The first maximum is associated with the closing of the $\Delta v = -1$ channel. The second one arises from the mixing, induced by the anharmonicity of the I_2 subunit, between the first resonance sustained by the $v = 33$ channel and the resonances of the $v = 32$ channel. As a consequence, there is an important contribution of the $v = 32$ channel in the predissociation of $v = 33$; and the small content of the $\Delta\nu = -2$ channel ($v = 33 \rightarrow 31$) is greatly enhanced by adding a $\Delta\nu = -1$ ($v = 32 \rightarrow 31$) contribution. It is then possible to describe the predissociation of $v = 33$ in two steps: the first involving intramolecular vibrational energy redistribution (IVR) and the second one producing vibrational predissociation . This effect has been recently born out in the Ar···Cl_2 complex [31].

The final rotational distribution of the diatomic fragment also provides a deep insight into the dynamics of the process and into the anisotropy of the intermolecular potential. The He···I_2 and Ne···I_2 complexes are not expected to produce highly excited rotational states of the I_2 fragment because the equilibrium configuration is T-shaped and the I_2 fragment is very heavy as compared with the He and Ne atoms, making effective angular momentum transfer very unfavorable. Presently, experimental data concerning the final state rotational distributions of the I_2 fragment is not available. However, the results obtained with different theoretical methods [18], quantal as well as quasiclassical, show very weak rotational excitation of the I_2 subunit after dissociation.

On the other hand, heteronuclear systems like He···ICl and Ne···ICl [28, 29, 32], which are not T-shaped, can produce fragments which are rotationally excited. In Figure 4 the experimental final state distribution of the ICl fragment in the vibrational predissociation of Ne···ICl is presented together with different theoretical calculations. The agreement between the experimental data and the

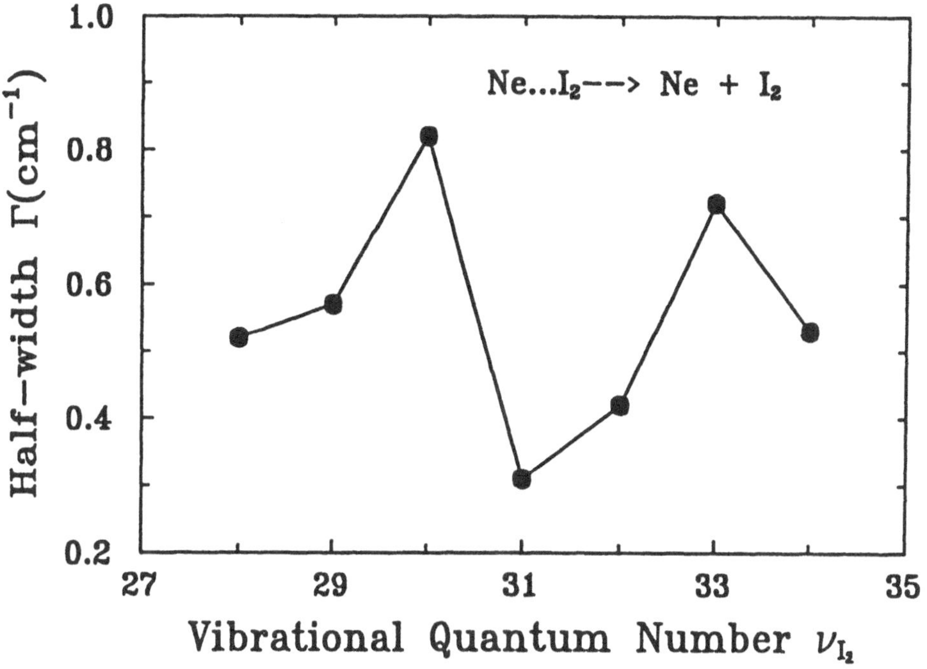

Figure 3. Vibrational predissociation linewidths (in cm^{-1}), of NeI_2, VD-RIOS Approximation.

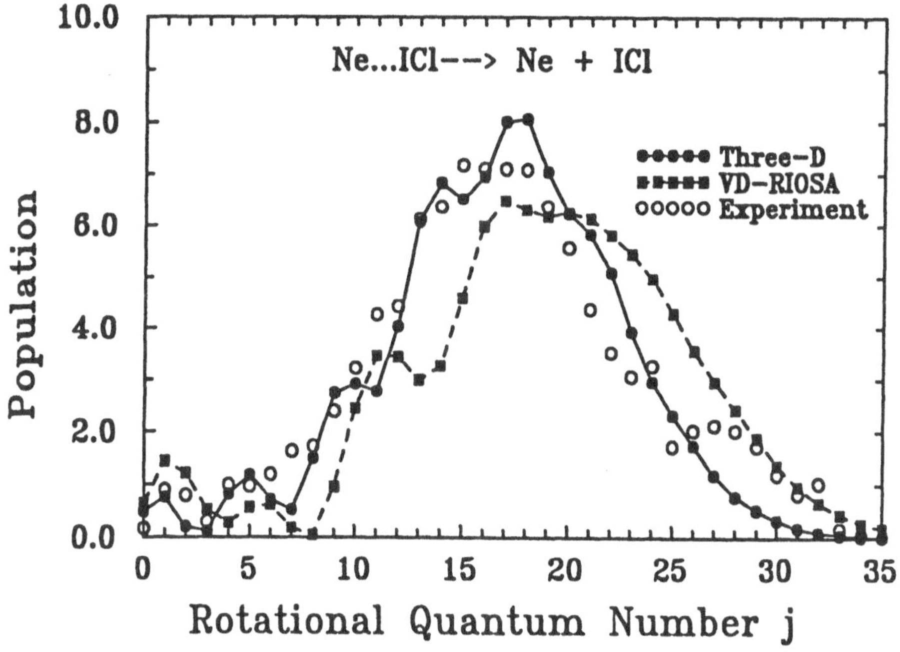

Figure 4. Final rotational distribution of the ICl after the vibrational predissociation .

results of "exact" calculations, as well as the ones using the VD approximation, is very good. On the other hand, the results obtained by the VD-IOS approximation are fairly good for low j states, but deviate from the exact results for high j values. This is a consequence of the neglect of the centrifugal barrier and this produces an overestimation of the population for high rotational values.

6. Rotational Predissociation

The rotational predissociation (RP) is another type of dissociation induced by internal energy which occurs in van der Waals complexes. In this process the complexes predissociate by using some of their internal rotational energy to break the van der Waals bond. This form of predissociation, as well as VP, can be viewed as simple unimolecular reaction in which the vdW bonds play the role of the intermediate, long-lived *activated state* yielding products with different internal states.

After separation of the center of mass of the whole system, the Hamiltonian for the nuclear motion of an X-BC (rigid-rotor) triatomic molecule can be written as

$$H = -\frac{\hbar^2}{2m}\frac{\partial^2}{\partial R^2} + \frac{\boldsymbol{\ell}^2}{2mR^2} + B_e\mathbf{j}^2 + V(R,\theta) \tag{52}$$

where R is the distance between the atom and the diatomic center of mass, m is the atom-diatom reduced mass, and θ is the relative orientation, between $\mathbf{R}$ and the diatomic. The j and $\boldsymbol{\ell}$ are the angular momentum operators associated to BC (with rotational constant B_e) and the centrifugal rotation, respectively, while V describes the intermolecular van der Waals interaction.

In the space-fixed representation, the wave function is expanded as

$$\Psi^{J(p)M}(\mathbf{R},\hat{r};E) = \sum_{\alpha} \chi_{\alpha}^{J(p)}(R;E)F_{\alpha}^{J(p)M}(\hat{R},\hat{r}) \tag{53}$$

where J is the total angular momentum $\mathbf{J} = \mathbf{j} + \boldsymbol{\ell}$, M its projection on the space-fixed z-axis, and $p = (-1)^{(J+j+l)}$ is the parity index. In Equation (53) $F_{\alpha}^{J(p)M}(\hat{R},\hat{r})$ denotes a complete orthonormal set of angular basis functions with $\alpha = (j,l)$.

Using the $F_{\alpha}^{J(p)M}(\hat{R},\hat{r})$ basis set in the Schrödinger equation results in the system of coupled equations

$$\left[-\frac{\hbar^2}{2m}\frac{\partial^2}{\partial R^2} + V_{\alpha\alpha}^{J(p)}(R) - (E - E_{\alpha k_\alpha})\right]\chi_{\alpha}^{J(p)}(R;E) = -\sum_{\substack{\alpha,\alpha' \\ (\alpha'\neq\alpha)}} V_{\alpha\alpha'}^{J(p)}(R)\chi_{\alpha'}^{J(p)}(R;E) \tag{54}$$

where the $V_{\alpha\alpha'}^{J(p)}(R)$ functions are defined by

$$V_{\alpha\alpha'}^{J(p)}(R) = \langle F_{\alpha}^{J(p)M} | \left(\frac{\ell^2}{2mR^2} + B_e \mathbf{j}^2 + V(R,\theta) \right) | F_{\alpha'}^{J(p)M} \rangle. \tag{55}$$

The equations (54) can be solved by standard techniques [33]. Once the relevant S matrix is obtained, the energy position and the linewidth of the resonances, as well as the final distribution of the fragments, can be determined.

7. Approximate Quantal Models for Rotational Predissociation

7.1. DIABATIC ROTATIONAL METHOD

This method has been applied to several systems [34, 35, 36]. It assumes that all nondiagonal $V_{\alpha\alpha'}$ elements in Equation (54) are negligibly small. For each α channel the following equation is then solved:

$$\left[-\frac{\hbar^2}{2m}\frac{\partial^2}{\partial R^2} + V_{\alpha\alpha}^{J(p)}(R) - E \right] \chi_{\alpha}^{J(p)}(R; E_{\alpha k_\alpha}) = E_{\alpha k_\alpha} \chi_{\alpha}^{J(p)}(R; E_{\alpha k_\alpha}) \tag{56}$$

where k_α is the quantum number associated with stretching in the vdW bond. In this approximation the wave function is written as a simple product

$$\Psi_{\alpha k_\alpha}^{J(p)M}(\mathbf{R}, \hat{r}) = \chi_{\alpha}^{J(p)}(R; E_{\alpha k_\alpha}) F_{\alpha}^{J(p)M}(\hat{R}, r). \tag{57}$$

Since asymptotically j becomes a good quantum number, this method can provide accurate resonance energies when the diatomic rotational spacing is large and the off-diagonal couplings $V_{\alpha\alpha'}$ are small compared to the diagonal elements $V_{\alpha\alpha}$. One of the systems to which this approximation has been applied is the He-HF complex [35]. A body-fixed frame was used so that the wave functions were written for each $J(p)M$ as

$$\Psi_{\alpha j\Omega}(\mathbf{R}, \hat{r}) = \phi_{\alpha j\Omega}(R) \langle \hat{R}, \hat{r} | j\Omega \rangle \tag{58}$$

where Ω is the "tumbling" angular momentum quantum number and corresponds to the component of J (total angular momentum) on the body-fixed z axis $\mathbf{R}$, α denotes either the vibrational quantum number associated with the vdW bond for the discrete wave functions (ν) or the continuum relative kinetic energy (ε) of the fragments for continuum eigenfunctions. Then $\phi_{\alpha j\Omega}$ is the solution of the Schrödinger equation

$$\left[-\frac{\hbar^2}{2m}\frac{\partial^2}{\partial R^2} + U_{j\Omega}(R) \right] \phi_{\alpha j\Omega} = E \phi_{\alpha j\Omega} \tag{59}$$

with

$$U_{j\Omega} = \langle j\Omega | V | j\Omega \rangle + \frac{\langle j\Omega | \hat{l}^2 | j\Omega \rangle}{2m\, R^2} + Bj(j+1). \tag{60}$$

The rotational angular momentum ℓ^2 is no longer diagonal in this representation: its nonzero matrix elements has been given in Equation (17).

TABLE II. Energies and total widths for RP of He-HF, $\Gamma_{j\Omega}$ from j = 1, 2 and 3. All values are given in cm^{-1}.

$J=0$	CC	Numerical	Analytical
E_{10}	34.75	35.76	35.17
Γ_{10}	0.13	0.12	0.14
E_{20}	120.74	120.71	121.15
Γ_{20}	0.20	0.25	0.284
E_{30}	248.82	248.70	249.56
Γ_{30}	0.11	0.16	0.180

The Schrödinger equation (59) obviously has discrete and continuum eigenfunctions. Now, in the simplest "golden rule" approach the width associated to the quasibound states is given by

$$\Gamma_{\nu j\Omega} = \pi \sum_{j',\Omega' \neq j,\Omega} |\langle \Psi_{nj\Omega} \mid H \mid \Psi_{\epsilon j'\Omega'} \rangle|^2 \tag{61}$$

where the continuum wave functions are energy normalized. The calculations implied in Equation (61) are restricted to open channels (i.e. with $\varepsilon > 0$). Hence, continuum-continuum and discrete-discrete couplings are completely neglected.

In Tables II and III the calculated partial and total widths for rotational predissociation of He-HF with $J = 0$ and $J = 1$ are presented. For $J = 0$, three different calculations have been performed: a close coupling (CC) calculation using Equations (54), a numerical diabatic one through Equation (61), and an analytical calculation using a fitting of the matrix elements [34]. The general agreement between the CC and these two approximate methods is very good both for the position of the resonances and their total linewidth.

The same theoretical techniques have been applied to the calculation of bound and quasibound levels of Ne-H_2 and Ar-H_2 complexes by Roncero *et al.* [36]. Their results are presented in Table 4. Due to symmetry selection rules, RP occurs with a change of the diatomic rotational state $\Delta j = -2, -4, \ldots$. Therefore levels with $j = 1$ cannot predissociate and become bound levels even if their energy is above the threshold for dissociation. Above $j = 3$ for Ne-H_2 and $j = 6$ for Ar-H_2, the corresponding effective diagonal potentials do not support discrete levels [36].

A calculation for Ar-H_2 ($J = 0, j = 2$) with the same potential energy surface but using a complex-coordinate coupled channel (CCCC) method has also been performed [37]. Good agreement was found between the results of the different theoretical calculations: $E = 346.172\ cm^{-1}$ and $\Gamma = 0.13 \times 10^{-1}\ cm^{-1}$ for the rotational diabatic while the CCCC values are $346.1433\ cm^{-1}$ and $0.127 \times 10^{-1}\ cm^{-1}$

TABLE III. Energies and total widths for RP processes in He-HF, $\Gamma_{j\Omega}$ from j = 1, 2 and 3 and with Ω = 0 and 1. All values are given in cm^{-1}.

$J = 1$	Numerical	Analytical
E_{10}	36.51	36.05
Γ_{10}	0.11	0.14
E_{11}	36.79	37.90
E_{20}	122.90	122.02
Γ_{20}	0.21	0.28
E_{21}	121.45	121.99
Γ_{21}	0.04	0.05
E_{30}	251.47	250.43
Γ_{30}	0.12	0.17
E_{31}	249.72	249.94
Γ_{31}	0.07	0.10

TABLE IV. Energies and half-widths (units are cm^{-1}) for Ne-H_2 and Ar-H_2 bound and metastable predissociating levels. They were obtained by numerical application of the diabatic rotational model with the potentials of Tang and Toennies [39]. A rotational diatomic constant of 56.987 cm^{-1} was used.

	Ne-H_2		Ar-H_2	
j	Energy	Width	Energy	Width
0	-5.31	···	-21.37	···
			-0.35	
1	109.35	···	92.26	···
			113.68	
2	339.14	0.102×10^{-1}	322.44	0.134×10^{-1}
3			667.17	0.287×10^{-2}
4			1126.69	0.650×10^{-3}
5			1700.96	0.151×10^{-3}

TABLE V. Energies and widths, in cm^{-1}, for shape resonances of X-H_2 (X = He, Ne, Ar) complexes. Except that marked with an asterisk, all of them correspond to overbarrier resonances [36].

	He-H_2		Ne-H_2		Ar-H_2	
j	Energy	Width	Energy	Width	Energy	Width
2	345.22	4.67	346.58	7.64	341.91	0.79
3			683.78*	0.13	684.75	2.81
4			1142.94	3.97	1142.07	4.59
5					1713.66	6.58

respectively. On the other hand, larger differences were found if the potential energy surface is modified. This illustrates the extreme sensitivity of these results to details in the intermolecular potentials and couplings.

Another type of quasibound states arises when the centrifugal barrier is sufficiently high to accommodate levels which dissociate by tunnelling. In Table V the energies and widths associated with shape resonances in He, Ne, Ar-H_2 complexes, are presented. They correspond to the first shape resonance found for each j channel. They are obtained by analysis of the asymptotic behaviour of the continuum wave function

$$\phi_{\varepsilon j\Omega}(R) \underset{R\to\infty}{\sim} A\sin\{kR + \delta_{j\Omega}(\varepsilon)\} \tag{62}$$

where $k = (2m\varepsilon)^{1/2}/\hbar$ is the wave number, A a normalization factor, and $\delta_{j\Omega}$ the phase shift. Near a resonance phase shift takes the well-known Breit–Wigner form [38].

$$\delta_{j\Omega}(\varepsilon) = \delta_{j\Omega}^{(0)}(\varepsilon) + \arctan\left\{\frac{\Gamma_{j\Omega}}{2(\varepsilon_{j\Omega} - \varepsilon)}\right\} \tag{63}$$

where $\varepsilon_{j\Omega}$ denotes the energy resonance while $\Gamma_{j\Omega}$ stands for its corresponding width. In Equation (63) $\delta_{j\Omega}^{(0)}$ is a background contribution which, for isolated narrow resonances, can be considered as a smooth function of the energy. The corresponding cross section is given by

$$\sigma_{j\Omega}(\varepsilon) = \frac{4\pi}{k^2}(2j+1)\sin^2\delta_{j\Omega}(\varepsilon) \tag{64}$$

and will have different functional forms depending on the local value of the background $\delta_{j\Omega}^{(0)}$ at the resonance energy. For the X-H_2 complexes there were found wide overbarrier resonances, except the one corresponding to $j = 3$ for Ne-H_2 which is supported by the barrier (i.e., it is an orbiting resonance). This narrow resonance appears when the effective potential has no bound levels. Its width is of the same order of magnitude as some orbiting resonances reported by Hutson and le Roy [40] using a very precise method.

Concerning Ar-H_2 only overbarrier resonances were found in the range of the j values considered. In order to check the validity of the diabatic method in this case, a CC calculation was performed. The position of the resonance was found with a difference of only 0.001 cm^{-1} and the width was 0.82 cm^{-1} as compared to 0.79 cm^{-1} obtained by the diabatic approximation.

The rotational diabatic approximation can be applied to systems presenting large rotational spacings in the diatomic subunit and low anisotropy. However, for other systems the method can be improved by introducing discrete-discrete couplings for instance.

7.2. CONFIGURATION INTERACTION MODELS

In order to improve the diabatic approximation the wave function can be written as a linear combination of diabatic functions:

$$\Psi_n^{J(p)M}(\mathbf{R}, \hat{r}) = \sum_{\alpha,k_\alpha} a_{\alpha k_\alpha}^{(n)} \chi_\alpha^{J(p)}(R; E_{\alpha k_\alpha}) F_\alpha^{J(p)M}(\hat{R}, \hat{r}) \tag{65}$$

where the summation extends over all the bound states supported by each potential $V_{\alpha,\alpha}$ defined by Equation (55). The problem can then be solved by diagonalization of the Hamiltonian matrix. This method is called the Diabatic Configuration Interaction (DCI) model.

It is possible to extend the linear combination by assuming the bound wave functions to be the product of a radial times an angular function and rewriting the wave function as

$$\Phi_n^{J(p)M}(\mathbf{R}, \hat{r}) = \sum_{\alpha,q} b_{\alpha q}^{(n)} \phi_q^{J(p)}(R) F_\alpha^{J(p)M}(\hat{R}, \hat{r}) \tag{66}$$

where the summation extends over the channel index α and a new index q spanning the radial basis. These radial functions $\{\phi_q\}$ are obtained as linear combinations of the previous χ functions, i.e.,

$$\phi_q^{J(p)}(R) = \sum_{\alpha' k'_\alpha} \beta_{\alpha' k'_\alpha}^{(q)} \chi_{\alpha'}^{J(p)}(R; E_{\alpha' k'_\alpha}) \,. \tag{67}$$

The problem is reduced once more to a diagonalization. This method may be expected to yield more accurate results than the previous ones, since bound states are here obtained from an enlarged basis. On the other hand, its success strongly depends on the quality of the original χ functions employed to describe the motion along the vdW vibrational coordinate. This method is called the Full Diabatic Configuration Interaction (FDCI) model.

The FDCI starts with diabatic radial functions to get the orthonormal radial basis set. We can thus search for more suitable basis functions and then repeat the above procedure. The He-CO complex [41] for instance, presents some features that make more efficient the use of an adiabatic angular approximation. Therefore, by fixing the angle θ for a number value in the range $[0, \pi]$, it is possible to choose

TABLE VI. Energy levels (in cm^{-1}) of bound and metastable states in He-CO system for a total angular momentum $J = 0$. The zero of energy is chosen to be that of the asymptotic rotor with $j = 0$.

j	D	DCI	FDCI	AAFCI	CC
0	-2.268	-2.295	-3.331	-3.560	
1	2.212	2.226	1.410	1.094	0.939
2	10.830	10.843	10.208	9.079	9.037
3			22.817	23.149	22.440

as radial functions the solutions of

$$\left[\frac{\hbar^2}{2m}\frac{\partial^2}{\partial R^2} + v(r,\theta_l)\right]\Psi_{k_l}(R;\theta_l) = E_{k_l}(\theta_l)\Psi_{k_l}(R;\theta_l). \tag{68}$$

The total wave function is written as in Equation (66) but now the ϕ_q radial functions are given by

$$\phi_q^{J(p)}(R) = \sum_{l,k_l} C_{lk_l}^{(q)}\Psi_{k_l}(R;\theta_l) \tag{69}$$

the k_l index being the corresponding stretching quantum number. This is the Angular Adapted FDCI (AAFDCI) model.

In Table VI the energies of some bound and quasibound levels of He-CO, obtained using these methods, are presented together with the results of a CC calculation. From an inspection of Table VI it is concluded that the diabatic approximation overestimates the energies by 1–2 cm^{-1}. The DCI method does not provide significant improvements due to the limited quality of the basis set. However, a qualitative and also quantitative improvement is obtained with the use of FDCI, and for this system, an even better agreement using the AAFDCI method.

In Table VII the resonance energies and halfwidths for the $J(p) = 0+, 1+$ states of He-CO obtained by a CC calculation are presented. It is important to note that these models are useful even in the cases where they are not expected to work, as they provide a simple physical picture which helps to understand the results obtained from rigorous CC calculations. The D model provides reasonable results when off-diagonal couplings are small and also gives a fairly good starting point for both DCI and FDCI approaches. The DCI model is a useful method for systems where the number of diabatic levels supported by the potential is large enough to ensure a better basis set as is the case for Ar-N_2 and Ar-O_2. The FDCI model includes a more efficient representation and should therefore improve on situations where the D method is already a good starting point. However, it is the AAFDCI that is the best of the models discussed here, and it appears to be

TABLE VII. Energies and widths at half maximum for rotational predissociation in the He-CO system.

$J(p)$	E (cm^{-1})	$\Gamma/2$ (cm^{-1})
0(+)	0.939	0.283
	9.037	0.690
	22.440	0.585
1(+)	0.301	0.040
	1.818	0.205
	8.512	0.580
	10.756	0.428
	21.059	0.962

particularly appropriate for complexes with slow rotational motion of part of the system and fast vibrational motion along vdW coordinates.

7.3. ADIABATIC EXPANSIONS

In the cases where the drastic separation implied by the diabatic model is not a realistic description of the internal motion within a vdW molecule, it appears more reasonable to construct the total wave functions for the bound and metastable levels of the complex as a product of a function which depends on only one of the variables, either the stretching or the bending, and another function which depends only weakly on that variable and more strongly on all others. This approach simply tries to take advantage of the physical differences which exist between fast and slow-motion [42]. In the following paragraphs we briefly review the models which can be used to exploit the adiabatic separation for treating the dynamical couplings within vdW complexes.

7.3.1. *Adiabatic Angular Expansion*

In this approximation the wave function is written as a product of a radial function which only weakly depends on orientation and a purely angular function,

$$\Psi_{s,l}(\mathbf{R}, \hat{r}) = \phi_s(R;\theta)F_{s,l}(\theta) \tag{70}$$

where s and l are the quantum numbers associated with the vdW stretching and librational motions, respectively. Assuming that the weak θ-dependence of the ϕ_s functions can be neglected, i.e.

$$\begin{aligned} \boldsymbol{\ell}^2\Psi_{s,l} &= \phi_s\boldsymbol{\ell}^2\, F_{s,l} \\ \mathbf{j}^2\Psi_{s,l} &= \phi_s\mathbf{j}^2 F_{s,l} \end{aligned} \tag{71}$$

the ϕ_s functions are then the solutions of

$$\left\{-\frac{1}{2m}\frac{\partial^2}{\partial R^2}+V(R;\theta)\right\}\phi_s(R;\theta)=W_s(\theta)\phi_s(R;\theta) \tag{72}$$

and the F function satisfies

$$\left\{-\frac{1}{4m}[A_s(\theta),\boldsymbol{\ell}^2]+B_e\mathbf{j}^2+W_s(\theta)\right\}F_{s,l}(\theta)=E_{s,l}F_{s,l}(\theta) \tag{73}$$

where

$$A_s(\theta)=\langle\phi_s(R;\theta)|\left(\frac{1}{R^2}\right)|\phi_s(R;\theta)\rangle \tag{74}$$

and $[A,B]=AB+BA$. In order to solve Equation (73), the F functions can be expanded as linear combinations of body-fixed (BF) free rotor functions with the coefficients being determined by diagonalization. This approach is very similar in spirit to the adiabatic approach discussed previously for vibrational predissociation processes (see Section 3.4).

It is possible to improve the above representation of discrete eigenstates of the complex by a configuration interaction method, i.e., the wave function is written as

$$\Psi_k(\mathbf{R},\hat{r})=\sum_{s,l}a^k_{s,l}\phi_s(R;\theta)F_{s,l}(\theta) \tag{75}$$

with the coefficients being determined by diagonalization of the full Hamiltonian [43].

7.3.2. *Adiabatic Stretching Expansion (Best Local)*

In Equation (70) we have assumed that the angular motion is slow and that the motion along the vdW radial coordinate is fast. It is of interest to examine the other possible situation, i.e., the fast motion along the librational coordinate with a relatively slow motion along the stretching coordinate. The equation to be solved in that case is

$$\left[V(R,\theta)+B_e\mathbf{j}^2+\frac{\boldsymbol{\ell}^2}{2mR^2}\right]g_l(\theta;R)=U_l(R)g_l(\theta;R) \tag{76}$$

where l represents the librational (angular) quantum number. The g's functions are then expanded in a BF angular basis set

$$g_l(\theta;R)=\sum_{j\Omega}G^l_{j\Omega}(R)\langle\hat{R},\hat{r}\mid JMj\Omega\rangle \tag{77}$$

and the coefficients are obtained by diagonalization.

The complete wave function for the discrete states is written down as a simple product

$$\Psi_{s,l}(\mathbf{R},\hat{r})=g_l(\theta;R)\phi_{s,l}(R) \tag{78}$$

and assuming the adiabatic approximation

$$\left(-\frac{1}{2m}\frac{\partial^2}{\partial R^2}\right)\Psi_{s,l} = g_l\left(-\frac{1}{2m}\frac{\partial^2}{\partial R^2}\right)\phi_{s,l} \tag{79}$$

the radial functions are the solutions of the equation

$$\left[-\frac{1}{2m}\frac{\partial^2}{\partial R^2} + U_l(R)\right]\phi_{s,l}(R) = E_{sl}\phi_{s,l}(R). \tag{80}$$

The goodness of this type of adiabatic separation is given by the degree of validity of Equation (79) i.e. to the amount of decoupling between the angular motion of the diatomic fragment and the stretching of the rare-gas atom in the complex. In any case this approximation can always be improved by a configuration interaction treatment using the basis set defined by Equation (78).

7.3.3. *Continuum Functions*

The continuum wave functions can be obtained by the use of the IOS approximation:

$$\Psi_{j,\Omega,\varepsilon,\bar{l}}(\mathbf{R},\hat{r}) = \phi_{\varepsilon,\bar{l}}(R;\theta)\langle\hat{R}\hat{r} \mid JMj\Omega\rangle \tag{81}$$

where ϕ are the energy normalized solutions of the θ-dependent equation

$$\left[-\frac{1}{2m}\frac{\partial^2}{\partial R^2} + V(R,\theta)\right]\phi_{\varepsilon,\bar{l}=0}(R;\theta) = \varepsilon\phi_{\varepsilon,\bar{l}=0}(R;\theta) \tag{82}$$

with $\bar{l}$ being an arbitrarily chosen value of the orbital angular momentum and ε the "on the energy-shell" kinetic energy. The RP halfwidth associated to a metastable state of the complex, labelled by the discrete quantum numbers $|s,l\rangle$ can now be calculated using

$$\Gamma_{s,l} = \pi\sum_{j\Omega}|V^{d\to c}_{sl\to j\Omega}|^2 \tag{83}$$

where

$$V^{d\to c}_{sl\to j\Omega} = \langle\Psi_{j\Omega,\varepsilon\bar{l}}|H|\Psi_{s,l}\rangle. \tag{84}$$

When the CI method is used, we have instead

$$\Gamma_k = \pi\sum_{j\Omega}|U^{d\to c}_{k\to j\Omega}|^2 \tag{85}$$

with

$$U^{d\to c}_{k\to j\Omega} = \langle\Psi_{j\Omega,\varepsilon\bar{l}}|H|\Psi_k\rangle = \sum_{sl}a^k_{s,l}\langle\Psi_{j\Omega,\varepsilon\bar{l}}|H|\Psi_{s,l}\rangle. \tag{86}$$

7.3.4. *Rotational Predissociation of He-AB (AB=CO, N_2, O_2).*

The adiabatic model was applied to the He-AB (AB=CO, N_2, O_2) complexes which present stretching frequencies in the range $\sim[10,20]$ cm^{-1} and rotational constants from 1 to 2 wave numbers (1.931 (CO), 1.998 (N_2), 1.445 (O_2)). Therefore, we may expect that for these systems the adiabatic decoupling between

TABLE VIII. Energies and full widths for RP (in cm^{-1}) using the adiabatic approximation and comparison with the exact results.

	Adiabatic	C.C.		Adiabatic	C.C.
$He-N_2$			He-CO		
$J(p) = 0(+)$			$J(p) = 1(+)$		
E_{10}	-1.728		E_{00}	-3.306	
E_{30}	20.383	21.053	E_{11}	0.094	0.301
$\Gamma_{30\to10}$	0.861	0.878	$\Gamma_{11\to00}$	0.098	0.040
$J(p) = 1(+)$			E_{10}	1.758	1.818
E_{11}	-5.067		$\Gamma_{10\to00}$	0.307	0.205
E_{10}	-1.047		E_{21}	8.485	8.512
E_{31}	18.615	19.321	Γ_{21}	0.832	0.580
Γ_{31}	0.666	0.857	E_{20}	10.860	10.756
E_{30}	22.376	22.923	Γ_{20}	0.547	0.428
Γ_{30}	0.246	0.597	E_{31}	20.987	21.059
			Γ_{31}	1.057	0.962
$He-O_2$			He-CO		
$J(p) = 0(+)$			$J(p) = 0(+)$		
E_{10}	-2.334		E_{00}	-3.818	
E_{30}	14.543	14.780	E_{10}	0.922	0.939
$\Gamma_{30\to10}$	0.623	0.581	$\Gamma_{10\to00}$	0.461	0.281
$J(p) = 1(+)$			E_{20}	9.301	9.037
E_{11}	-4.759		Γ_{20}	1.016	0.691
E_{10}	-1.549		E_{30}	22.477	22.440
E_{31}	12.740	13.049	Γ_{30}	0.651	0.585
Γ_{31}	0.535	0.572			
E_{30}	16.709	16.796			
Γ_{30}	0.515	0.403			

rotation and vdW bond vibrations is valid. Table VIII lists the results obtained with the adiabatic angular approximation and those obtained by a CC calculations. The best results are obtained for O_2 and the worse for N_2. This is expected due to the increase of the rotational constant of the diatom. The bound and quasibound levels of the Ne-HF have also been studied and in this case the adiabatic (and also the AAFDCI) approximation provides results very close to the CC ones (accuracy of the order of 0.01 cm^{-1}) [44, 45].

8. Vibrational Predissociation of a Complex with More than One vdW Bond

The systems discussed above have been restricted to one rare gas atom bound to a diatomic molecule. It is also interesting to study systems formed of two or more rare gas atoms bound to a diatomic molecule. This allows the study of energy flow from the excited diatom to each one of the weak vdW bonds, i.e., the branching ratios among the different dissociation channels together with its dependence on the initial excitation in the chemical bond. Some experimental [46] and theoretical [47–50] work have been performed on the dissociation dynamics of vibrational excited complexes of I_2 with two or more rare gas atoms.

A full quantal treatment of the four-body system becomes prohibitively difficult. On the other hand, quasiclassical trajectory calculations [47] as well as approximate quantal ones [48] on the He-I_2^*Ne have been performed. Since the perpendicular configuration is the most probable for each triatomic XI_2^* molecule, X = He, Ne and the rotational constant associated to I_2^*(Be $\sim$ 0.04 cm^{-1}), NeI_2^* ($\sim$ 0.05 cm^{-1}), and HeI_2^* ($\sim$ 0.3 cm^{-1}) are much lower than the stretching vibrational energies, a particular configuration in which the He and Ne atoms are restricted to move on a perpendicular plane to the axis of the (nonrotating) I_2 molecule was assumed. In this model, four motions are accounted for within the complex: the stretching of I_2, the vibrations of He and Ne with respect to the center of mass of I_2 and also the oscillation (bending) of the angle formed by both weak bonds γ.

After separation of the center of mass motion of the whole system, the Hamiltonian for the X-BC-Y complex, where X, Y the rare gas atoms are restricted to move on a perpendicular plane to the nonrotating BC molecule, may be written as

$$H = \frac{P_r^2}{2m} + \frac{\mathbf{P}_1^2}{2\mu_1} + \frac{\mathbf{P}_2^2}{2\mu_2} + \frac{\mathbf{P}_1 \cdot \mathbf{P}_2}{M} + V(r, R_1, R_2, \gamma) \tag{87}$$

where r is the BC bondlength with conjugate momentum P_r, while $\mathbf{R}_1, \mathbf{R}_2$ are the vectors going from the center of mass of BC to the X and Y atoms, respectively, with conjugate momenta $\mathbf{P}_1$, and $\mathbf{P}_2$. In Equation (87) $M = m_B + m_C$ is the diatomic mass, while $m = m_B m_C / M$, $\mu_1 = m_X M/(m_X + M)$ and $\mu_2 = m_Y M/(m_Y + M)$ are reduced masses. Finally V describes the potential energy function defined by

$$V = V_{BC}(r) + V_{X-BC}(r, R_1) + V_{Y-BC}(r, R_2) + V_{X-Y}(R_1, R_2, \gamma). \tag{88}$$

The Hamiltonian (87) can be rewritten quantum mechanically as

$$\begin{aligned} H \;=\; & -\frac{1}{2\mu}\left(\frac{\partial^2}{\partial r^2} + \frac{2}{r}\frac{\partial}{\partial r}\right) - \frac{1}{2\mu_1}\left(\frac{\partial^2}{\partial R_1^2} + \frac{2}{R_1}\frac{\partial}{\partial R_1}\right) \\ & - \frac{1}{2\mu_2}\left(\frac{\partial^2}{\partial R_2^2} + \frac{2}{R_2}\frac{\partial}{\partial R_2}\right) + \frac{\mathbf{j}_1^2}{2\mu_1 R_1^2} + \frac{\mathbf{j}_2^2}{2\mu_2 R_2^2} \\ & + \frac{1}{M}\left(-\frac{\partial^2}{\partial R_1 \partial R_2}\hat{R}_1 \cdot \hat{R}_2 - \frac{1}{R_1 R_2}\nabla_{\theta_1\varphi_1} \cdot \nabla_{\theta_2\varphi_2}\right. \end{aligned}$$

$$- \frac{1}{R_1}\hat{R}_2 \cdot \nabla_{\theta_1\varphi_1}\frac{\partial}{\partial R_2} - \frac{1}{R_2}\hat{R}_1 \cdot \nabla_{\theta_2\varphi_2}\frac{\partial}{\partial R_1}\Bigg) + V \tag{89}$$

where $\hat{R}_k$ is the unit vector in the $\mathbf{R}_k$ direction, $\nabla_{\theta_k\varphi_k}/R_k$ is the tangential component of the gradient, while $\mathbf{J}_k = -i\hat{R}_k \times \nabla_{\theta_k\varphi_k}$ is the angular momentum associated to $\mathbf{R}_k$, $k = 1, 2$. The last term in the kinetic part is divided by the total diatomic mass M. This term may be neglected for heavy diatomic molecules at first approximation. In order to solve the Schrödinger equation

$$H\Psi(r, \mathbf{R}_1, \mathbf{R}_2) = E\Psi(r, \mathbf{R}_1, \mathbf{R}_2) \tag{90}$$

we take advantage of the usual large vibrational spacing of BC to write the total wave function in the diabatic vibrational approximation

$$\Psi(r, \mathbf{R}_1, \mathbf{R}_2) = \Psi_v(r_1, \mathbf{R}_1, \mathbf{R}_2) = \frac{1}{r}\,\chi_v(r)\phi^{(v)}(R_1, R_2) \tag{91}$$

where $\chi_v(r)$ is an eigenfunction of the BC intramolecular Schrödinger equation, and $\phi^{(v)}(\mathbf{R}_1, \mathbf{R}_2)$ satisfies the equation

$$\Bigg[-\frac{1}{2\mu_1}\left(\frac{\partial^2}{\partial R_1^2} + \frac{2}{R_1}\frac{\partial}{\partial R_1}\right) - \frac{1}{2\mu_2}\left(\frac{\partial^2}{\partial R_2^2} + \frac{2}{R_2}\frac{\partial}{\partial R_2}\right)$$

$$+ \frac{\mathbf{j}_1^2}{2\mu_1 R_1^2} + \frac{\mathbf{j}_2^2}{2\mu_2 R_2^2} + \frac{1}{M}\Bigg(-\frac{\partial^2}{\partial R_1 \partial R_2}\cos\gamma - \frac{1}{R_1 R_2}\nabla_1\nabla_2$$

$$- \frac{1}{R_1}\hat{R}_2 \cdot \nabla_1\frac{\partial}{\partial R_2} - \frac{1}{R_2}\hat{R}_1 \cdot \nabla_2\frac{\partial}{\partial R_1}\Bigg)$$

$$+ V_{vv}(R_1, R_2, \gamma)\Bigg]\phi^{(v)}(\mathbf{R}_1, \mathbf{R}_2)$$

$$= (E - E_{BC}(v))\phi^{(v)}(\mathbf{R}_1, \mathbf{R}_2) \tag{92}$$

where we have replaced the subscripts θ_k, φ_k in the tangential gradient components by $k = 1, 2$ and where $V_{vv} = \langle\chi_v(r)[V(r, R_1, R_2, \gamma)]\chi_v(r)\rangle$ is the averaged potential

$$V_{vv} = V_{X-BC}^{(v,v)}(R_1) + V_{Y-BC}^{(v,v)}(R_2) + V_{X-Y}. \tag{93}$$

Since the frequency of the diatomic subunit $I_2 \sim 125\ \mathrm{cm}^{-1}$ is large as compared to the other frequencies of the molecule, diabatic separation should be appropriate. Moreover, due to the difference existing between the stretching frequencies associated to Ne-I_2 and He-I_2 ~ 25 and $16\ \mathrm{cm}^{-1}$, respectively and to He-Ne $\sim 5\ \mathrm{cm}^{-1}$, which is directly related to the oscillation in the angle γ, the function $\phi^{(v)}$ in the adiabatic angular approximation can be written as

$$\phi^{(v)}(\mathbf{R}_1, \mathbf{R}_2) = \frac{1}{R, R_2}\varphi_v(R_1, R_2; \gamma)F^{(v)}(\hat{R}_1, \hat{R}_2) \tag{94}$$

where $F^{(v)}$ is a pure rotational function while φ_v is a radial function that depends parametrically on γ and is assumed to be transparent to the action of the angular operators. It is then given by the solution of

$$\left[-\frac{1}{2\mu_1}\frac{\partial^2}{\partial R_1^2}-\frac{1}{2\mu_2}\frac{\partial^2}{\partial R_2^2}-\frac{\cos\gamma}{M}\frac{\partial^2}{\partial R_1\partial R_2}+V_{vv}(R_1,R_2;\gamma)\right]$$
$$\times\,\varphi_v(R_1,R_2;\gamma)=\varepsilon\varphi_v(R_1,R_2;\gamma) \tag{95}$$

In order to solve Equation (95) the wave function φ_v^{dis} can be expanded, with an additional quantum number k (necessary to specify the two-dimensional stretching problem), in terms of products of triatomic stretching functions, i.e.,

$$\varphi_{vk}^{\mathrm{dis}}(R_1,R_2;\gamma)=\sum_{mn}a_{mn}^{(v,k)}(\gamma)\xi_m^{(v)}(R_1)\eta_n^{(v)}(R_2) \tag{96}$$

where ξ is the discrete solution of

$$\left[-\frac{1}{2\mu_1}\frac{\partial^2}{\partial R_1^2}+V_{X-BC}^{(v,v)}(R_1)\right]\xi_m^{(v)}(R_1)=E_{X-BC}^{v,m}\xi_m^{(v)}(R_1) \tag{97}$$

and η the solution of a similar equation with $Y-BC$ instead of $X-BC$.

We now consider, for each value of the angle γ, the continuum states which describe the rare gas atom X plus a triatomic fragment $Y-BC$ in a given stretching n state, where the diatomic subunit is in a lower $v'<v$ vibrational level, as a product

$$\varphi_{v'\varepsilon n'}^{\mathrm{cont},X}(R_1,R_2;\gamma)=\xi_\varepsilon^{(v',n')}(R_1;\gamma)\eta_{n'}^{v'}(R_2) \tag{98}$$

where η is defined in Equation (97) and ξ is the solution of

$$\left[-\frac{1}{2\mu_1}\frac{\partial^2}{\partial R_1^2}+V_{X-BC}^{(v',v')}\right.$$
$$\left.+\langle\eta_{n'}^{(v')}(R_2)[V_{X-Y}(R_1,R_2,\gamma)]\eta_{n'}^{(v')}(R_2)\rangle-\varepsilon_1\right]\xi_\varepsilon^{(v',n')}(R_1;\gamma)=0 \tag{99}$$

with ε being the relative kinetic energy of the fragments. Continuum states for each angle γ corresponding to Y plus $X-BC$ can be described in a similar way. Using the Fermi's golden rule, two different γ-dependent halfwidths for VP of $X-BC(v)-Y$ in a given k state can be calculated. For breaking up the $Y-BC$ bond we have

$$\begin{aligned}\Gamma_{vk}^{Y}(\gamma) &= \pi\sum_{v'<v}\sum_{m'}|\langle\varphi_{vk}^{\mathrm{dis}}|\langle\chi_v|v|\chi_{v'}\rangle|\varphi_{v\,\varepsilon\,m'}^{\mathrm{cont},Y}\rangle|^2\\ &= \pi\sum_{v'<v}\sum_{m'}|\sum_{m,n}a_{mn}^{(v,k)}(\gamma)[\langle\xi_m^{(v)}\mid\xi_{m'}^{(v')}\rangle\langle\eta_n^{(v)}\mid V_{Y-BC}^{(v,v')}\mid\eta_\varepsilon^{(v',m')}\rangle\\ &\quad+\langle\eta_n^{(v)}|\eta_\varepsilon^{(v',m')}\rangle\langle\xi_m^{(v)}|V_{X-BC}^{(v,v')}|\xi_{m'}^{(v')}\rangle]|^2\end{aligned} \tag{100}$$

with

$$\varepsilon=(E_{BC}^{v}-E_{BC}^{v'})+W_k^{(v)}(\gamma)-E_{X-BC}^{(v',m')}. \tag{101}$$

Therefore in the frame of the angular adiabatic approximation applied here the half-widths for VP of $X-BC-Y$ will be functions of the angle γ. For this the rotational functions $F^{(v)}(\hat{R}_1, \hat{R}_2)$ of Equation (95) that are now labelled by the quantum numbers v and k associated to the stretching motions, and a third quantum number l associated to the bending motion in γ, have to be evaluated. They can be obtained by diagonalization of the angular Hamiltonian which has $W_k^{(v)}(\gamma)$ as an effective potential [48]. For a level specified by the (v, k, l) quantum numbers one can estimate the associated VP rates by the following averages:

$$\Gamma_{vkl}^X = \langle F_{kl}^{(v)}(\hat{R}_1, \hat{R}_2) \mid \Gamma_{vk}^X(\gamma) \mid F_{kl}^{(v)}(\hat{R}_1, \hat{R}_2)\rangle \tag{102}$$

$$\Gamma_{vkl}^Y = \langle F_{kl}^{(v)}(\hat{R}_1, \hat{R}_2) \mid \Gamma_{vk}^Y(\gamma) \mid F_{kl}^{(v)}(\hat{R}_1, \hat{R}_2)\rangle \tag{103}$$

and the total VP of the complex becomes

$$\Gamma_{vkl} = \Gamma_{vkl}^X + \Gamma_{vkl}^Y \tag{104}$$

In this treatment the dissociation of the complex is considered sequentially, i.e., either X dissociates first or Y does. The possibility of simultaneous dissociation of the two rare gas atoms is not accounted for. The following processes could occur:

$$X - BC - Y \longrightarrow X - Y + BC \tag{105}$$

$$X - BC - Y \longrightarrow X + Y + BC \tag{106}$$

A calculation for HeI_2Ne was recently done [49] and the main conclusion was that the sequential mechanism dominates and contributes to its VP as high as 99%, the rest corresponding to dissociation into two diatomic fragments since the complete, double continuum, fragmentation, Equation (106), becomes almost forbidden. These contributions may dramatically change for other complexes like Ne_2Cl_2.

The potential energy surface used for HeI_2Ne was a simple addition of atom-atom interaction represented by Morse functions. For details of the calculations see Reference [47].

In the range of $25 \leq v \leq 35$ the T-shaped molecules He-I_2 and Ne-I_2 show two and five energy levels, respectively. For the vibrational level $v = 25$, the effective angular potentials $W_k^{v=25}(\gamma)$ with $k = 0$, has the minimum at $\gamma = \frac{\pi}{4}$ and its value in cm^{-1} is

$$W_0^{25}(\pi/4) = -98,89 \simeq -18.37 - 71.24 - D_{\text{He-Ne}}(= 9.94) = 99.55.$$

For $k = 1$ the behavior is similar.

The ground level measured from the diatomic vibrational energy $E_{I_2}(25)$ is $E_{25,00} = -94.28$ cm^{-1} and roughly corresponds to the simple sum of the ground triatomic energies plus the energy of the single level supported by the He-Ne potential. It is worth noting that the zero-point bending level $l = 0$ of the first stretching excitation $k = 1$ ($E_{25,1,0} = -73.83$ cm^{-1}), is found above the sixth bending excitation $l = 5$ of $k = 0$ ($E_{25,0,5} = -76.97$ cm^{-1}). This provides a justification for the use of the angular adiabatic approximation. In addition, the stretching excitation

$$E_{25,1,0} - E_{25,0,0} = 20.76\,\mathrm{cm}^{-1}$$

is very close to the value found by Kenny *et al.* [46] in their fluorescence excitation spectrum (20.09 cm^{-1}) suggesting that the second observed peak, denoted by $w' = 1$ in their work, may correspond to a stretching excitation in the HeI_2 bond. On the other hand the zero-point energy levels $E_{v,0,0}$ range from -94.28 cm^{-1} for $(v = 25)$ up to -93.10 cm^{-1} for $(v = 35)$ showing an increase in the blue shift for higher diatomic excitation. This is in qualitative, and even quantitative, agreement with the experimental findings (for example, the difference in positions for the first level between $v' = 26$ and $v' = 16$ in Table I of Kenny *et al.* [46] 0,87 cm^{-1} compares favorably with the theoretical result of 94.28 - 93.10 = 1.18 cm^{-1} found in the range [25–35]).

We can represent the functions $\Gamma^{He}_{25.0}(\gamma)$ and $\Gamma^{Ne}_{25.0}(\gamma)$ in the range $\gamma\epsilon[0,\pi]$ (the corresponding values for $\gamma\epsilon[\pi, 2\pi]$ are obviously symmetric), and are also represented. Despite the fact that these angular dependent rates are physically meaningless, we can note that the Ne rate is always higher than the He one. In particular this holds in the region of the well where the corresponding values are the most important for obtaining (meaningful) averages. The same happens for $v = 26, 27$ and 28at every angle. However for $v = 29$, due to the anharmonicity of the I_2 potential, the $v - 1$ channel for dissociation of Ne becomes energetically closed at $\gamma = \pi/4$ and the corresponding rate $\Gamma^{Ne}_{29.0}(\pi/4) = 0,86 \times 10^{-3}\ \mathrm{cm}^{-1}$ is smaller than $\Gamma^{He}_{29.0}(\pi/4) = 0,27\ \mathrm{cm}^{-1}$. As v increases the angular region for which only the $v - 2$ channel contributes to Γ^{Ne} is wider and, finally, for $v \geq 34$ at all the angles $\gamma \geq \frac{\pi}{4}$, the Ne-$I_2$bonds needs two vibrational quanta to be broken. This behavior is reflected in the variation of the averaged rates in the ground state $k = l = 0$ vs. the initial diatomic excitation v_{I_2}. In the range $v\epsilon[25, 28]$, $\Gamma^{Ne}_{v,0,0}$ is roughly two times $\Gamma^{He}_{v,0,0}$ and both show a superlinear increasing law as v increases. Hence, the total rate $\Gamma_{v,0,0} = \Gamma^{He}_{v,0,0} + \Gamma^{Ne}_{v,0,0}$ that could be compared with possible experiments follows also a superlinear increasing law with v. $\Gamma^{He}_{v,0,0}$ continues its tendency in the full range studied here, but $\Gamma^{Ne}_{v,0,0}$ decreases sharply for $v > 28$ and consequently the total rate does. At $v = 29$, $\Gamma^{Ne}_{29,0,0} = 0{,}32\ \mathrm{cm}^{-1}$ is still higher but very close to $\Gamma^{He}_{29,0,0} = 0,26\ \mathrm{cm}^{-1}$, however at $v = 30$ the situation changes and $\Gamma^{Ne}_{30,0,0} = 0,20\ \mathrm{cm}^{-1}$ is smaller than $\Gamma^{He}_{30,0,0} = 0,29\ \mathrm{cm}^{-1}$. From here on, the He atom clearly dissociates first. For $v \geq 30$ the very small contributions of the two quanta channels to Γ^{Ne} yield a total rate that practically coincides with Γ^{He}.

A quasiclassical calculation for the VP of HeI_2Ne has also been performed [49]. The agreement between quasiclassical and quantal rates is fairly good.

9. Electronic Predissociation and Nonadiabatic Transitions in van der Waals Complexes

When a van der Waals complex is prepared in an excited electronic state, nonadiabatic mixing between different electronic configurations may be induced by the van der Waals interaction. This can be viewed as the analog of electronic nonadiabatic interactions induced by the intermolecular potential in collisions. It is the purpose of this section to present some examples of this type of effect which have been studied theoretically, and experimentally, in simple van der Waals molecules.

9.1. NONADIABATIC TRANSITIONS IN THE PHOTODISSOCIATION OF I_2-X COMPLEXES

It is well established that photoexcitation of a "free" iodine molecule above the dissociation limit of $I_2(B)$ leads to the production of atoms with a quantum yield of unity. On the other hand, when a I_2-M (M= rare gas atom) van der Waals is excited at the same total energy, fluorescence from a large number of vibrational levels of $I_2(B)$ is observed [51–53]. Thus excitation of the van der Waals complex with energy up to 1000 cm^{-1} in excess of the dissociation limit of the B state, can lead to

$$I_2 - M \xrightarrow{h\nu} I_2(B, v', j') + M + \Delta E \tag{107}$$

with a rather broad distribution of vibrational levels ($\Delta v' \sim 10$) around $v' = 35$. It should be noted at this point, that the dissociation of I_2 into

$$I_2 - M \xrightarrow{h\nu} 2I + M + \Delta E \tag{108}$$

can be present but has not been measured, nor have other channels, involving other electronic states of I_2 or the formation of IM. However, even if the branching ratio for reaction (107) is very small, it is rather surprising to observe a process in which such a large amount ($\sim$ 1500 cm^{-1}) of energy is going into relative translational energy of the products. In the VP of very high vibrationally excited levels of $I_2(B)$, the prominent channels involve the lost of a few vibrational quanta (see Section 3), i.e., of the order of 100 cm^{-1}.

Another piece of information comes from experiments performed in gases and liquids [54, 55]. The I_2 photodissociation quantum yields in a variety of nonpolar gases was found to drop below unity at surprisingly low densities, namely one or two orders of magnitude below liquid densities. Otto, Schroeder, and Troe [54] proposed an explanation for the observed drop in the quantum yield of I_2 photodissociation based on the experimental findings of Saenger, McClelland, and Herschbach [51], Valentini and Cross [52], and Philippoz, and van den Bergh [53]. They proposed that the concentration of I_2-solvent complexes is large enough and that these complexes undergo fragmentation into an I_2 vibrationally excited in its ground electronic state.

Another important set of experimental results [56] shows that collisions between a large variety of nonpolar solvent particles and $I_2(B)$ lead to dissociation of the I_2 (collision induced electronic predissociation). So, collisions lead to dissociation while fragmentation of an I_2-M complex leads to I_2 fragments.

Two mechanisms have been proposed in the literature to explain the occurrence of the reaction (107). The first involved "caging" produced by a single solvent particle [52]. Quasiclassical trajectory calculations of this process was performed by Noorbatcha, Raff and Thompson on I_2-M, M = Ar, Kr, Xe [57]. These authors have found that stabilization of the I_2 molecule occurs by a combination of direct impulsive energy transfer from near collinear geometries and an anchoring effect from near 90 degrees geometries which results in long-lived trajectories. In all three cases studied however, the results are that the energy transfer to the rare gas atom is small and there combined I_2 is found in very high lying vibrational states in contrast to the experimental findings.

The second mechanism proposed to explain reaction (107) was based on a nonadiabatic electronic transition induced by the presence of the rare gas atom [58].

9.1.1. *Electronic Nonadiabatic Model*

It is well known [59] that in addition to the B state there is another state ($^1\Pi_{1u}$) that contributes significantly to the absorption cross section in the λ = 500–450 nm spectral region. For example, at λ = 480 nm the $^1\Pi_{1u}$ state accounts for up to 50% of the total cross section. The $^1\Pi_{1u}$ state is repulsive and is known to contribute to the spontaneous and magnetic field induced $B \to ^1 \Pi_{1u}$ predissociation [60]. The $^1\Pi_{1u}$ and the $B^3\Pi_{0_u^+}$ states are weakly coupled by hyperfine interactions in the free I_2 molecule. They are likely to be more strongly coupled by the presence of a solvent particle. It has been suggested [4] that indeed the $^1\Pi_{1u}$ state may be the preferential route for collision induced predissociation of low lying vibrational levels of $I_2(B)$ in collisions with helium.

In the electronically nonadiabatic recombination model of Beswick *et al.* [58] it is assumed that the complexes that have been photoexcited to the repulsive $^1\Pi_{1u}$ state have a probability of undergoing a nonadiabatic electronic transition down to vibrational levels of the B state, the energy difference being transformed to relative kinetic energy of the solvent particle. According to this model the energy loss will be governed by the I_2 vibrational matrix elements of the electronic nonadiabatic coupling. In the vibrational diabatic approximation (see Section 3.2), the total vibrational wave function for the initially excited $^1\Pi_{1u}$ state of the I_2-M complex can be written as

$$\psi^{(1u)}(R,r) = \chi_E^{(1u)}(r)\phi_\ell^{(1u)}(R) \tag{109}$$

where $\chi_E^{(1u)}(r)$ is the continuum wave function at energy E for the I_2 motion on the repulsive potential energy curve of the $^1\Pi_{1u}$ state, while $\phi_\ell^{(1u)}(R)$ describes the relative bound motion of M with respect to the center of mass of I_2. Similarly, the final state after the electronically nonadiabatic transition can be written as

$$\psi^{(B,v')}(R,r) = \chi_{v'}^{(B)}(r)\phi_{\epsilon'}^{(B)}(R) \tag{110}$$

where now $\chi_{v'}^{(B)}(r)$ is the bound wave function of the v' vibrational level of $I_2(B)$, while $\phi_{\epsilon'}^{(B)}(R)$ describes the relative dissociative motion of M with respect to the center of mass of I_2, ϵ' being the relative kinetic energy of the two fragments.

The probability for an electronically nonadiabatic transition between these two states will be given, in the first-order Born approximation [38], by

$$P_{1u\to B,v'} = 4\pi^2|\langle\psi^{(1u)}|V(r,R)|\psi^{(B,v')}\rangle|^2 \tag{111}$$

where V is the nonadiabatic electronic coupling operator. From Equations (108)–(110), it follows that

$$P_{1u\to B,v'} = 4\pi^2|\langle\phi_{\ell}^{(1u)}(R)|V_{1uE,Bv'}(R)|\phi_{\epsilon'}^{(B)}\rangle|^2 \tag{112}$$

where $V_{1uE,Bv'}(R)$ is given by

$$V_{1uE,Bv'}(R) = \langle\chi_E^{(1u)}(r)|V(r,R)|\chi_{v'}^{(B)}(r)\rangle. \tag{113}$$

The probability for recombination of I_2 depends crucially on this matrix element. For the time being, the nature of the coupling operator V is unknown. Besides Coriolis forces, which are most probably small in this case, there are two possible origins for this coupling: spin-orbit and "non Born-Oppenheimer" interactions. In the latter, the coupling is proportional to the velocity operator d/dr. In the case of spin-orbit coupling, one usually assumes that $V(r,R)$ is a slowly varying function of the coordinates, and therefore

$$V_{1uE,Bv'}(R) = V\langle\chi_E^{(1u)}(r)|\chi_{v'}^{(B)}(r)\rangle. \tag{114}$$

If in the other hand, one takes the "non-Born–Oppenheimer" coupling between the two states, then

$$V_{1uE,Bv'}(R) \propto \langle\chi_E^{(1u)}(r)|\frac{\partial}{\partial r}\chi_{v'}^{(B)}(r)\rangle. \tag{115}$$

Equations (112) and (113) have been evaluated [58] with the result that the predicted dependence of the product recoil energy with respect to the excitation wave length is in fairly good agreement with the experiment. The agreement appears to be better with the "non-Born–Oppenheimer" model, Equation (113), in the case of the heavier and more polarizable rare gas atoms. Since the two curves $1u$ and B are almost parallel in the region of interest, both models lead to a very narrow final vibrational distribution in the $I_2(B)$ fragment. However, final state dynamics (i.e., the dynamical interaction between the receding rare gas atom and the I_2 molecule) will broaden this distribution. Further dynamical calculations are required to elucidate this point. The nonadiabatic electronic transition model seems to provide a possible mechanism for the production of bound I_2 in the photodissociation of I_2-M complexes.

Alternatively, it appears that there is a correlation between the experimental energy loss for these systems and the matrix elements of the nonadiabatic operator d/dr between the I_2 vibrational wave functions of the $^1\Pi_{1u}$ and B states.

It is important to note that the coupling between the B and the $^1\Pi_{1u}$ states induced by the presence of a rare gas atom should be important in two other processes at energies below the dissociation limit of the B state. The first has already been mentioned above, the collision-induced predissociation of $I_2(B)$ by collisions with rare gas atoms [56]. The second one corresponds to electronic predissociation of $I_2(B)$-M van der Waals complexes when excited below the dissociation threshold of the B state [61]. All these processes are intimately related and it may well be that the interactions and electronic states involved are the same.

9.2. FINE-STRUCTURE ELECTRONIC PREDISSOCIATION IN VAN DER WAALS MOLECULES

The fine-structure electronic relaxation of atoms perturbed by collisions with rare gas atoms or molecules has been extensively studied in the past [62]. Collisional fine-structure transitions can be represented as

$$A(^{2S+1}L_{J_A}) + M \longrightarrow A(^{2S+1}L_{J'_A}) + M \tag{116}$$

where $A(^{2S+1}L_{J_A})$ is an atom in the electronic state specified by the total spin S, the orbital angular momentum L, and the atomic total angular momentum J_A. In Equation (114) M denotes an atomic or molecular perturber.

Another type of experiment which also addresses fine-structure electronic relaxation has been developed by Jouvet and Soep [62]. It was conducted as follows: a van der Waals complex A-M is first formed in a supersonic expansion and then excited in the spectral region corresponding to the transition from the ground state to the $^{2S+1}L_{J_A}$ state of the atom A. For photon energies below the dissociation threshold the complex can be excited to bound levels of the $^{2S+1}L_{J_A}$ state (see Figure 5). These levels are coupled to the $^{2S+1}L_{J'_A}$ levels by nonadiabatic and rotational couplings. Therefore, if they lie above the dissociation threshold of the $^{2S+1}L_{J'_A}$ state of the complex, they will predissociate according to

$$A(^{2S+1}L_{J_A}) - M \longrightarrow A(^{2S+1}L_{J'_A}) + M \tag{117}$$

which can be viewed as the "half-collision" analog of Equation (114).

It should be noted, however, that collisional relaxation occurs necessarily above the $A(^{2S+1}L_{J_A})$ threshold, while predissociation corresponds to total energies below that threshold (see Figure 5). Although the potential energy curves and couplings are the same, the relative importance of the different interaction terms as well as the critical regions on the potential energy surfaces where the process occurs, may be different in "half" and "full" collisions. In addition to Equation (115), other processes may also occur following excitation of the van

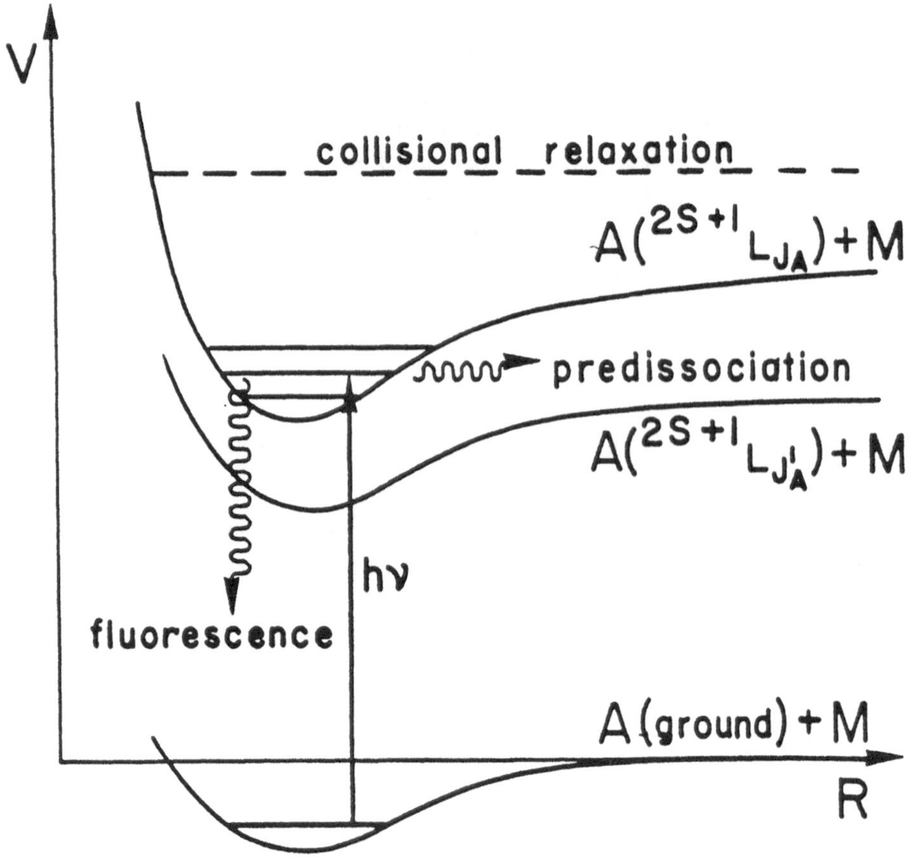

Figure 5. Schematic potential energy curves for fine-structure predissociation and relaxation in the atom-atom case.

der Waals molecule:

1. Electronic predissociation producing ground electronic state atoms,

$$A(^{2S+1}L_{J_A}) - M \longrightarrow A(\text{ground}) + M \tag{118}$$

where if M is a molecule, it can be vibrationally excited. This will be the half-collision analog of collisional electronic to vibrational relaxation.

2. Direct excitation of the $A(^{2S+1}L_{J'_A}) + M$ continuum:

$$A(\text{ground}) - M + h\nu \longrightarrow A(^{2S+1}L_{J'_A}) + M. \tag{119}$$

The probability for this process to occur depends on the electronic dipole selection rules as well as on Franck–Condon factors. In some very favorable cases direct dissociation and predissociation may occur simultaneously and give rise to interference effects.

3. If the photon is such that the complex is excited above the $A(^{2S+1}L_{J_A})$ threshold (see Figure 5), the system may dissociate directly leading to both

$A(^{2S+1}L_{J_A})$ and $A(^{2S+1}L_{J'_A})$ atoms. In this case, the potential energy surfaces and the total energy reached by photon excitation are the same as those involved in the collisional experiment. For the complex, however, the impact parameter and the orientation are restricted to a small range of values while in collisions an average over all impact parameter and orientation is performed.

9.2.1. *Study of Fine-Structure Predissociation in Hg-M Complexes*

In this section we shall discuss the fine-structure electronic predissociation in relation with the experiments performed on Hg-M complexes. The reaction channels (116) and (117) are neglected. This is likely to be a good approximation. In the Hg-N_2 case, for instance, the electronic relaxation to the ground state in collisions is known to be rather inefficient [62] and it is expected the same is true in predissociation. In addition, since the transition from Hg(ground) to Hg($^3P_0$) is forbidden for free Hg atoms, we expect the direct excitation of the $^3P_0$ state to be negligibly small in the complex.

Several mechanisms have been invoked to describe fine-structure electronic relaxation in collisions [64]:

1. Electrostatic coupling within the fine-structure manifold. This mechanism is efficient for large internuclear distances where the electrostatic interactions is of the order of the spin-orbit coupling. Partial decoupling between **L** and **S** then occurs leading to fine-structure transitions. For atom-atom collisions this coupling is often referred to as the "radial" coupling since it is induced by the relative motion of the two atoms. When M is a molecule, additional coupling effects appear. The removal of the $C_{\infty v}$ symmetry may give rise to couplings between states which in the atom-atom case would be uncoupled. In addition, the existence of vibrational and rotational degrees of freedom in the M fragment opens new channels for relaxation. In the case of Hg($^3P_1$)+N_2, for instance, the vibrational excitation of N_2 was invoked to explain the enhancement of the relaxation cross section with respect to the Hg-Ar case [62].
2. Coriolis coupling. This arises from the rotation of the collision complex. The component of the electronic angular momentum on the intermolecular axis (the vector going from A to the center of mass of M) is a good quantum number only in the nonrotating frame. For fast rotation the electronic motion partially decouples from the internuclear axis rotation and this induces mixing between different fine-structure states. This mechanism will then be most efficient for small internuclear distances where the rotation is faster.
3. Coupling to other electronic states. In particular the interaction with the ionic surface should be considered when the ionization potential of one of the collisional partners is low (intermediate ionic complex model). On the other hand, their importance in fine-structure relaxation has not yet been determined with certainty.

In the case of fine-structure electronic predissociation in van der Waals complexes, one may also consider similar mechanisms. Jouvet and Beswick [65] have studied fine-structure predissociation in van der Waals complexes induced by electrostatic and Coriolis interactions within the fine-structure manifold and neglecting couplings to other electronic states. In particular, they have neglected the influence of the intermediate ionic complex. Since in van der Waals molecules the transition occurs in the region of the well, they were implicitly assuming that the crossing with the ionic surface occurs at much shorter internuclear distances. The total Hamiltonian for an atom A interacting with a diatomic molecule BC can be written, after separation of the center of mass motion, as

$$H = \frac{\hbar^2}{2m}\left(-\frac{\partial^2}{\partial R^2} + \frac{\boldsymbol{\ell}^2}{R^2}\right) + \frac{\hbar^2}{2\mu}\left(-\frac{\partial^2}{\partial r^2} + \frac{\mathbf{j}^2}{r^2}\right) + H_{\text{el}} + H_{\text{so}} \tag{120}$$

where H_{el} is the nonrelativistic electronic Hamiltonian while H_{so} is the spin-orbit interaction operator. All other quantities in Equation (118) have been defined above (see Section 3.1).

The atom A is described by a total electronic orbital angular momentum $\mathbf{L}$ and a total spin angular momentum $\mathbf{S}$, with projections on the z axis equal to Λ and Σ, respectively. The total electronic angular momentum of atom A is $\mathbf{J}_A = \mathbf{L}+\mathbf{S}$ with projection on the z axis $\Omega_A = \Lambda + \Sigma$. Thus, the electronic basis set $|L, S, J_A, \Omega_A\rangle$ (denoted by $|J_A, \Omega_A\rangle$ for simplicity) diagonalize H_{so}.

The electronic Hamiltonian H_{el} can be written as

$$H_{\text{el}} = H_{\text{el}}^{(0)} + H_{\text{el}}' \tag{121}$$

where $H_{\text{el}}^{(0)}$ is the asymptotic nonrelativistic electronic Hamiltonian, while $H_{\text{el}}'$is the intermolecular interaction which goes to zero as $R \to \infty$. The following simplifying assumptions are made:

1. Only one electronic state of BC with total angular momentum equal to zero ($^1\Sigma$ state) is considered.
2. The atom A is described by only two electronic manifolds (one corresponding to the ground electronic state and the other being an electronically excited manifold) characterized by well-defined quantum numbers L and S.

Thus asymptotically, there will be $(2L + 1)(2S + 1)$ degenerate eigenstates of H_{el} in each one of the manifolds. This degeneracy is lifted by the spin-orbit interaction and is obtained $2\min(L, S) + 1$ states $|J_A, \Omega_A\rangle$ with energies: $E_{J_A} + E_{BC}(r)$, where E_{J_A} are the fine-structure atomic energies of A, while $E_{BC}(r)$ is the BC diatomic potential energy.

For finite values of R these states are coupled by the intermolecular electronic interaction H_{el}'. For the atom-atom case this is often referred to as the"radial" coupling. An additional coupling between those states is induced by the centrifugal term in Equation (118), i.e., by $\boldsymbol{\ell}^2/2mR^2$. This "rotational" (also named Coriolis) coupling is important in many cases.

It is convenient to choose the body-fixed frame of reference in which the z axis lies along the **R** vector and the x axis lies on the plane of the molecule. With this choice the matrix elements of $H'_{\rm el}$ will depend only on R, r, and θ (the angle between **R** and **r**). Denoting by $|\phi_{nL\Lambda S\Sigma}\rangle$ the eigenfunctions of the asymptotic electronic Hamiltonian $H^{(0)}_{\rm el}$, the matrix elements of $H'_{\rm el}$ in this basis set will be given by

$$\langle \phi_{nL\Lambda S\Sigma}|H'_{\rm el}|\phi_{nL\Lambda' S\Sigma'}\rangle = \delta_{\Sigma\Sigma'} \sum_K V^{(K)}_{\Lambda\Lambda'}(R,r) Y_{K(\Lambda'-\Lambda)}(\theta,0) \tag{122}$$

with $V^{(K)}_{-\Lambda-\Lambda'} = (-1)^{\Lambda'-\Lambda} V^{(K)}_{\Lambda'\Lambda} = V^{(K)}_{\Lambda\Lambda'}$. As discussed above, the actual basis set used in the analysis is defined by Hund's case (c) wave functions

$$|J_A, \Omega_A\rangle = \sum_{\Lambda\Sigma} C(L,\Lambda,S,\Sigma; J_A,\Omega_A)|\phi_{nL\Lambda S\Sigma}\rangle \tag{123}$$

where $C(j_1,m_1,j_2,m_2;j_3,m_3)$ are Clebsh–Gordan coefficients. Using Equation (121) we get the matrix elements of $H'_{\rm el}$ in the $|J_A,\Omega_A\rangle$ basis set as

$$\begin{aligned} &\langle J_A,\Omega_A|H'_{\rm el}|J'_A,\Omega'_A\rangle \\ &\quad= \sum_{\Lambda\Lambda'}\sum_{\Sigma}\sum_K C(L,\Lambda,S,\Sigma;J_A,\Omega_A) \\ &\quad\quad\times C(L,\Lambda',S,\Sigma;J'_A,\Omega'_A) V^{(K)}_{\Lambda\Lambda'}(R,r) Y_{K(\Lambda'-\Lambda)}(\theta,0). \end{aligned} \tag{124}$$

Introducing now the total angular momentum $\mathbf{J} = \mathbf{J}_A + \boldsymbol{\ell} + \mathbf{j}$ and the body-fixed total angular momentum wave functions

$$\begin{aligned} &|J,M,\Omega,j,J_A,\Omega_A\rangle \\ &\quad= \left(\frac{2J+1}{4\pi}\right)^{1/2} D^{J*}_{M\Omega}(\phi_R,\theta_R,\phi) Y_{j,\lambda=\Omega-\Omega_A}(\theta,0)|J_A,\Omega_A\rangle \end{aligned} \tag{125}$$

with Ω, the projection of **J** on the body-fixed axis, being equal to the sum of the projections of **j** and$\mathbf{J}_A$. The Euler angles ϕ_R, θ_R, and ϕ are the same as those used in Section 2 (see Equation (16)). From Equations (122) and (123) we obtain

$$\begin{aligned} &\langle J,M,\Omega,j,J_A,\Omega_A|H'_{\rm el}|J,M,\Omega,j',J'_A,\Omega'_A\rangle \\ &\quad= \sum_{\Lambda\Lambda'}\sum_{\Sigma}\sum_K (-1)^K \sqrt{\frac{2K+1}{4\pi}} V^{(K)}_{\Lambda\Lambda'}(R,r) C(L,\Lambda,S,\Sigma;J_A,\Omega_A) \\ &\quad\quad\times C(L,\Lambda',S,\Sigma;J'_A,\Omega'_A) C(j,0,K,0;j',0) \\ &\quad\quad\times C(j',\Omega-\Omega'_A,K,\Omega'_A-\Omega_A;j,\Omega-\Omega_A). \end{aligned} \tag{126}$$

Moreover, as in Section 2, it is convenient to work with wave functions of definite parity with respect to total inversion through the space fixed origin:

$$\begin{aligned} &|J,M,\Omega,j,J_A,\Omega_A,p\rangle \\ &\quad= C_{\Omega,\Omega_A}\{|J,M,\Omega,j,J_A,\Omega_A\rangle \end{aligned}$$

$$+ p(-1)^{J+J_A+L}|J, M, -\Omega, j, J_A, -\Omega_A\rangle\} \tag{127}$$

where $p = \pm 1$ for even/odd parity, and $C_{\Omega,\Omega_A} = 1/\sqrt{2(1+\delta_{\Omega 0}\delta_{\Omega_A 0})}$. It should be noticed that in the case $\Omega = \Omega_A = 0$, only the states with parity $p = (-1)^{J+J_A+L}$ exist.

The other important matrix elements which are needed in the problem are those of $\boldsymbol{\ell}^2$. They can be written similarly to Equation (17) as

$$\begin{aligned}\langle J, M, \Omega, j, J_A, \Omega_A|\boldsymbol{\ell}^2|J, M, \Omega', j, J_A, \Omega'_A\rangle &= \delta_{\Omega\Omega'}\delta_{\Omega_A\Omega'_A} \\ &\times [(J+1) + j(j+1) + J_A(J_A+1) - 2\Omega^2 - 2\Omega_A^2 + 2\Omega\Omega_A] \\ &- \langle J, M, \Omega, j, J_A, \Omega_A|(J_+j_- + J_-j_+ + J_+J_{A-} + J_-J_{A+} \\ &- j_+J_{A-} - j_-J_{A+})|J, M, \Omega', j, J_A, \Omega'_A\rangle\end{aligned} \tag{128}$$

with the raising and lowering operators defined as usual:

$$\begin{aligned}J_\pm|J,\Omega\rangle &= \sqrt{J(J+1) - \Omega(\Omega \mp 1)}|J, \Omega \mp 1\rangle \\ j_\pm|J,\lambda\rangle &= \sqrt{j(j+1) - \lambda(\lambda \pm 1)}|j, \lambda \pm 1\rangle \\ J_{A\pm}|J_A,\Omega_A\rangle &= \sqrt{J_A(J_A+1) - \Omega_A(\Omega_A \pm 1)}|J_A, \Omega_A \pm 1\rangle.\end{aligned} \tag{129}$$

It should be noted at this point that in the atom-atom case, v and j will not appear in the above equations. In addition, $\Omega = \Omega_A$ since in this case the projection of the total angular momentum on the internuclear axis is equal to the projection of the electronic angular momentum. The multipole expansion in Equation (120) reduces to only one term, $K = 0$, in this case. Thus

$$\langle\phi_{nL\Lambda S\Sigma}|H'_{\text{el}}|\phi_{nL\Lambda' S\Sigma'}\rangle = \delta_{\Sigma\Sigma'}\delta_{\Lambda\Lambda'}V^{(0)}_{\Lambda\Lambda}(R) \tag{130}$$

and the nonzero electronic coupling elements between rotational wave functions become:

$$\begin{aligned}&\langle J, M, \Omega_A, J_A, \Omega_A, p|H'_{\text{el}}|J, M, \Omega_A, J'_A, \Omega_A, p\rangle \\ &= \sum_{\Lambda\Sigma}(4\pi)^{-1/2}V^{(0)}_{\Lambda\Lambda}(R) \\ &\times C(L, \Lambda, S, \Sigma; J_A, \Omega_A)C(L, \Lambda, S, \Sigma; J'_A, \Omega_A).\end{aligned} \tag{131}$$

Finally, the rotational coupling induced by $H_{\text{rot}} = \hbar^2\boldsymbol{\ell}^2/2mR^2$ term in Equation (118) has the following off-diagonal matrix elements:

$$\begin{aligned}&\langle J, M, \Omega_A, J_A, \Omega_A, p|H_{\text{rot}}|J, M, \Omega'_A, J_A, \Omega'_A, p\rangle \\ &= \frac{\hbar^2}{2mR^2}\frac{1}{2\sqrt{(1+\delta_{\Omega_A 0})(1+\delta_{\Omega'_A 0})}}\delta_{\Omega_A\Omega'_A\pm 1} \\ &\times\sqrt{J(J+1) - \Omega'_A(\Omega'_A \mp 1)}\sqrt{J_A(J_A+1) - \Omega'_A(\Omega'_A \mp 1)}.\end{aligned} \tag{132}$$

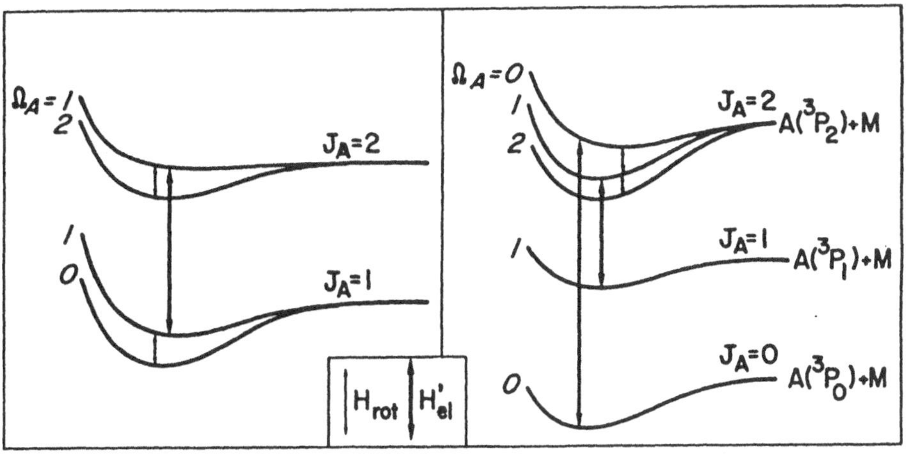

Figure 6. Schematic potential energy curves and couplings for the atom-atom case: $A^*(L = 1, S = 1) - M$. The spin-orbit coupling parameter g is assumed to be positive and larger than the binding energy of the complex. H_{rot} and H'_{el} refer to the rotational and the nonrelativistic electronic Hamiltonians, respectively. (a) $p = (-1)^J$ parity states; (b) $p = -(-1)^J$ parity states.

From Equations (129) and (130) it can be seen that H'_{el} couples states with the same component Ω_A and different J_A, while H_{rot} couples states with Ω_A differing by ± 1 but with the same J_A. As an example, let us consider the case $L = 1$ and $S = 1$ (as in Hg(^{3}P)). Thus J_A can take the values 0, 1, and 2 with asymptotic energies $-g$, $-g/2$, and $+g$, respectively. The energy curves corresponding to these states are represented schematically on Figure 6 for $g > 0$ and the two parities. The coupling scheme according to Equations (129) and (130) is also indicated on Figure 6.

Several interesting conclusions can be drawn. The $|1, 0\rangle$ state, for instance, cannot predissociate at all due to the different total symmetry of the $|1, 0\rangle$ and $|0, 0\rangle$ states. In general, it is expected that relaxation from the $^3P_1$ manifold will be less efficient than that from the $^3P_2$. The $^3P_1$ state is coupled to the lower fine-structure state $^3P_0$ only indirectly via the $^3P_2$. On the other hand, the $|2, 1\rangle$ and $|2, 0\rangle$ will predissociate fast because they are coupled directly to the lower lying states.

We now turn to the atom-diatom case. The H'_{el} matrix elements now has $K = 0$ (isotropic) and $K > 0$ (anisotropic) terms. The coupling between states with same Ω_A but different J_A is provided by the full H'_{el} Hamiltonian, while the coupling between states which differ both in J_A and Ω_A, are induced by the anisotropy terms ($K > 0$). They were denoted $H'_{el}(\theta)$ by Jouvet and Beswick [65]. This additional nonadiabatic interaction caused by the anisotropy of the potential will strongly enhance the relaxation of atom-diatom molecules as compared to the atom-atom case.

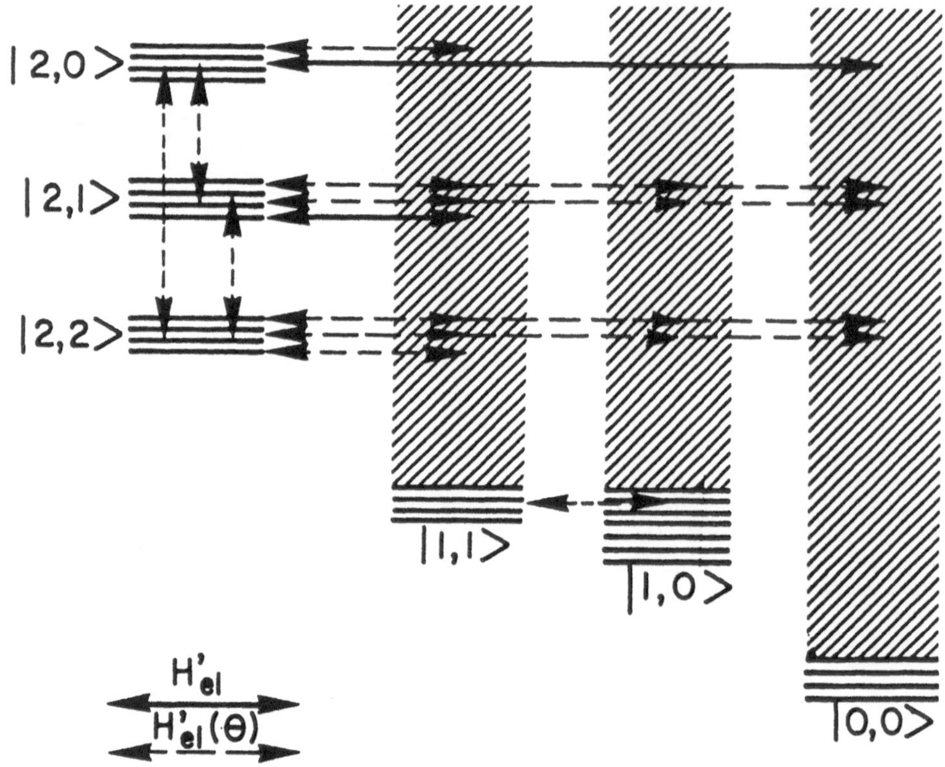

Figure 7. Energy levels and couplings for the atom-diatom case: $A^*(L = 1, S = 1) - M$. The spin-orbit coupling parameter g is assumed to be positive and larger than the binding energy of the complex. H'_{el} is the total nonrelativistic electronic Hamiltonian, while $H'\mathrm{el}(\theta)$ is the anisotropic part of H'_{el}. For simplicity the rotational couplings (assumed to be small) have been omitted.

Consider the case $L = 1$ and $S = 1$. In Figure 7 the energy level scheme and couplings are represented. The new couplings induced by the anisotropy $H'_{\mathrm{el}}(\theta)$ are drawn as dotted lines. We note for instance that the $|2,2\rangle$ state is now coupled to all other states while in the atom-atom case it was only coupled to the $|2,1\rangle$ by Coriolis coupling. Thus the $|2,2\rangle$ state can predissociate readily into the $|1,1\rangle$, $|1,0\rangle$, and $|0,0\rangle$ channels. The $|2,1\rangle$ state, which in the atom-atom case was coupled by H'_{el} only to the $|1,1\rangle$ continuum, is now coupled by $H'_{\mathrm{el}}(\theta)$ to the $|1,0\rangle$ and $|0,0\rangle$ channels as well. The $|2,0\rangle$ state, which in the atom-atom case was coupled only to $|0,0\rangle$, can now predissociate into the $|1,1\rangle$ directly via $H'_{\mathrm{el}}(\theta)$. On the other hand, for symmetry reasons, it is not coupled to the $|1,0\rangle$ channel.

We now turn to the $J_A = 1$ manifold which has been studied experimentally [63]. In the atom-atom case the two sublevels $|1,1\rangle$ and $|1,0\rangle$ were not coupled directly to the $|0,0\rangle$ channel (see Figure 6). In the atom-diatom case, the $|1,1\rangle$ state is now coupled to the $|0,0\rangle$ continuum by the anisotropy of the potential via the $|2,2\rangle$ channel. Therefore, it is expected to predissociate much faster in the atom-diatom case. This is exactly what was observed in the experiment [63]. The same situation prevails for the $|1, 0\rangle$ state.

References

[1] A. van der Avoird, P. Wormer, F. Mulder and R. Berns: *Top. Curr. Chem.* **93** (1980) 1.
[2] J.H. van Lenthe, J.G.C.M. van Duijneveldt-van deRijdt and F.B. van Duijneveldt: *Adv. Chem. Phys.* **69** (1987) 521.
[3] A.D. Buckingham, P.W. Fowler and J.M. Hutson: *Chem. Rev.* **88** (1988) 963.
[4] J.A. Beswick and J. Jortner: *Adv. Chem. Phys.* **47** (1981) 363.
[5] D.H. Levy: *Adv. Chem. Phys.* **47** (1981) 323.
[6] K.C. Janda: *Adv. Chem. Phys.* **60** (1985) 201.
[7] "Van der Waals molecules", *Faraday Discuss. Chem. Soc.* **73** (1982).
[8] A. Weber (Ed.): *Structure and Dynamics of Weakly Bound Molecular Complexes*, NATO ASI Series C: Mathematical and Physical Sciences, Vol. 212, Reidel, Dordrecht (1987).
[9] N. Halberstadt and K.C. Janda (Eds.): *Dynamics of Polyatomic van der Waals Complexes*, Vol. B, Plenum Publishing Corporation, New York (1989).
[10] R.E. Smalley, D.H. Levy and L. Wharton: *J. Chem. Phys.* **64** (1976) 3266; M.S. Kim, R.E. Smalley, L. Wharton and D.H. Levy: *J. Chem. Phys.* **65** (1978) 1216; K.E. Johnson, L. Wharton and D.H. Levy: *J. Chem. Phys.* **69** (1978) 2719; W. Sharfin, K.E. Johnson, L. Wharton and D.H. Levy: *J. Chem. Phys.* **71** (1979) 1283; J.A. Blazy, B.M. De Koven, T.D. Russell and D.H. Levy: *J. Chem. Phys.* **72** (1980) 2439.
[11] W. Sharfin, P. Kroger and S.C. Wallace: *Chem. Phys. Lett.* **85** (1982) 81.
[12] J.A. Beswick, G. Delgado-Barrio and J. Jortner: *J. Chem. Phys.* **70** (1979) 3895.
[13] D.M. Willberg, M. Gutmann, J.J. Breen and A.H. Zewail, *J. Chem. Phys.* **96** (1992) 198.
[14] J.I. Cline, D.D. Evard, B.P. Reid, N. Sivakumar, F. Thommen and K.C. Janda, in *Structure and Dynamics of Weakly Bound Molecular Complexes*, NATO ASI Series C: Mathematical and Physical Sciences, Vol. 212, A. Weber (Ed.), Reidel, Dordrecht (1987).
[15] N. Halberstadt, J.A. Beswick and K.C. Janda: *J. Chem. Phys.* **87** (1987) 3966; R.L. Waterland, M.I. Lester and N. Halberstadt: *J. Chem. Phys.* **92** (1990) 4261.
[16] U. Fano: *Phys. Rev.* **124** (1961) 1866.
[17] R.T. Pack: *J. Chem. Phys.* **60** (1974) 633; G.C. Schatz and A. Kupperman: *J. Chem. Phys.* **65** (1976) 4642; R.N. Zare: in *Angular Momentum: Understanding Spatial Aspects in Chemistry and Physics*, J. Wiley, New York (1988).
[18] G. Delgado-Barrio, S. Serna, S. Miret-Artés, O. Roncero, Campos-Martínez and P. Villarreal: *Laser Chemistry* **12** (1992) 103.
[19] G. Delgado-Barrio, P. Mareca, P. Villarreal, A.M. Cortina and S. Miret-Artés: *J. Chem. Phys.* **84** (1986) 4268.
[20] J.A. Beswick and G. Delgado-Barrio: *J. Chem. Phys.* **73** (1980) 3653.
[21] M. Aguado, P. Villarreal, G. Delgado-Barrio, P. Mareca and J.A. Beswick: *Chem. Phys. Lett.* **102** (1983) 227.
[22] D. Secrest: *J. Chem. Phys.* **62** (1975) 710; L.W. Hunter: *J. Chem. Phys.* **62** (1975) 2855; L. Eno and G.G. Balint-Kurti: *J. Chem. Phys.* **71** (1979) 1447.
[23] D.L. Bunker and N.C. Blais: *J. Chem. Phys.* **41** 2377 (1964); D.L. Bunker: *Methods in Computational Physics*, Academic, New York (1971), Vol. 10, p. 287; M.D. Pattengill, in *Atomic-Molecule Collision Theory*, R.B. Bernstein (Ed.), Plenum, New York (1979); R.N. Porter and L.M. Raff, in *Modern Theoretical Chemistry*, W.H. Miller (Ed.), Plenum, New York (1976), Vol. 1, Part B.
[24] G. Delgado-Barrio, P. Villarreal, P. Mareca and G. Albelda: *J. Chem. Phys.* **78** (1983) 280.
[25] J.A. Beswick, G. Delgado-Barrio, P. Villarreal and P. Mareca: *Faraday Discuss. Chem. Soc.* **73** (1982) 406.
[26] R.N. Porter, L.M. Raff and W.H. Miller: *J. Chem. Phys.* **63** (1975) 2214.
[27] S.B. Woodruff and D.L. Thompson: *J. Chem. Phys.* **71** (1979) 376.
[28] O. Roncero, J.A. Beswick, N. Halberstadt, P. Villarreal and G. Delgado-Barrio: *Bull. Soc.*

Roy. des Sciences de Liège **3-4** (1989) 227.
[29] O. Roncero, J.A. Beswick, N. Halberstadt, P. Villarreal and G. Delgado-Barrio: *J. Chem. Phys.* **92** (1990) 3348.
[30] P. Villarreal and G. Delgado-Barrio: in *Mode Selective Chemistry*, J. Jortner, R.D. Levine and B. Pullman (Eds.), The Jerusalem Symposia on Quantum Chemistry and Biochemistry, Vol. 24, Kluwer Academic Publishers, Dordrecht (1991).
[31] D.D. Evard, C.R. Bieler, J.I. Cline, N. Sivakumar and K.C. Janda: *J. Chem. Phys.* **89** (1988) 2829; N. Halberstadt, J.A. Beswick, O. Roncero and K.C. Janda: *J. Chem. Phys.* **96** (1990) 2404; N. Halberstad, S. Serna, O. Roncero and K.C. Janda: *J. Chem. Phys.* **97** (1992) 341.
[32] J.C. Drobits, J.M. Skene and M.I. Lester: *J. Chem. Phys.* **84** (1986) 2896; J.M. Skene, J.C. Drobits and M.I. Lester: *J. Chem. Phys.* **85** (1986) 2329; J.C. Drobits and M.I. Lester: *J. Chem. Phys.* **88** 1988) 120; J.C. Drobits and M.I. Lester: *J. Chem. Phys.* **89** (1988) 4716; R.L. Waterland, M. Lester and N. Halberstadt: *J. Chem. Phys.* **92** (1990) 4261.
[33] F.A. Gianturco: *The Transfer of Molecular Energy by Collisions*, Springer-Verlag, Berlin (1979).
[34] J.A. Beswick and A. Requena: *J. Chem. Phys.* **72** (1980) 3018.
[35] F.A. Gianturco, A. Palma, P. Villarreal and G. Delgado-Barrio: *Chem. Phys. Lett.* **111** (1984) 399.
[36] O. Roncero, S. Miret-Artés, G. Delgado-Barrio and P. Villarreal: *J. Chem. Phys.* **85** (1986) 2084.
[37] S.I. Chu and K.K. Datta: *J. Chem. Phys.* **76** (1982) 5307.
[38] M.S. Child, *Molecular Collision Theory*, Academic Press, London (1974).
[39] K.T. Tang and J.P. Toennies: *J. Chem. Phys.* **68** (1978) 5501; K.T. Tang and J.P. Toennies: *J. Chem. Phys.* **74** (1981) 1148.
[40] J.M. Hutson and R.J. Le Roy: *J. Chem. Phys.* **83** (1985) 1197.
[41] P. Villarreal, G. Delgado-Barrio, O. Roncero, F.A. Gianturco and A. Palma: *Phys. Rev.* **A36** (1987) 617.
[42] F.A. Gianturco, A. Palma, P. Villarreal, G. Delgado-Barrio and O. Roncero: *J. Chem. Phys.* **87** (1987) 1054.
[43] F.A. Gianturco, G. Delgado-Barrio, O. Roncero and P. Villarreal: *Int. Rev. Phys. Chem.* **7** (1988) 1.
[44] S. Miret-Artés, O. Roncero, P. Villarreal and G. Delgado-Barrio: *J. Phys. Chem.* **91** (1987) 5623.
[45] F.A. Gianturco, G. Delgado-Barrio, O. Roncero and P. Villarreal: *Int. J. Quantum Chem. Symposium* **21** (1987) 389.
[46] J.E. Kenny, K.E. Johnson, W. Sharfin and D.H. Levy: *J. Chem. Phys.* **72** (1980) 1109; D.M. Willberg, M. Gutmann, J.J. Breen and A.H. Zewail: *J. Chem. Phys.* bf 96 (1992) 198; M. Gutmann, D.M. Willberg and A.H. Zewail: *J. Chem. Phys.* bf 97 (1992) 8048; E.D. Potter, Q. Lin and A.H. Zewail: *Chem. Phys. Lett.* bf 200 (1992) 605; D.M. Willberg, M. Gutmann, E.E. Nikitin and A.H. Zewail: *Chem. Phys. Lett.* bf 201 (1993) 506.
[47] G.C. Schatz, V. Buch, M.A. Ratner and R.B. Gerber: *J. Chem. Phys.* **79** (1983) 1808; R.H. Bisseling, R. Kossloff, R.B. Gerber, M.A. Ratner, L. Gibson and C. Cerjan: *J. Chem. Phys.* **87** (1987) 2760; G.C. Schatz, R.B. Gerber and M.A. Ratner: *J. Chem. Phys.* **88** (1988) 3709; R.B. Gerber and M.A. Ratner: *Adv. Chem. Phys.* **70** (1988) 97; N. Martin, G. Delgado-Barrio, P. Villarreal, P. Mareca and S. Miret-Artés: *J. Mol. Struct.* **142** (1986) 501.
[48] P. Villarreal, A. Varade and G. Delgado-Barrio: *J. Chem. Phys.* **90** (1989) 2684.
[49] P. Villarreal, S. Miret-Artés, J. Campos-Martinez and G. Delgado-Barrio: in *Dynamics of Polyatomic van der Waals Complexes*, Vol. 27B, N. Halberstadt and K.C. Janda (Eds.), Plenum Publishing Corporation, New York, (1990), p. 517.
[50] G. Delgado-Barrio, P. Villarreal, A. Varade, N. Martin and A. Garcia-Vela: in *Structure and Dynamics of Weakly Bound Molecular Complexes*, NATO ASI Series C: Mathematical and Physical Sciences, Vol. 212, A. Weber (Ed.), Reidel, Dordrecht (1987), p. 573.
[51] K.L. Saenger, G.M. McClelland and D.R. Herschbach: *J. Phys. Chem.* **85** (1981) 3333.
[52] J.L. Valentini and J. Cross: *J. Chem. Phys.* **77** (1982) 572.
[53] J.-M. Philippoz, H. van der Bergh and R. Monot: *J. Phys. Chem.* **91** (1987) 2545.

[54] B. Otto, J. Schroeder and J. Troe: *J. Chem. Phys.* **81** (1984) 202.
[55] J.-C. Dutoit, J.M. Zellweger and H. van der Bergh: *J. Chem. Phys.* **78** (1983) 1825.
[56] J.I. Steinfeld and W. Klemperer: *J. Chem. Phys.* **42** (1965) 3475.
[57] I. Noorbatcha, L.M. Raff and D.L. Thompson: *J. Chem. Phys.* **81** (1984) 5658.
[58] J.A. Beswick, R. Monot, J.-M. Philippoz and H. van der Bergh: *J. Chem. Phys.* **86** (1987) 3965.
[59] J. Tellinghuisen: *J. Chem. Phys.* **76** (1982) 4736.
[60] J. Vigué, M. Broyer and J.C. Lehmann: *J. Phys.* **42** (1981) 937, 949, 961.
[61] G. Kubiak, P.S.H. Fitch, L. Wharton and D.H. Levy: *J. Chem. Phys.* **68** (1978) 4477; K.E. Johnson, W. Sharfin and D.H. Levy: *J. Chem. Phys.* **74** (1981) 163.
[62] J.M. Mestdagh, P. de Pujo, J. Cuvellier, A. Binet and J. Berlande: *J. Phys.* **B15** (1982) 663; G. Pipper, J.E. Velazco and D.W. Setser: *J. Chem. Phys.* **59** (1973) 3323; H.J. Yuh and P.J. Dagdigian: *Phys. Rev.* **A28** (1983) 63; H. Horiguchi and S. Tsuchiya: *J. Chem. Soc. Faraday Trans.* **71** (1975) 1164.
[63] C. Jouvet and B. Soep: *J. Chem. Phys.* **80** (1984) 2229.
[64] E.E. Nikitin and B.M. Smirnov: *Sov. Phys. Usp.* **21** (1978) 95; V. Aquilanti, P. Casavecchia, G. Grossi and A. Lagana: *J. Chem. Phys.* **73** (1980) 1173; A. Hickman: *J. Phys.* **B15** (1982) 3005.
[65] C. Jouvet and J.A. Beswick: *J. Chem. Phys.* **86** (1987) 5500.

DYNAMICS OF NON-RIGID MOLECULES:
THE EXPLORATION OF PHASE SPACE VIA RESONANT AND SUB-RESONANT COUPLING

DAVID E. WEEKS* AND RAPHAEL D. LEVINE
Fritz Haber Research Centre for Molecular Dynamics
The Hebrew University,
Jerusalem,
Israel 91904

Abstract. In systems of non-linear coupled oscillators, it is possible to specify initial distributions of excitation energy that cause the zero order frequencies of the oscillators to be in resonance. In the presence of coupling, these resonances mediate the exchange of energy between the zero order modes. The magnitude of this energy exchange is a measure of the width of the resonance. We show that for higher order resonances there exist sub-resonances that in other parts of phase space correspond to lower order resonances. These sub-resonances provide a mechanism for large amplitude, high frequency oscillatory excursions in phase space. They also induce sub-resonant modulation of the resonance widths. To analyse the dynamics of coupled non-linear oscillators, conventional methods typically treat the coupling as a small perturbation and "average" over rapidly oscillating non-resonant terms. This averaging neglects the effect of the sub-resonances which, for physically realistic couplings can have amplitudes as large or larger than the resonance. We discuss an alternative approach for the perturbative analysis of coupled non-linear oscillators designed to incorporate the effect of sub-resonances on the dynamics. We then show that for small values of coupling when first order perturbation theory is valid, the sub-resonances lead to new, nearly separable modes. However, for large coupling, sub-resonant effects cannot be transformed away and dominate phase space dynamics.

* Now at the Engineering Physics Department, Air Force Institute of Technology, Wright-Patterson AFB, OH, 45433, U.S.A.

Y.G. Smeyers (ed.), Structure and Dynamics of Non-Rigid Molecular Systems, 249–306.

1. Introduction

Ever since the pioneering work by Fermi, Pasta, and Ulam [1], vibrational energy transfer between modes as described by classical mechanics has been known to be a complex and not necessarily ergodic process. It is now well established [2–17] that the dynamics of this energy transfer are to a large degree governed by the existence of internal resonances. These resonances are conveniently discussed using zero order action-angle variables of an uncoupled Hamiltonian [18]. The resonances occur because the frequencies of the uncoupled oscillators depend only on the zero order actions, $\omega_i = \omega_i(\mathbf{I})$. It is therefore possible to select appropriate resonance actions $\mathbf{I} = \mathbf{I}_r$ so that the frequencies $\omega(\mathbf{I}_r)$ satisfy the resonance condition,

$$\mathbf{m}_r \bullet \omega(\mathbf{I}_r) = \sum_i m_{ri}\omega_i(\mathbf{I}_r) = 0. \tag{1}$$

The vectors $\mathbf{m}_r$ are called resonant Fermi vectors [19] and their components m_{ri} are integers of which at least one must be negative and another positive. An arbitrary vector $\mathbf{n}$ with integer components n_i that do not satisfy the resonance condition is a non-resonant Fermi vector. For every resonant Fermi vector, Equation (1) defines a resonance surface in action space. From the zero order Hamiltonian, itself a function of the actions only, $H_0 = H_0(\mathbf{I})$, a constant energy (CE) surface is defined. The intersection of the $\mathbf{m}_r$th resonance surface with the CE surface determines the location of the $\mathbf{m}_r$th resonance in action space for a particular energy $H_0 = E$. For any choice of energy E, an infinite number of resonance surfaces intersect the CE surface and form the so called Arnold web. This well-known construction will be rederived and extended as part of our introductory technical remarks in Section 2.

In the presence of coupling, energy transfer occurs between resonant zero order modes. This resonant energy transfer is a principal mechanism for intramolecular vibrational energy redistribution (IVR). Using well-known standard perturbative techniques [2–5], the motion of the system under the influence of a single resonance can be described by a pendular Hamiltonian. This description is obtained by first determining the Fourier series expansion of the coupling Hamiltonian, $H_1(\mathbf{I},\theta) = \Sigma A_{\mathbf{n}}(\mathbf{I})\cos\{\mathbf{n} \bullet \theta\}$, where in the absence of coupling, $\theta_i(t) = \omega_i t + \theta_{io}$. For an $\omega(\mathbf{I})$ that approximately satisfies the resonance condition in Equation 1, the $\mathbf{m}_r$th component in the Fourier expansion of $H_1(\mathbf{I},\theta)$ is a slowly varying function of time since $d\{\mathbf{m}_r \bullet \theta(\mathbf{I})\}/dt \approx 0$. It can therefore be isolated from the other terms in the expansion of H_1 that are rapidly varying since their corresponding Fermi vectors do not satisfy the resonance condition. The contribution of these rapidly varying terms is then assumed to "average out", thereby reducing the infinite Fourier series of H_1 to the single, slowly varying, resonant term, $H_1(\mathbf{I},\theta) \approx A_{\mathbf{m}_r}(\mathbf{I})\cos\{\mathbf{m}_r \bullet \theta\}$. For this simplified form of H_1 it is possible to make a canonical transformation to a new set of action-angle variables where one of the new actions corresponds to the oscillations of a pendulum, and the remaining actions are constants of the motion. In terms of the zero order modes this pendular oscillation describes a periodic exchange of energy between resonant modes with an amplitude bounded by the so-called "resonance width". This familiar result will be validated here for the lowest order resonance. For higher order resonances there exist sub-resonant coupling terms in the Fourier series of H_1 that correspond to lower order resonances in other regions of phase space. While these terms do indeed vary rapidly with time, their amplitude, for realistic choices of H_1, is exponentially larger than the amplitude of the resonant term [20, 21]. The central point of this paper is that due to their large amplitudes, the sub-resonant terms can play a major, even dominant, role in the dynamics of coupled non-linear oscillators.

The process of averaging employed in the standard perturbative approach [4, 15] neglects the effect of these large amplitude sub-resonant terms [22]. We therefore discuss an alternative approach in Section 3 for the perturbative analysis of coupled non-linear oscillators. This alternative approach is specifically designed to incorporate the effect of sub-resonant terms on the dynamics of coupled anharmonic oscillators, and differs from the standard approach in that no average over fast non-resonant terms is performed. Instead, using a type two generating function, a canonical transformation is constructed that *transforms* to new variables in which the sub-resonances are removed to first order in the coupling.

As discussed in Section 4, the principal effect of these sub-resonant terms near a higher order resonance is to induce large amplitude, high frequency oscillations that are superimposed on the much lower frequency canonical "pendular" behaviour of the zero order variables. The amplitude of these sub-resonant excursions is, for physically realistic values of the coupling, typically much larger than the amplitude of the pendular oscillation. The sub-resonant terms also induce a large amplitude high frequency modulation of the higher order resonance widths [23]. For low values of coupling, found for example between the local modes of H_20 [14], the effect of the sub-resonant coupling can to first order, be transformed away with a canonical transformation using the methods outlined in Section 3. This gives rise to three sets of action-angle variables: $\mathbf{I}$, θ which corresponds to the zero order description: $\mathbf{J}$, Ψ which corresponds to the first order description in which the sub-resonant coupling has been eliminated: and K, ϕ which corresponds to the pendular description. Even for weak coupling, the $\mathbf{I}$, θ variables exhibit the large amplitude, high frequency oscillations caused by the sub-resonant terms. These excursions are eliminated to first order by the transformation to the new action-angle variables $\mathbf{J}$, Ψ. Hence while the motion is still separable, it is separable in new modes which are not the uncoupled zero order modes. Thus it is the $\mathbf{J}$, Ψ variables, and not the zero order $\mathbf{I}$, θ variables, that provide a more physically appropriate approximate description of the system of coupled oscillators. For realistic but larger couplings, as found for example [24–26] in linear C_2H_2, perturbation theory can no longer be applied, and large sub-resonant excursions will be generic for any choice of modes. In this case, discussed in Section 5, the zero order $\mathbf{I}$, θ variables are as good as any other choice in the large coupling limit, and the sub-resonant excursions dominate the dynamics of phase space trajectories.

To illustrate the central ideas concerning sub-resonant coupling, we will first discuss the zero order Hamiltonian $H_0(\mathbf{I})$ and the construction of the Arnold web. Two, three and N Morse oscillators will be considered as specific examples. A realistic coupling Hamiltonian $H_1(\mathbf{I}, \theta)$ quadratic in the momenta will then be introduced and its effects on $H_0(\mathbf{I})$ examined.

2. The Zero Order Hamiltonian: $H_0(\mathbf{I})$

The Hamiltonian of $n = 2, 3, N$ coupled Morse oscillators in internal coordinates and momenta $\mathbf{q}$, $\mathbf{p}$ is given by,

$$H(\mathbf{p}, \mathbf{q}) = H_0(\mathbf{p}, \mathbf{q}) + H_1(\mathbf{p}, \mathbf{q}), \tag{2}$$

where H_0 is separable and H_1 is a coupling term. The zero order Hamiltonian H_0 is given by,

$$H_0(\mathbf{p}, \mathbf{q}) = \sum_{i=1}^{n} \frac{1}{2} g_{ii} p_i^2 + V(q) = \sum_{i=1}^{n} \frac{p_1^2}{2\mu_i} + D_i(1 - \exp\{-\alpha_i q_i\})^2, \tag{3}$$

where the $g_{ii} = 1/\mu_i$ are diagonal elements of the $\mathbf{G}$ matrix and μ_i, D_i and α_i are respectively the reduced mass, Morse dissociation energy and anharmonicity of the ith oscillator. The canonical transformation from the p_i, q_i variables to the angle-action variables I_i, θ_i of the ith Morse oscillator is,

$$\begin{aligned} q_i &= \alpha_i^{-1} \ln \left\{ \frac{1 - \eta_i \cos \theta_1}{\lambda_i^2} \right\} \\ p_i &= \frac{\Omega_i \lambda_i \mu_i}{\alpha_i} \left\{ \frac{\eta_i \sin \theta_i}{1 - \eta_i \cos \theta_i} \right\} \\ \eta_i &= \{1 - \lambda_i^2\}^{1/2} \\ \lambda_i &= 1 - \frac{2I_i}{\kappa_i}, \end{aligned} \tag{4}$$

where Ω_i is the frequency of the ith Morse oscillator in the harmonic limit, and κ_i is twice the maximum bound action. Substitution of Equation 4 into Equation 3 yields the transformed zero order Hamiltonian,

$$H_0(\mathbf{I}) = \sum_{i=1} \Omega_i I_i (1 - I_i / k_i). \tag{5}$$

The surface corresponding to $H_0(\mathbf{I})$ given by Equation 5 for $n = 2$ Morse oscillators is an inverted paraboloid. The maximum height of the paraboloid above the I_1, I_2 plane is the sum of the dissociation energies $E^{\Sigma D}$ and occurs for the values $\mathbf{I} = \{\kappa_1/2, \kappa_2/2\}$. In an uncoupled system, $H_0 = E$ is a first integral of the motion and is therefore conserved. For constant values of E, Equation 5 defines a set of concentric ellipses in the I_1, I_2 plane with principle axes parallel to the I_1, I_2 axes. As shown in Figures 1a and b, the set of these ellipses correspond to contours of the inverted paraboloid, and are the CE surfaces of the $n = 2$ Morse oscillators. As shown in Figure 2, for $n = 3$ the CE surfaces are concentric ellipsoids with a common centre at $\mathbf{I} = \{\kappa_1/2, \kappa_2/2, \kappa_3/2\}$. For $n > 3$ oscillators, the CE surface corresponds to a set of concentric hyper-ellipsoids.

By inverting Equation 5 it is possible to obtain the equation $I_i = I_i(E, I_j, I_k \ldots)$, for the $n-1$ dimensional CE surface in action space. For $n = 2$ and $n = 3$ Morse oscillators, $I_2 = I_2(E, I_1)$ and $I_2 = I_2(E, I_1, I_3)$ are given by,

$$I_2 = \frac{\kappa_2}{2} \pm \frac{\kappa_2}{2}\left\{1 - \frac{4(\Omega_1 I_1^2 - \kappa_1\Omega_1 I_1 + \kappa_1 E)}{\kappa_1\kappa_2\Omega_2}\right\}^{1/2} \Leftarrow n = 2 : n = 3 \Downarrow$$

$$I_2 = \frac{\kappa_2}{2} \pm \frac{\kappa_2}{2} \times$$

$$\left\{1 - \frac{4(\Omega_1\kappa_3 I_1^2 + \Omega_3\kappa_1 I_3^2 - \kappa_3\kappa_1\Omega_1 I_1 - \kappa_1\kappa_3\Omega_3 I_3 + \kappa_1\kappa_3 E)}{\kappa_1\kappa_2\kappa_3\Omega_2}\right\}^{1/2} . \quad (6)$$

Two constraints must be imposed on the CE surface equations in order to limit them to those actions that correspond to physically bound states; (i) I_1, I_2 and I_3 cannot exceed the maximum bound actions $\kappa_1/2$, $\kappa_2/2$, and $\kappa_3/2$ which limits Equation 6 to the negative branch, and (ii) I_1, I_2 and I_3 must be positive. The physically bound portion of the CE surfaces for $n = 2$ Morse oscillators in Figure 1 are drawn with a solid line. For $n = 3$ Morse oscillators the physically bound portion of the CE surfaces are shown in Figure 2 and correspond to truncated concentric ellipsoids. For $n = 2, 3$ Morse oscillators, Equation 5 has no real solutions for energies greater than the sum of the dissociation energies $E^{\Sigma D}$, and therefore no CE surfaces exist above this energy. The point $\mathbf{I} = \{\kappa_i/2, \kappa_j/2, \ldots\}$ corresponding to $E^{\Sigma D}$, and the origin $\mathbf{I} = \{0, 0, \ldots\}$, sit in opposite corners of a volume of action space bounded by the requirements that the actions are positive, and less than the maximum bound state actions $\kappa_i/2$. The volume enclosed corresponds to the physically relevant bound actions of the system and is illustrated by the highlighted square in Figure 1 for $n = 2$ Morse oscillators, and by the highlighted cube in Figure 2 for $n = 3$ Morse oscillators.

The gradient of the CE surfaces is given by,

$$\omega(\mathbf{I}) = \vec{\nabla} H_0(\mathbf{I}) = \sum_{i=1}^{n} \mathbf{e}_i \frac{\partial H_0 \mathbf{I})}{\partial I_i} , \quad (7)$$

where $\mathbf{e}_i$ are unit vectors that span action space. From Hamilton's equations, $\omega_i(\mathbf{I}) = \partial H/\partial I_i$, components of the gradient correspond to the zero order frequencies evaluated at $\mathbf{I}$. Thus, by definition, $\omega(\mathbf{I})$ is everywhere perpendicular to

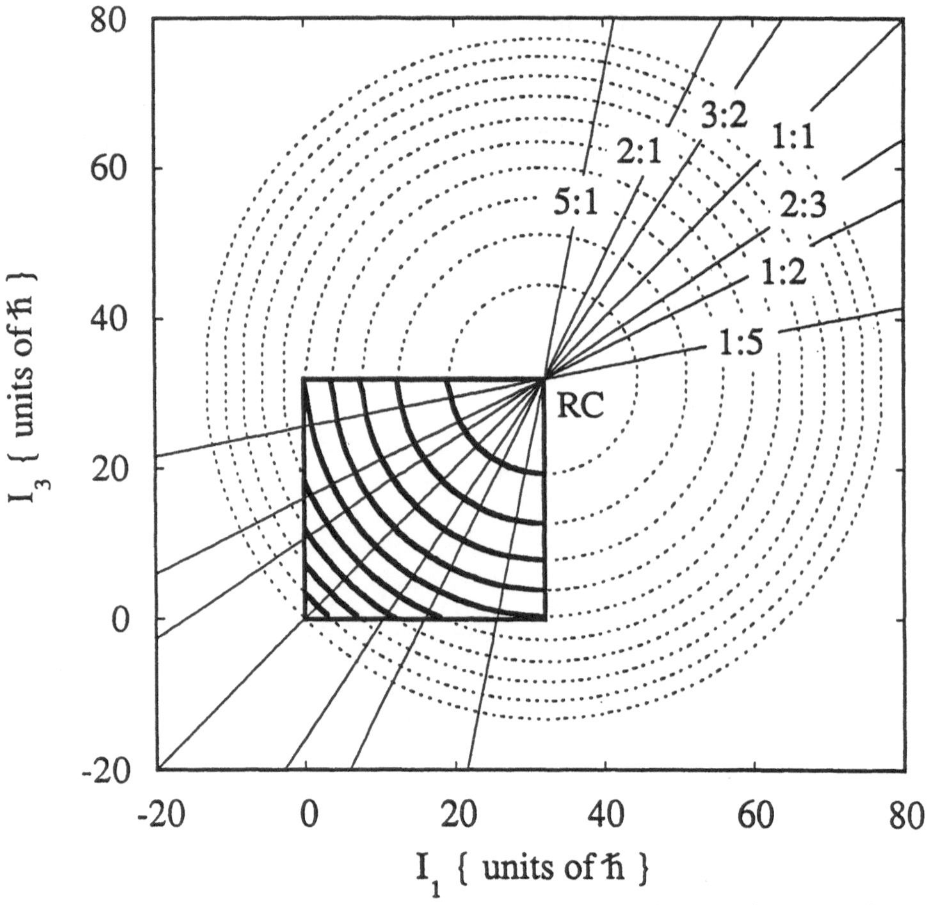

Figure 1a. CE surfaces and resonance surfaces for $n = 2$ Morse oscillators are shown in (a) for two C–H oscillators and (b) for C–H and C–C oscillators. The actions I_i correspond to Morse oscillators parametrized by acetylinic constants listed in the ith column in Table I. In both (a) and (b), the dotted lines correspond to complete CE surfaces as given by Equation 6 for total energies E that vary from 0.0 cm^{-1} (corresponding to the outermost, largest ellipse) to 90540 cm^{-1} (corresponding the innermost, smallest ellipse) in steps of 10060 cm^{-1} . Portions of the CE surfaces inside the highlighted square in (a) and rectangle in (b) are drawn with a thick solid line and correspond to physically realistic bound state actions. The thin solid straight lines are the resonance surfaces determined by Equation 8 and all intersect at the resonance centre RC which for (a) occurs at an energy $E^{\Sigma D} = 98192$ cm^{-1}, and for (b) at $E^{\Sigma D} = 101306$ cm^{-1}.

the CE surface. From the resonance condition in Equation 1, it follows that the resonant Fermi vectors are tangent to the CE surface at the point **I**. By inverting Equation 1, it is possible to obtain an equation for the $\mathbf{m}_r$th resonance surface. For $n = 2, 3$ Morse oscillators, these equations are,

$$I_2 = \left\{\frac{\kappa_2}{2} + \frac{\kappa_2 m_1 \Omega_1}{2m_2\Omega_2}\right\} - \left\{\frac{\kappa_2 m_1 \Omega_1}{\kappa_1 m_2 \Omega_2}\right\} I_1 \Leftarrow n = 2 : n = 3 \Downarrow$$

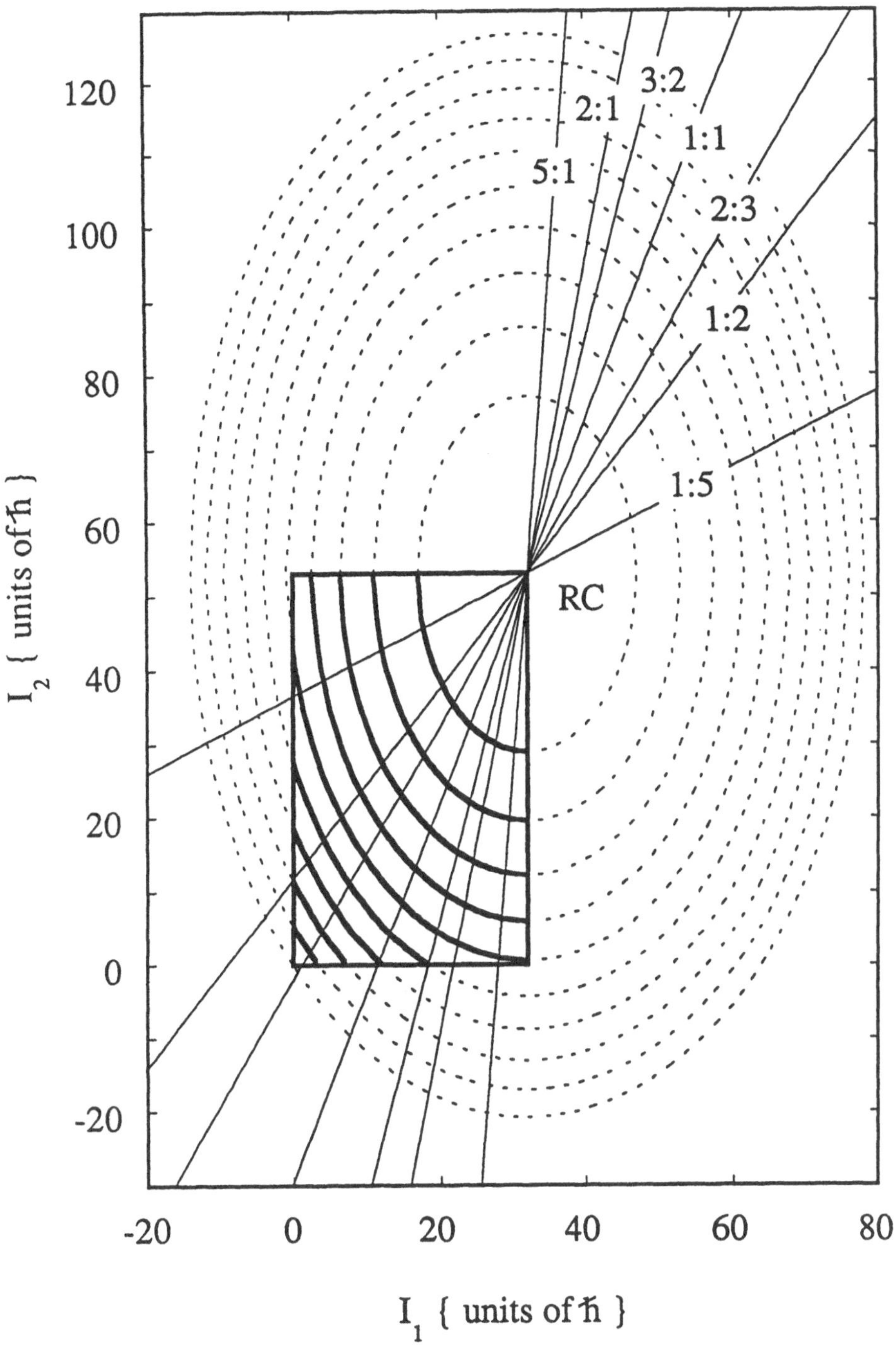

3:2
2:1
1:1
5:1
2:3
1:2
1:5
RC
120
100
80
60
40
20
0
-20
I_2 { units of ħ }
-20
0
20
40
60
80
I_1 { units of ħ }

Figure 1b.

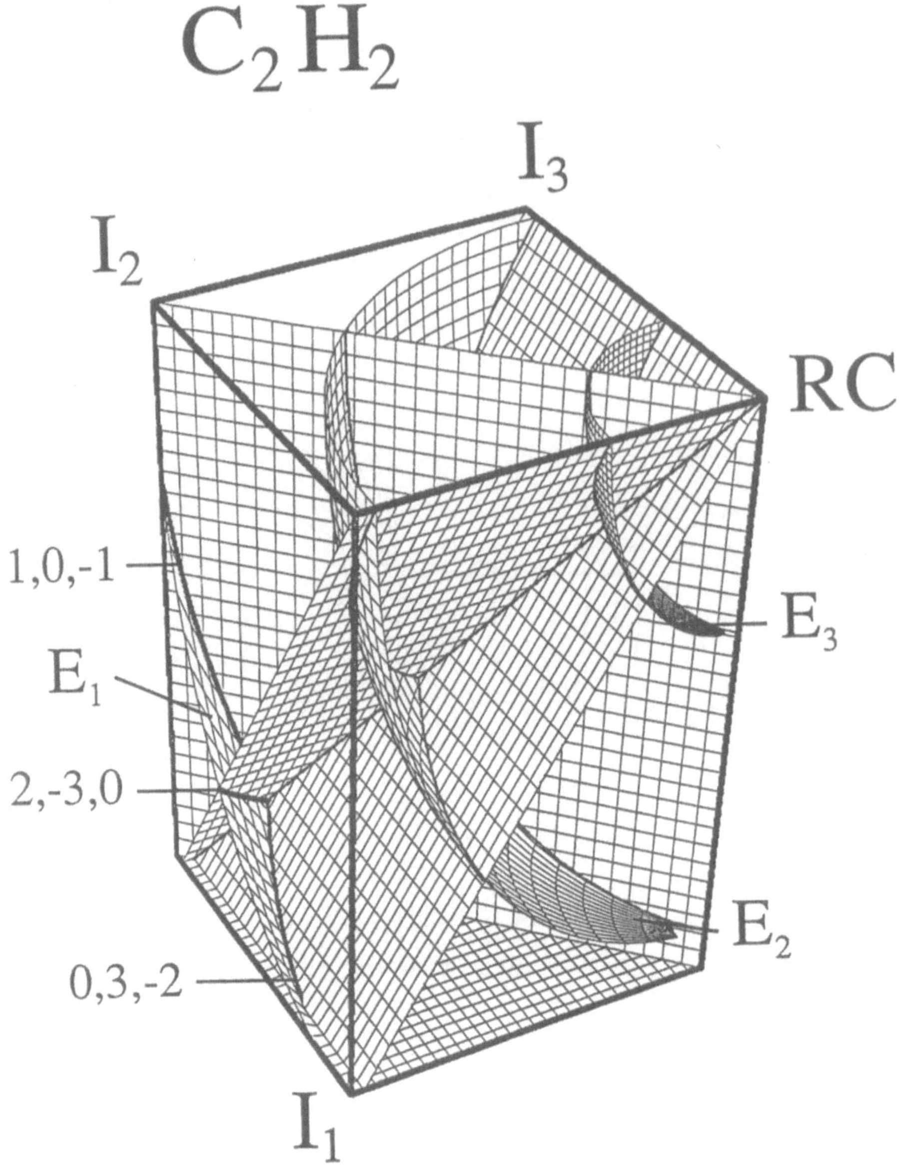

Figure 2. CE surfaces and resonance surfaces for $n = 3$ Morse oscillators. The actions I_i correspond to Morse oscillators parametrized by acetylinic constants listed in the ith column in Table I. Three CE surfaces are illustrated with energies $E_1 = 47740$ cm^{-1}, $E_2 = 106089$ cm^{-1}, and $E_3 = 143220$ cm^{-1}. Each surface corresponds to an ellipsoid given by Equation 6, and is truncated to lie within the highlighted cube of physically realistic bound actions. The $m_r = \{1, 0, \text{-}1\}$, $m_r = \{2, \text{-}3, 0\}$, and $m_r = \{0, 3, \text{-}2\}$ resonance surfaces determined by Equation 8 are also shown and intersect at the resonance centre RC which corresponds to an energy $E^{\Sigma D} = 150407$ cm^{-1}. The resonance surfaces are identified by their intersections with the E_1 CE surface.

$$I_2 = \left\{ \frac{\kappa_2}{2} + \frac{\kappa_2 m_1 \Omega_1}{2 m_2 \Omega_2} + \frac{\kappa_2 m_3 \Omega_3}{2 m_2 \Omega_2} \right\} - \left\{ \frac{\kappa_2 m_1 \Omega_1}{\kappa_1 m_2 \Omega_2} \right\} I_1 - \left\{ \frac{\kappa_2 m_3 \Omega_3}{\kappa_3 m_2 \Omega_2} \right\} I_3, \quad (8)$$

where the $\mathbf{m}_i$ are components of the resonance vector m_r. For $n = 2$ Morse oscillators the resonance surface is a line and $\mathbf{m}_r$ is a two dimensional resonant Fermi vector where by definition the ratio of components $m_1/m_2 < 0$. As a result, all resonance lines will have a positive slope proportional to $|m_1/m_2|$. Possible values of the slope range from limits of zero to infinity. The gradient of the CE surfaces vanishes at the extremum of $H_0(\mathbf{I})$ and the corresponding point $\mathbf{I} = \{\kappa_1/2, \kappa_2/2\}$ defines what we call the resonance centre (RC). The CE surface for this value of $\mathbf{I}$ is a single point at the centre of all the ellipses. At the resonance centre, $\omega(\mathbf{I}) = \mathbf{0}$ and therefore the resonance condition is satisfied by all $\mathbf{m}_r$. As a result, all resonance surfaces intersect at the resonance centre and are tangent to the CE surface. Several resonance surfaces for $n = 2$ Morse oscillators are shown in Figure 1. This generalizes to $n = 3$ Morse oscillators where the planes defined by Equation 8 intersect at the centre of the ellipsoids as shown in Figure 2.

The conclusion that for bilinear Hamiltonians all resonances intersect at the resonance centre while obvious is an important one in its implications for IVR. At or near the resonance centre there will be a significant amount of resonance width overlap even for very weak coupling. As a result it is reasonable to assume that there will exist very facile IVR with rapid energy exchange between modes followed by dissociation. For Hamiltonians of the form given by Equation 5, the energy of the resonance centre is the sum of the individual dissociation energies $E^{\Sigma D}$ and is therefore well above the dissociation energy of any one bond. For a total energy equal to the dissociation energy of a single bond, a much larger coupling is therefore required for the facile mixing of all the modes. The lower the energy, the higher the required coupling strength. Complete mode–mode mixing will therefore not necessarily occur for a total energy on the order of the dissociation energy of a single bond.

For each resonance vector $\mathbf{m}_r$, there will be a resonant surface. The $n - 2$ dimensional intersections in action space of the resonant surfaces and the constant energy surface form the so-called Arnold web. It should be noted that the constant energy surface, the resonant surfaces and the Arnold web can be constructed using H_0 only, with no reference made to the coupling H_1. Examples of the constant energy surface, and Arnold web for C_2H_2 and HOCl are shown in Figures 3a and b where, for three or more dimensions, it is possible for a sub-set of all resonances to intersect at a point in action space other than the resonance centre. These intersections exist for arbitrarily small energy E and, in the presence of coupling, provide a mechanism for exploring phase space.

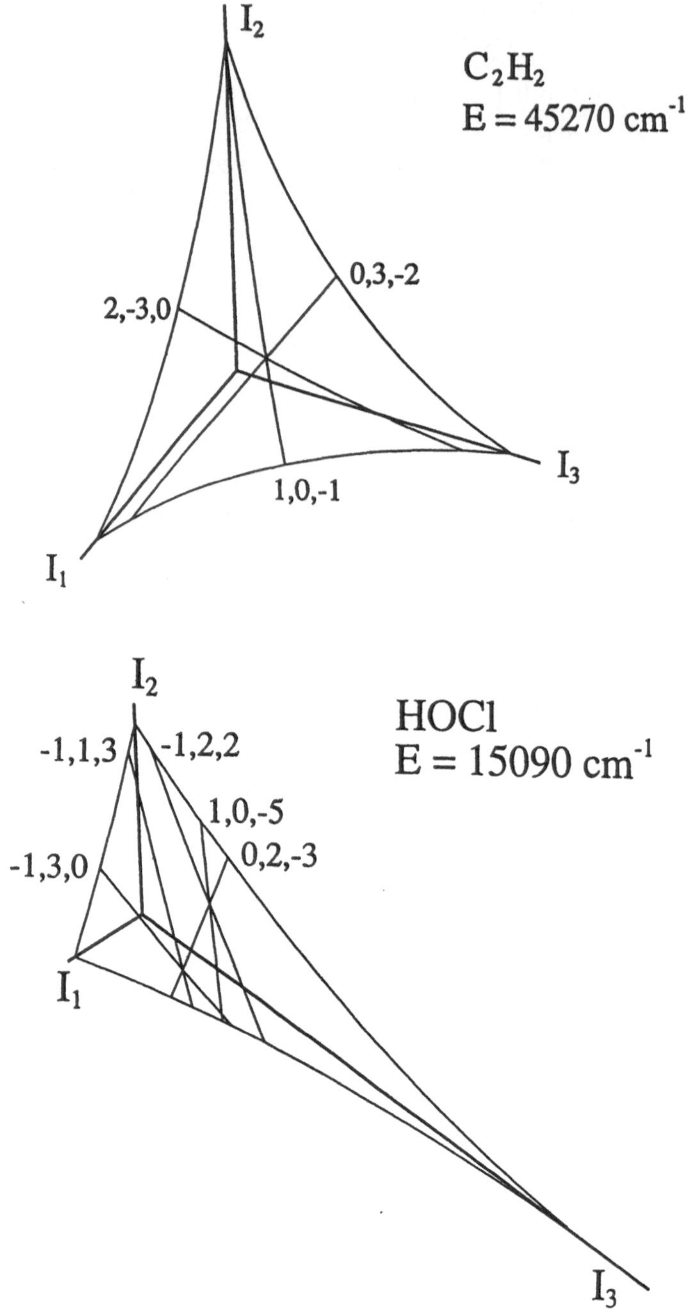

Figure 3. A single CE surface and part of the Arnold web for (a) C_2H_2 and (b) HOCl. In (a) and (b), the actions I_i correspond to Morse oscillators parametrized by the ith column in Tables I and II respectively. Although drawn with a slightly lower energy, the C_2H_2 CE surface and Arnold web in (a) correspond to the E_1 surface and resonance intersections labelled in Figure 2.

3. The Coupling Hamiltonian $H_1 = H_1(\mathbf{I}, \theta)$

To analyse the dynamics of coupled non-linear oscillators, conventional methods typically treat the coupling as a small perturbation and "average" over rapidly oscillating non-resonant terms [4, 15]. This approach works well when treating the lowest order "one to one" type resonance. However, near higher order resonances, in the presence of physically realistic values of the coupling, the process of averaging neglects sub-resonant terms that can exert an exponentially larger effect on the dynamics than the resonant term [22]. We therefore develop an alternative approach for the perturbative analysis of coupled non-linear oscillators designed to incorporate the effect of sub-resonances on the dynamics. We then apply this alternative approach to $n = 2, 3, N$ Morse oscillators coupled by a perturbation H_1, in order to demonstrate the origin and effects of sub-resonant coupling. As a specific example for H_1, we will consider quadratic coupling in the momentum of the form,

$$H_1(\mathbf{p}, \mathbf{q}) = \sum_{i,j>i}^{n} g_{ij} p_i p_j = \sum_{i,k>i}^{n} \varepsilon_{ij} \frac{p_i p_j}{\mu_{ij}}, \tag{9}$$

where the coupling parameters g_{ij} between the ith and jth oscillators are off diagonal elements of the G matrix given by,

$$\begin{aligned} g_{ij} &= \frac{\varepsilon_{ij}}{\mu_{ij}} \\ \varepsilon_{ij} &= \cos(\sigma_{ij}). \end{aligned} \tag{10}$$

The bond angle between, and the mass common to, the ith and jth oscillators are denoted σ_{ij} and μ_{ij} respectively. By fixing the bond angle σ_{ij} at various values between $\pi/2$ to π for a constant value of μ_{ij}, the magnitude of g_{ij} will range from a minimum of zero, corresponding to an uncoupled orthogonal arrangement of the ith and jth oscillators, to a maximum of $-1/\mu_{ij}$ corresponding to a maximally coupled linear configuration. The coupling strength is conveniently parametrized by the unitless quantity ε_{ij} which varies from zero for the uncoupled $\sigma_{ij} = \pi/2$ configuration to -1 for the maximally coupled $\sigma_{ij} = \pi$ configuration. The essential conclusions obtained for quadratic kinetic coupling are expected to be similar for potential coupling.

For small but non-zero coupling, the trajectory of the system will no longer be a fixed point on the CE surface since the variables **I** are not action variables of the coupled system. Instead, there will be both motion along the surface and across the surface. Motion across the surface occurs because the uncoupled energy is not conserved and is a measure of the size of the coupling energy which, for example in linear C_2H_2, can be quite large. To apply the techniques of first-order perturbation theory in the small coupling limit, H_1 is typically expanded in a Fourier series in the angle variables. In the case of quadratic coupling, this is a double Fourier series given by the product of the individual expansions of p_i and

p_j. The origin of sub-resonant coupling is found in the Fourier series expansion of p_i given by Equation 4, where,

$$
\begin{aligned}
p_i &= \sum_{n_i=1}^{\infty} b_{n_i}(I_i)\sin\{n_i\theta_i\} \\
b_{\eta_i} &= \frac{2\Omega_i\lambda_i\mu_i}{\alpha_i}\left\{\frac{I_i}{\kappa_i - I_i}\right\}^{n_i/2}
\end{aligned}
\qquad (11)
$$

and the n_i are restricted to the set of positive integers. The Fourier coefficients b_{η_i} are obtained by contour integration around the unit circle in the complex plane that encompasses a single first-order pole on the real axis at $r^- = 1/\eta_i - (1/\eta_i^2 - 1)^{1/2}$. Since we are considering bound states, the quantity $I_i/(\kappa_i - I_i)$ is less than one. It therefore follows that,

$$n_i < n_i' \Rightarrow b_{ni} > b_{n'i}. \qquad (12)$$

where $b_{\eta_i'}$ is exponentially smaller than b_{η_i}. This is illustrated in Figure 4 where $b_{\eta_i}\alpha_i/\Omega_i\mu_i$ is plotted as a function of $2I_i/\kappa_i$ for $n = 1$ to 10. It is also evident from Figure 4 that the ratio $b_{\eta_i}/b_{\eta_i'}$ is much larger for smaller values of $2I_i/\kappa_i$. Figure 4 can also be used to estimate the number of terms in the Fourier series of p_i that are required for a reasonable fit to the exact expression given by Equation 4. For example, for $2I_i/\kappa_i = 0.1$, approximately three terms will be sufficient. This is indeed the case as shown in Figure 5, where a comparison is made between the exact solution of Equation 4, and the Fourier series expansion in Equation 11 truncated at $n = 3$. From the plots it is clear that for $2\ I_i/\kappa_i = 0.1$, three terms in Equation 11 are sufficient, for $2I_i/\kappa_i = 0.5$ more than three terms will be required, and for $2I_i/\kappa_i = 0.9$, three terms just barely reflect the major features of the trajectory. The main point here is that for all values of I_i, the largest contribution to the Fourier series expansion of p_i always comes from the $n = 1$ term and decreases exponentially with increasing n. We will show that this is the origin of sub-resonant coupling.

As a quick check of our calculations, it is useful to expand Equations 4 and 11 using a Taylor's series, in the harmonic limit of small I_i. For a harmonic oscillator,

$$p_i^{\text{harmonic}} = \{2\Omega_i I_i \mu_i\}^{1/2}\sin\theta_i. \qquad (13)$$

A Taylor's series expansion around $I_i = 0$ of the exact expression for the momentum given in Equation 4 yields,

$$p_i^{\text{Morse}} = \{2\Omega_i I_i \mu_i\}^{1/2}\sin\theta_i + \vartheta(I_i). \qquad (14)$$

A similar expansion around $I_i = 0$ of the Fourier coefficients b_{η_i} in Equation 11 gives,

$$b_{\eta_i} = \left\{ \begin{array}{ll} \{2\Omega_i I_i\mu_i\}^{1/2} + \vartheta(I_i) & n = 1 \\ \vartheta(I_i) & n \geq 2 \end{array} \right\}, \qquad (15)$$

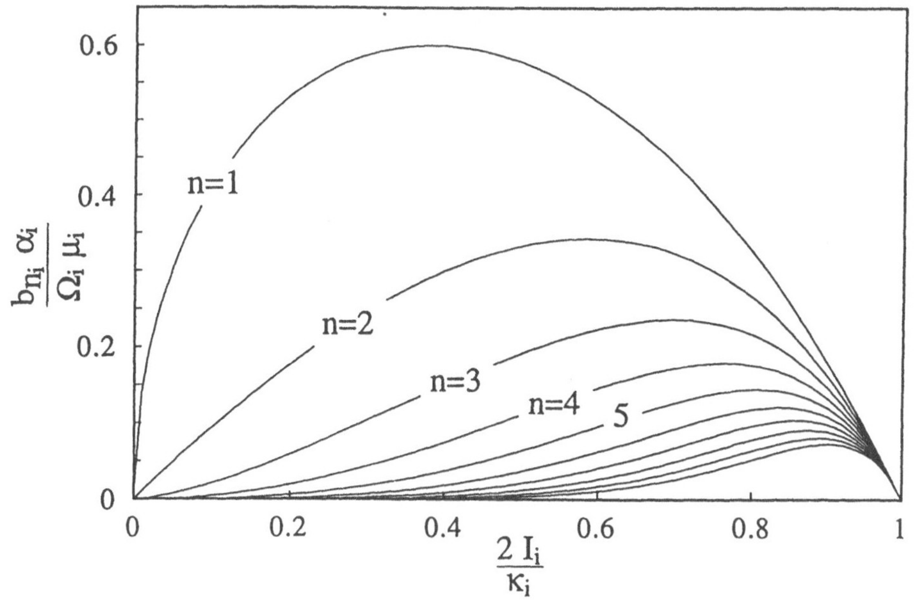

Figure 4. A plot of the unitless quantity $b_{ni}\alpha_i/(\Omega_i\mu_i)$ as a function of the unitless parameter $2I_i/\kappa_i$. The functional dependence of the Fourier coefficients b_{ni} on the action I_i given by Equation 11 is illustrated, and shows that for all values of bound action, $b_{n(i)} > b_{n(i+1)}$.

which when substituted into the Fourier series yields,

$$p_i = \sum_{n=1}^{\infty} b_{\eta_i} \sin\{n\theta_i\} = \{2\Omega_i I_i \mu_i\}^{1/2} \sin\theta_i + \vartheta(I_i). \tag{16}$$

The Fourier series expansion of the coupling Hamiltonian H_1 is obtained from the product of the expansions of p_i and p_j where,

$$\begin{aligned} H_1(\mathbf{I},\theta) &= \sum_{i,j>i}^{n} \varepsilon_{ij} \frac{p_i(\mathbf{I},\theta)p_j(\mathbf{I},\theta)}{\mu_{ij}} \\ &= \sum_{i,j>i} \varepsilon_{ij} \frac{\beta_{ij}\lambda_i\lambda_j\eta_i\eta_j \sin(\theta_i)\sin(\theta_j)}{\{1-\eta_i\cos(\theta_i)\}\{1-\eta_j\cos(\theta_j)\}} \\ &= \sum_{i,j>i} \varepsilon_{ij} \sum_{\mathbf{n}_{ij}} A_{\mathbf{n}_{ij}} \cos(\mathbf{n}_{ij} \bullet \theta) \end{aligned} \tag{17}$$

and the Fourier coefficients A_{nij} are,

$$\begin{aligned} A_{\mathbf{n}_{ij}} &= \frac{\zeta_{\mathbf{n}_{ij}}}{2\mu_{ij}} \frac{-n_j}{|n_j|} b_{n_i} b_{n_j} = 2\zeta_{\mathbf{n}_{ij}} \beta_{ij}\lambda_i\lambda_j \frac{-n_j}{|n_j|} \Gamma_i^{n_i/2}\Gamma_j^{|n_j|/2} \\ \beta_{ij} &= \frac{\Omega_i\Omega_j\mu_i\mu_j}{\mu_{ij}\alpha_i\alpha_j} \end{aligned}$$

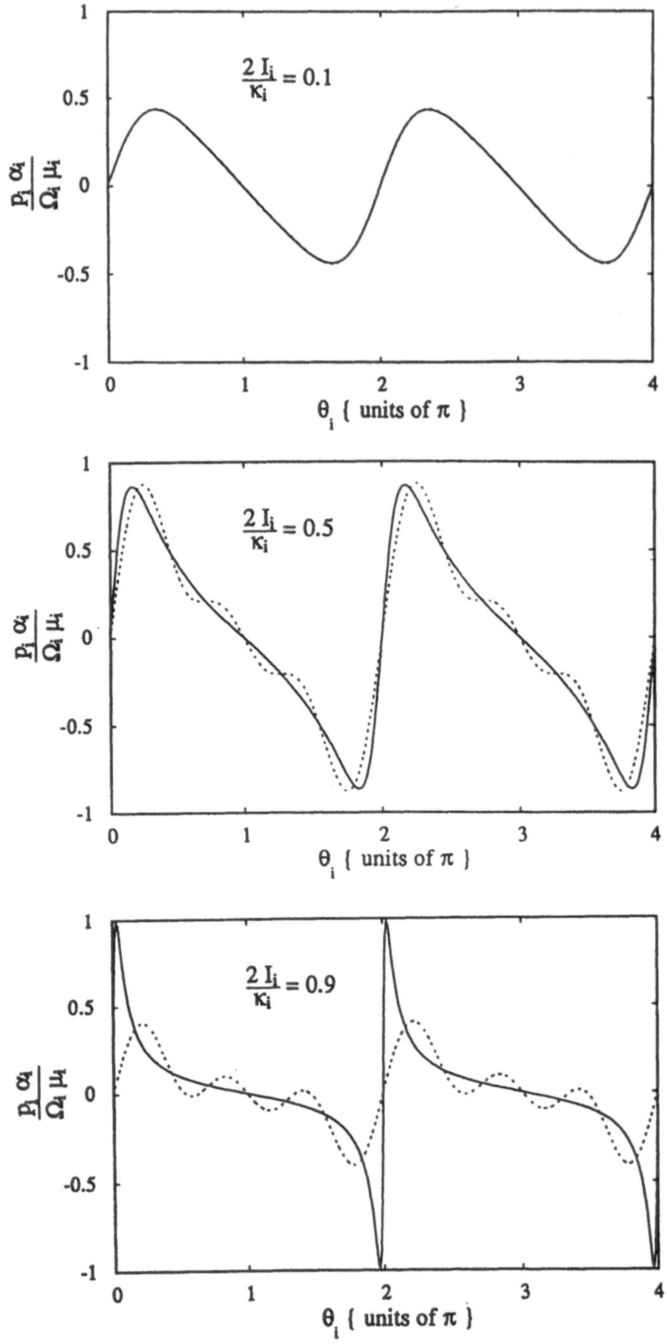

Figure 5. A plot of the unitless quantity $p_i\alpha_i/(\Omega_i\mu_i)$ as a function of angle θ_i for (a) $2I_i/\kappa_i = 0.1$, (b) $2I_i/\kappa_i = 0.5$, and (c) $2I_i/\kappa_i = 0.9$. The solid line is the exact solution given by Equation 4 and is compared with the dashed line given by the Fourier series expansion in Equation 11 truncated at three terms.

$$\Gamma_i = \frac{I_i}{\kappa_i - I_i}. \tag{18}$$

The coefficients $\zeta_{\mathbf{n}_{ij}}$ in Equation 18 are artificial and included for use in later sections as a mechanism to selectively set to zero various $A_{\mathbf{n}_{ij}}$ in order to determine their relative contributions to the dynamics of the coupled oscillators where,

$$\zeta_{\mathbf{n}_{ij}} = \begin{cases} 0 \Rightarrow A_{\mathbf{n}_{ij}} = 0 & \{\text{coefficient turned off}\} \\ 1 \Rightarrow A_{\mathbf{n}_{ij}} = 2\beta_{ij}\lambda_i\lambda_j \frac{-n_j}{|n_j|} \Gamma_i^{n_i/2}\Gamma_j^{|n_j|/2} & \{\text{coefficient turned on}\} \end{cases} \tag{19}$$

The dimension n of the vectors labelled $\mathbf{n}_{ij}$ in Equations 17, 18, and 19 is equal to the number of Morse oscillators where $n-2$ components are zero and the remaining two ith and jth components are the integers n_i and n_j where $\mathbf{n}_{ij} = \{\ldots, n_i, \ldots, n_j, \ldots\}$. The components of $\mathbf{n}_{ij}$ are used as indices for the double summation over $\mathbf{n}_{ij}$, as well as to evaluate $A_{\mathbf{n}_{ij}}$ and $\mathbf{n}_{ij} \bullet \theta$. In obtaining Equation 17, a trigonometric substitution is employed which introduces the factor of $-n_j/\{2|n_j|\}$ in Equation 18, and extends the sum over n_j to negative as well as positive integers. It therefore follows that the vectors $\mathbf{n}_{ij}$ can be identified as both resonant and non-resonant Fermi vectors. This identification is made clearer by considering the time derivative of the argument of the cosine term in Equation 17,

$$\frac{d}{dt}\{n_{ij} \bullet \theta\} = n_{ij} \bullet \omega, \tag{20}$$

which for $\mathbf{n}_{ij} = \mathbf{m}_r$ evaluates to zero and is precisely the resonance condition given by Equation 1.

The coefficients $A_{\mathbf{n}_{ij}}$ in Equations 17 and 18 are proportional to the product of b_{ni} and b_{nj} and will therefore exhibit a generalization of Equation 12. Hence for Fermi vectors $\mathbf{n}_{ij} = \{\ldots n_i, \ldots, n_j, \ldots\}$ and $\mathbf{m}_{ij} = \{\ldots, m_i, \ldots, m_j, \ldots\}$ for which the modulus $|\mathbf{n}_{ij}| = \sqrt{\{n_i^2 + n_j^2\}}$ is less than the modulus $|\mathbf{m}_{ij}| = \sqrt{\{m_i^2 + m_j^2\}}$, the coefficient $A_{\mathbf{n}_{ij}}$ will be exponentially larger than $A_{\mathbf{m}_{ij}}$. If $\mathbf{m}_r$ satisfies the resonance condition in Equation 1, then Fermi vectors $\mathbf{n}_{ij}$ for which $|\mathbf{n}_{ij}| < |\mathbf{m}_r|$ are called sub-resonant Fermi vectors and the corresponding $A_{\mathbf{n}_{ij}}$ are referred to as the sub-resonant coefficients. In a similar fashion, Fermi vectors $\mathbf{m}_{ij}$ for which $|\mathbf{m}_{ij}| > |\mathbf{m}_r|$ are called super-resonant and the corresponding $A_{\mathbf{m}_{ij}}$ are referred to as the super-resonant coefficients. The standard perturbative approach isolates the slowly oscillating resonant term $A_{\mathbf{m}_r}$ in the Fourier series of H_1 given by Equation 17, by assuming that the rapidly oscillating non-resonant terms "average out". The infinite Fourier series of H_1 is thereby reduced to the single slowly oscillating resonant term, $H_1(\mathbf{I}, \theta) \approx A_{\mathbf{m}_r}(\mathbf{I}) \cos\{\mathbf{m}_r \bullet \theta\}$. This process of averaging is equivalent to neglecting both the super-resonant *and* sub-resonant terms in Equation 17. Since the super-resonant coefficients $A_{\mathbf{m}_{ij}}$ in Equation 17 are exponentially smaller than the resonant coefficient $A_{\mathbf{m}_r}$ it is legitimate, and in agreement with the established standard procedure [4, 12–15, 27], to approximate to H_1 by neglecting the super-resonant terms $A_{\mathbf{m}_{ij}} < A_{\mathbf{m}_r}$. However, with the exception of the lowest order resonances $\mathbf{m}_r = \{\ldots, 1, \ldots, -1, \ldots\}$ for which no sub-resonant Fermi vectors exist, it is *not* legitimate to approximate H_1 by neglecting the sub-resonant terms $\mathrm{A}_{\mathbf{n}_{ij}}$ because the sub-resonant coefficients $A_{\mathbf{n}_{ij}}$ are exponentially *larger* than the resonant term $A_{\mathbf{m}_r}$.

An alternative approach is to isolate the $\mathbf{m}_r = \{\ldots, r, \ldots, s, \ldots\}$ resonant term in Equation 17 using a canonical transformation from the angle-action variables $\mathbf{I}$, θ of the Morse oscillator to a new set of canonical variables $\mathbf{J}$, Ψ. The transformation is local in that it is restricted to regions on the $\mathbf{I}$, θ CE surface near the $\mathbf{m}_r$th resonance. Formally this transformation is given by a type two generating function,

$$S(\mathbf{J}, \theta) = \mathbf{J} \bullet \theta + S_1(\mathbf{J}, \theta), \tag{21}$$

where S_1 is expanded in a Fourier series,

$$S_1 = \sum_{i,j>i} \varepsilon_{ij} \sum_{\mathbf{n}_{ij} \neq p\mathbf{m}_r} \frac{B_{\mathbf{n}_{ij}} \mathbf{n}_{ij} \bullet \omega}{} \sin\{\mathbf{n}_{ij} \bullet \theta\}. \tag{22}$$

To avoid singularities introduced by the resonance condition, the summation in Equation 22 excludes the set of resonant Fermi vectors $p\ \mathbf{m}_r = \{\ldots, pr, \ldots, ps, \ldots\}$, where p consists of all the positive integers. If the components s and r of the resonant Fermi vector $\mathbf{m}_r$ are relatively prime then $\mathbf{m}_r$ labels a *fundamental* Fermi resonance. Harmonic Fermi resonances are labelled by the vectors $p\mathbf{m}_r$ for $p > 1$. Typically, $A_{\mathbf{n}_{ij}}$ corresponding to harmonic Fermi resonances are negligible. Using Equations 21 and 22, the old actions $\mathbf{I} = \mathbf{I}(\mathbf{J}, \theta)$ are given by,

$$\mathbf{I} = \frac{\partial S}{\partial \theta} = \mathbf{J} + \frac{\partial S_1}{\partial \theta} = \mathbf{J} + \sum_{i,j>i} \varepsilon_{ij} \sum_{\mathbf{n}_{ij} \neq p\mathbf{m}_r} \mathbf{n}_{ij} \frac{B_{\mathbf{n}_{ij}}}{\mathbf{n}_{ij} \bullet \omega} \cos\{\mathbf{n}_{ij} \bullet \theta\}. \tag{23}$$

Thus the difference between the old and new actions is,

$$\mathbf{I} - \mathbf{J} = \sum_{i,j>i} \varepsilon_{ij} \sum_{\mathbf{n}_{ij} \neq p\mathbf{m}_r} \mathbf{n}_{ij} \frac{B_{\mathbf{n}_{ij}}}{\mathbf{n}_{ij} \bullet \omega} \cos\{\mathbf{n}_{ij} \bullet \theta\}. \tag{24}$$

Expanding H_0 given by Equation 5, in a Taylor's series around J, and using the result in Equation 24 gives,

$$\begin{aligned} H_0(\mathbf{I}) &= H_0(\mathbf{J}) + \left.\frac{\partial H_0}{\partial \mathbf{I}}\right|_{\mathbf{I}=\mathbf{J}} (\mathbf{I} - \mathbf{J}) + \ldots \\ &= H_0(\mathbf{J}) + \sum_{i,j>i} \varepsilon_{ij} \sum_{\mathbf{n}_{ij} \neq p\mathbf{m}_r} B_{\mathbf{n}_{ij}} \cos\{\mathbf{n}_{ij} \bullet \theta\} + \ldots \end{aligned} \tag{25}$$

where to zero order in the coupling, $\omega(\mathbf{I}) = \omega(\mathbf{J})$.

Combining the approximate H_0 given by Equation 25 and the exact Fourier series expansion of H_1 given by Equation 17, yields an approximate total Hamiltonian $H_t = H_0 + H_1$,

$$H_t = H_0(\mathbf{J} + \sum_{i,j>i} \varepsilon_{ij} \sum_{\mathbf{n}_{ij} \neq p\mathbf{m}_r} B_{\mathbf{n}_{ij}} \cos\{\mathbf{n}_{ij} \bullet \theta\} + \ldots$$

$$+\sum_{i,j>i}\varepsilon_{ij}\sum_{\mathbf{n}_{ij}}A_{\mathbf{n}_{ij}}\cos\{\mathbf{n}_{ij}\bullet\theta\}. \tag{26}$$

By identifying the Fourier coefficients of S_1 in Equation 22 to be,

$$B_{\mathbf{n}_{ij}} = -A_{\mathbf{n}_{ij}}, \tag{27}$$

the Hamiltonian in Equation 26 becomes,

$$\begin{aligned} H_t(\mathbf{J},\Psi) &= H_0(\mathbf{J}) + \sum_{i,j>i}\varepsilon_{ij}\sum_{p\mathbf{m}_r}A_{p\mathbf{m}_{r_{ij}}}\cos(p\mathbf{m}_{r_{ij}}\bullet\Psi) \\ &= H_0(\mathbf{J})\varepsilon_{ij}\sum_{p\mathbf{m}_r}A_{\mathbf{m}_r}\cos(p\mathbf{m}_r\bullet\Psi) \\ &= H_0(\mathbf{J}) + \varepsilon_{ij}A_{\mathbf{m}_r}\cos(\mathbf{m}_r\bullet\Psi) \end{aligned} \tag{28}$$

where to zero order in the coupling,

$$\cos(p\mathbf{m}_{r_{ij}}\bullet\theta) = \cos(p\mathbf{m}_{r_{ij}}\bullet\Psi), \quad \text{and} \quad A_{p\mathbf{m}_{r_{ij}}}(\mathbf{I}) = A_{p\mathbf{m}_{r_{ij}}}(J). \tag{29}$$

The summation over $i, j > i$ in Equation 28 includes the possibility that resonances between different pairs of modes can, in systems of three or more oscillators, simultaneously co-exist [17]. This corresponds to the situation shown in Figure 3 where for three or more oscillators resonances between different pairs of modes intersect. The subscripts i, j used to label the $\mathbf{m}_{r_{ij}}$ in the first line of Equation 28 and in Equation 29, are used to distinguish between the resonant Fermi vectors that correspond to simultaneously extant resonances. For a single isolated resonance the summation over i, j may be dropped as in the second line of Equation 28, and the final form of the transformed Hamiltonian is given in the last line of Equation 28 where coefficients labelled with harmonic Fermi vectors are neglected. Using Equations 21, 22, 27 and 29, the transformations $\mathbf{I} = \mathbf{I}(\mathbf{J},\Psi)$ and $\theta = \theta(\mathbf{J},\Psi)$ are given by,

$$\begin{aligned} \mathbf{I} &= \frac{\partial S}{\partial\theta} = \mathbf{J} + \frac{\partial S_1}{\partial\theta} \\ &= \mathbf{J} - \sum_{i,j>i}\varepsilon_{ij}\sum_{\mathbf{n}_{ij}\neq p\mathbf{m}_r}\mathbf{n}_{ij}\frac{A_{\mathbf{n}_{ij}}}{\mathbf{n}_{ij}\bullet\omega}\cos\{\mathbf{n}_{ij}\bullet\Psi\} \\ \theta &= \Psi - \frac{\partial S_1}{\partial\mathbf{I}} \\ &= \Psi + \sum_{i,j>i}\varepsilon_{ij}\sum_{\mathbf{n}_{ij}\neq p\mathbf{m}_r}\left\{\frac{\partial(A_{\mathbf{n}_{ij}}/\mathbf{n}_{ij}\bullet\omega)}{\partial\mathbf{J}}\right\}\sin\{\mathbf{n}_{ij}\bullet\Psi\} \end{aligned} \tag{30}$$

and to first order in the coupling, the inverse transformations are,

$$\begin{aligned} \mathbf{J} &= \mathbf{I} - \frac{\partial S}{\partial\theta} \\ &= \mathbf{I} + \sum_{i,j>i}\varepsilon_{ij}\sum_{\mathbf{n}_{ij}\neq p\mathbf{m}_r}\mathbf{n}_{ij}\frac{A_{\mathbf{n}_{ij}}}{\mathbf{n}_{ij}\bullet\omega}\cos\{\mathbf{n}_{ij}\bullet\theta\} \end{aligned}$$

$$\begin{aligned}
\Psi &= \frac{\partial S}{\partial \mathbf{J}} = \theta = \frac{\partial S_1}{\partial \mathbf{I}} \\
&= \theta - \sum_{i,j>i} \varepsilon_{ij} \sum_{\mathbf{n}_{ij} \neq p\mathbf{m}_r} \left\{ \frac{\partial (A_{\mathbf{n}_{ij}}/\mathbf{n}_{ij} \bullet \omega)}{\partial \mathbf{I}} \right\} \sin\{\mathbf{n}_{ij} \bullet \theta\}. \qquad (31)
\end{aligned}$$

The central point of this paper is that the Fourier coefficients $A_{\mathbf{n}_{ij}}$ of H_1 in Equation 17, labelled with non-resonant Fermi vectors are *transformed* away. Thus, when discussing the dynamics of a system governed by the Hamiltonian in Equation 2, it is crucial to remember that Equations 30 and 31 are in general not the unit transformation and therefore the dynamics of the variables $\mathbf{I}$, θ and $\mathbf{J}$, Ψ can in general be quite different. The transformations in Equations 30 and 31 exclude coefficients labelled by the fundamental Fermi vector $\mathbf{m}_r$ and its harmonics $p\mathbf{m}_r$. It should be stressed that this in no way excludes terms labelled with either sub-resonant or super-resonant Fermi vectors $\mathbf{n}_{ij}$. It is easy to show that for a given resonant Fermi vector $\mathbf{m}_r = \{\ldots, m_i, \ldots, m_j, \ldots\}$ there is an equivalence class of approximately resonant Fermi vectors $\mathbf{n}_e$ such that $\mathbf{n}_e \bullet \omega$ is a minimum given by the constant value $\pm 1/m_j$. In general, both sub-resonant and super-resonant Fermi vectors are elements of this equivalence class. However, since the $A_{\mathbf{n}_e}$ are exponentially smaller for super-resonant $\mathbf{n}_e$, and for elements of the equivalence class $\mathbf{n}_e \bullet \omega$ is a minimum given by the constant value $\pm 1/m_j$, all super-resonant terms $A_{\mathbf{n}_e/\mathbf{n}_e} \bullet \omega = \pm m_j A_{\mathbf{n}_e}$ are exponentially smaller than the corresponding sub-resonant terms. Of course, this does not guarantee that the series in Equations 22, 30, and 31 converge. It does however indicate that for the purpose of obtaining asymptotic solutions, it is reasonable and convenient to neglect the super-resonant terms. For example, near an isolated lowest order $\mathbf{m}_r = \{\ldots, 1, \ldots, -1, \ldots\}$ resonance, there are no sub-resonant Fermi vectors and therefore by neglecting the terms $A_\mathbf{n}/\mathbf{n}\bullet\omega$ labelled with super-resonant Fermi vectors, Equations 30 and 31 are indeed approximately the unit transformation [15, 27]. This, however, is not the case for higher order resonances. For example, in an expansion about an isolated $\mathbf{m}_r = \{\ldots, 2, \ldots, -3, \ldots\}$ Fermi resonance, the transformations in Equations 30 and 31 will include terms labelled with the sub-resonant vectors $\mathbf{n}_{ij} = \{\ldots, 1, \ldots, -2, \ldots\}$, $\mathbf{n}_{ij} = \{\ldots, 1, \ldots, -1, \ldots\}$, $\mathbf{n}_{ij} = \{\ldots, 1, \ldots, 1, \ldots\}$ and $\mathbf{n}_{ij} = \{\ldots, 2, \ldots, -1, \ldots\}$. For all of these terms, $A_{\mathbf{n}_{ij}} > A_{\mathbf{m}_r}$ and each one will make a significant contribution to the transformation. In fact, with the single exception of the lowest order $\mathbf{m}_r = \{\ldots, 1, \ldots, -1, \ldots\}$ resonances, expansions about all higher-order resonances will include significant sub-resonant contributions in the transformation given by Equations 30 and 31. As the order of the Fermi resonance increases, the number of sub-resonant Fermi vectors increases, and therefore the total sub-resonant contribution also increases. It should be noted that the above discussion is limited to initial actions that correspond to regions near isolated resonances that are far from the resonance intersections on the CE surface, shown for example in Figure 3. Near these regions of resonance intersection, there will be a resonance vector $\mathbf{m}_{r_{ij}}$ for each intersecting resonance between the ith and jth oscillators. The corresponding set of sub-resonant contributions will be included in the transformations in Equations 30 and 31 by an additional summation over i and j. The transformed Hamiltonian will be given by the top line in Equation 28, which for two intersecting resonances corresponds to an anisotropic Foucault pendulum when the $A_{p\mathbf{m}rij}$ labelled with harmonic Fermi vectors are ignored.

To further examine the nature of the transformation in Equations 30 and 31 it is useful to first complete the transformation of the intermediate Hamiltonian in the last line of Equation 28 to the standard pendular Hamiltonian. The transformation to pendular variables $\mathbf{K}$, ϕ is canonical and near the $\mathbf{m}_r$ resonance is given by [17],

$$\begin{aligned}
\mathbf{K} &= (\mathbf{m}^{-1})^{\top}[\mathbf{J} - \mathbf{J}_r] \\
\phi &= \mathbf{m}\Psi \\
\mathbf{J} &= \mathbf{m}^{\top}\mathbf{K} + \vec{\mathbf{J}}_r \\
\psi &= \mathbf{m}^{-1}\phi
\end{aligned} \tag{32}$$

where J_r labels the resonant tori of the uncoupled Morse oscillators in Equation 28. For quadratic coupling, the Fermi vectors $\mathbf{n}_{ij}$ used to label the Fourier series expansion have only two non-zero components n_i and n_j. Therefore, near an isolated resonance $\mathbf{m}_{rij} = \{\ldots, r, \ldots, s, \ldots\}$, the Hamiltonian in Equation 28 will be cyclic in all angles except for Ψ_i, and Ψ_j. For n coupled Morse oscillators, the original Hamiltonian $H = H(\mathbf{I}, \theta)$ is thereby transformed to $H = H(\mathbf{J}, \Psi)$ with $n - 2$ constants of the motion, and a coupling term between the ith and jth variables. An additional constant of the motion is obtained by transforming from $\mathbf{J}$, Ψ to $\mathbf{K}$, ϕ using Equation 32 where the transformation matrix $\mathbf{m}$ can be chosen as,

$$\mathbf{m} = \left\{ \begin{array}{cc} r & s \\ 0 & 1 \end{array} \right\} \begin{array}{l} \leftarrow \text{Resonance vector} \\ \leftarrow \text{Arbitrary vector} \end{array} . \tag{33}$$

The canonical transformation in Equation 32 is then given by,

$$\begin{aligned}
\phi_1 &= r\Psi_i + s\Psi_j \qquad \Psi_i = \frac{\phi_1}{r} - \frac{s}{r}\phi_2 \\
\phi_2\phi_2 &= \psi_j \qquad \Psi_j = \phi_2 \\
K_1 &= \frac{1}{r}(J_i - J_{ir}) \qquad J_i = rK_i + J_{1r} \\
K_2 &= -\frac{s}{r}(J_i - j_{ir}) + (J_j - J_{jr}) \qquad J_j = K_2 + sK_1 + J_{2r}..
\end{aligned} \tag{34}$$

Substituting transformations on the right hand side of Equation 34 into Equation 28 and keeping terms to second order in K_1 and to first order in K_2, yields the

TABLE I. Uncoupled C_2H_2 Morse oscillator parameters [17]. Labels 1, 2, and 3 correspond respectively to the H–C stretch, C–C stretch, and C–H stretch.

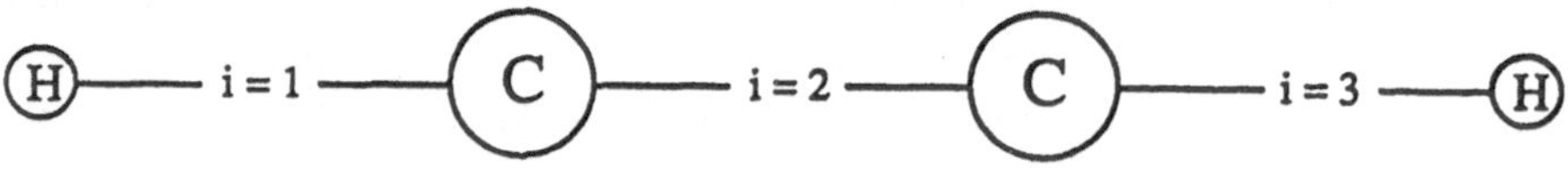

$\mu_i =$	0.153 (x 10^{-23}g)	0.996 (x 10^{-23}g)	0.153 (x 10^{-23}g)
$\alpha_i =$	1.621 (x 10^{8}cm^{-1})	2.572 (x 10^{8}cm^{-1})	1.621 (x 10^{8}cm^{-1})
$D_i =$	9.761 (x 10^{-12}erg)	10.38 (x 10^{-12}erg)	9.761 (x 10^{-12}erg)
$\Omega_i =$	5.785 (x 10^{14}s^{-1})	3.713 (x 10^{14}s^{-1})	5.785 (x 10^{14}s^{-1})
$\kappa_i =$	64.00 ($\hbar$)	106.0 ($\hbar$)	64.00 ($\hbar$)

final pendular Hamiltonian,

$$\begin{aligned} H\{\mathbf{K},\phi\} &= \frac{1}{2}\frac{\partial^2 H_0}{\partial K_1^2}\bigg|_0 K_1^2 + \varepsilon_{ij} A_{\mathbf{m}_r}\cos\{\phi_1\} + K_2\omega_{2r} \\ &= -\{r^2\Omega_1/\kappa_1 + s^2\Omega_2/\kappa_2\}K_1^2 \\ &\quad + \varepsilon_{ij} A_{\mathbf{m}_r}\cos\{\phi_1\} + K_2\omega_{2r}, \end{aligned} \tag{35}$$

where ϕ_2 is cyclic, K_2 is constant, and the coefficient $A_{\mathbf{m}_r}$ is evaluated at resonance.

The pendular Hamiltonian given by Equation 35 can be separated into the free particle motion of K_2, ϕ_2, and the pendular motion of K_1, ϕ_1,

$$\begin{aligned} H\{K_1,\phi_1\} &= GK_1^2 - F\cos\{\phi_1\} \\ G &= \{r^2\Omega_1/\kappa_1 + s^2\Omega_2/\kappa_2\} \\ F &= \varepsilon_{ij} A_{\mathbf{m}_r}. \end{aligned} \tag{36}$$

A pendular phase space portrait of the Hamiltonian in Equation 36 is shown in Figure 6 near the $\mathbf{m}_r = \{1,-1\}$ resonance between two coupled Morse oscillators used to model local C–H stretch modes. Acetylinic constants used to parametrize the Morse oscillators are listed in columns 1 and 3 of Table I, and a value of $\varepsilon_{12} = -0.1$ is used to couple the oscillators. The pendular phase space is separated into two regions by the separatrix where trajectories inside the separatrix

TABLE II. Uncoupled HOCl Morse oscillator parameters [31]. Labels 1, 2, and 3 correspond respectively to the H–O stretch, the H–O – Cl bend, and the O–Cl stretch. Morse oscillator parameters are obtained for the bending bond by transforming to pendular angle-action variables, expanding the action as a function of energy, inverting, and keeping terms to 2nd order.

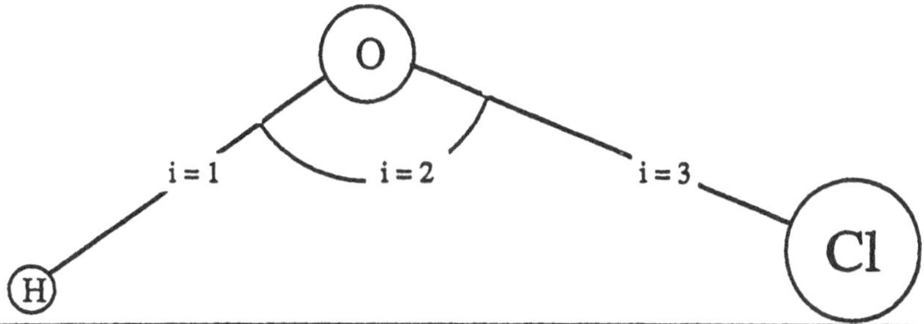

μ_i =	0.156 (x 10^{-23}g)	Undefined	1.823 (x 10^{-23}g)
α_i =	2.37 (x 10^{8}cm^{-1})	Undefined	2.87 (x 10^{8}cm^{-1})
D_i =	6.948 (x 10^{-12}erg)	25.02 (x 10^{-12}erg)	2.987 (x 10^{-12}erg)
Ω_i =	7.069 (x 10^{14}s^{-1})	2.065 (x 10^{14}s^{-1})	1.643 (x 10^{14}s^{-1})
κ_i =	37.28 ($\hbar$)	459.5 ($\hbar$)	68.95 ($\hbar$)

correspond to librational motion and trajectories outside the separatrix correspond to rotational motion. The width of the $\mathbf{m}_r$th resonance is computed by inverting Equation 36 and solving for K_1 with $\phi_1 = \pi$, and a total energy $E = F$,

$$W_{\mathbf{m}_r} = 2K_1^{\max} = \left\{ \frac{8|F|}{G} \right\}^{1/2} . \tag{37}$$

Trajectories with initial conditions that correspond to points on the separatrix will exhibit an exchange of action between modes that corresponds to the full resonance width. For example, using initial conditions corresponding to point B in Figure 6, numerical integration of the full Hamiltonian $H(\mathbf{I}, \theta) = H_0(\mathbf{I}) + H_1(\mathbf{I}, \theta)$ where H_0 and H_1 are respectively given by Equations 5 and 17, results in the trajectories of $I_1(t)$ and $I_3(t)$ shown in Figure 7b. As can be seen from the trajectory, there is a symmetric exchange of action between the two oscillators given by $W_{\mathbf{m}_r}$. Using initial conditions that correspond to the points A and C in Figure 6, new trajectories of I_1, and I_3 are obtained and are shown respectively in Figures 7a

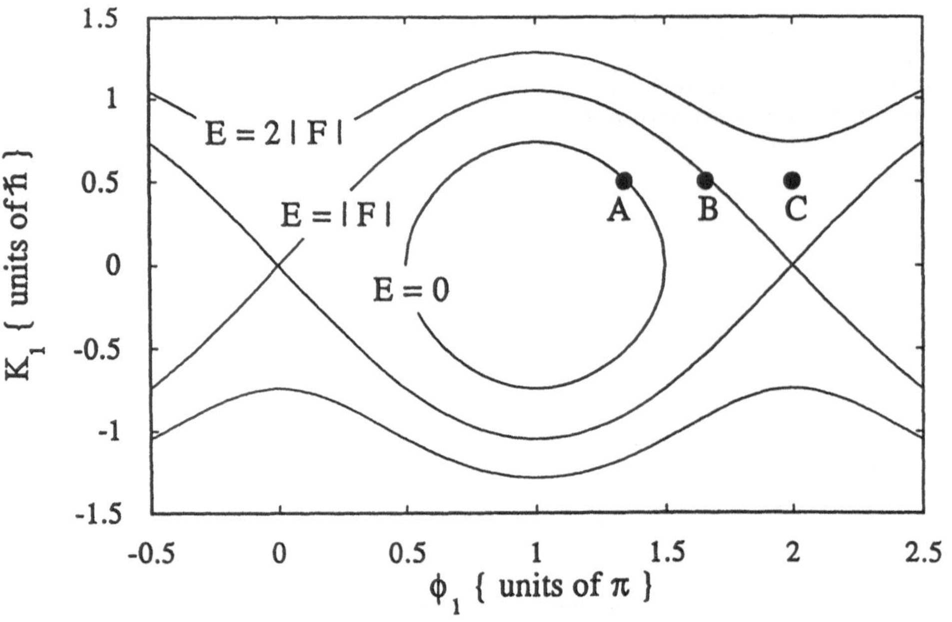

Figure 6. The pendular phase space portrait of the Hamiltonian in Equation 36 around the $m_r = \{1, -1\}$ resonance for two C–H oscillators . Contours are drawn for three energies $E = 2|F|$, $E = |F|$, and $E = 0$ corresponding respectively to rotational motion, the separatrix, and librational motion. The $E = 0$ contour is at the middle of the potential well given by $|F| \cos(\phi_1)$. The points labelled A, B and C lie on the line $K_1 = 0.5h/2\pi$, and are initial conditions for the trajectories shown in Figure 7a, b, and c respectively.

and 7c. In Figure 7a, the actual amount of action exchanged between modes is less than the maximum given by the resonance width. In general, the amount of action exchanged is a function of both initial actions and angles, and can vary between a maximum of $W_{\mathbf{m}_r}$ to a minimum of zero. Using the choice of initial conditions labelled A, B, and C in Figure 6, the dependence of the trajectories on the initial value of ϕ_1 is illustrated. By increasing the initial angle ϕ_1 from 1.34π in Figure 7a to 1.66π in Figure 7b, the amount of action exchanged between the C–H oscillators increases, and for the initial value of $\phi_1 = 2\pi$, the trajectory in Figure 7c exhibits *rotational* behaviour. Thus, to make the distinction between "normal" mode or librational behaviour, and "local" mode or rotational behaviour, knowledge of *both* the initial actions *and* angles is required.

The maximum resonance width given by Equation 37 is proportional to $\varepsilon_{ij}^{1/2}\Gamma_i^{|n_i|/2}\Gamma_j^{|n_j|/2}$. Thus, for large enough values of the coupling and resonance actions, it is possible for resonance widths that correspond to different resonances, to overlap [2]. When significantly large amounts of resonance overlap occur, the classical dynamics of the system will exhibit chaotic behaviour. The values of the coupling and resonance actions needed for the onset of this chaotic behaviour are approximately determined by Chirikov's overlap criteria [2–4]. As illustrated

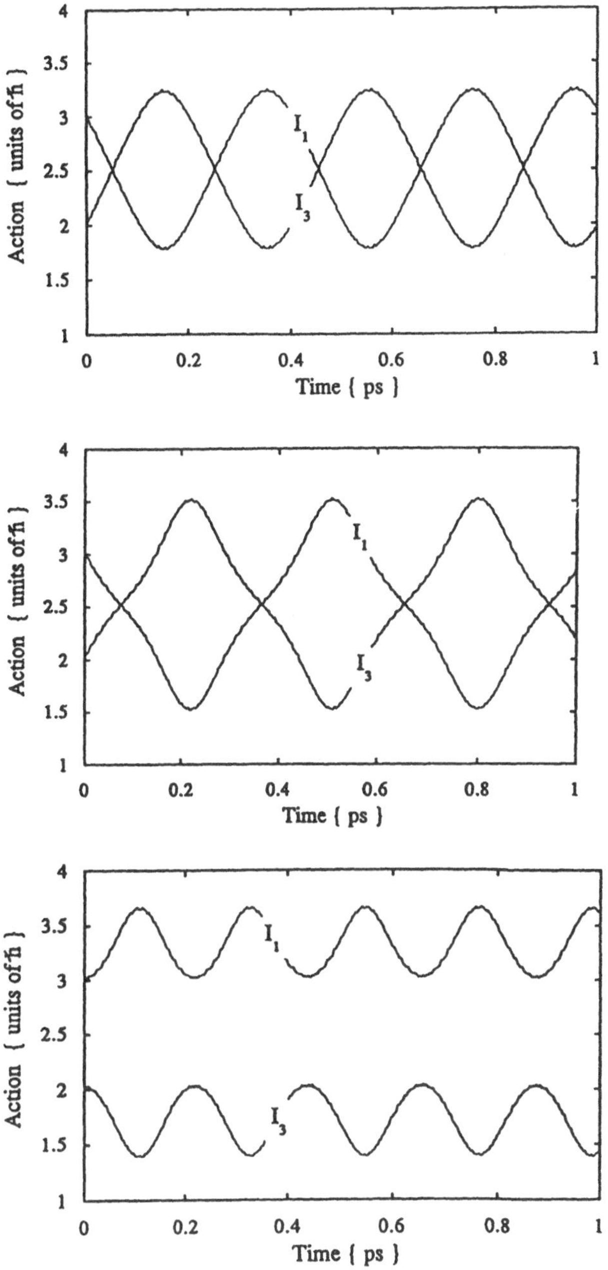

Figure 7. Trajectories of $I_1(t)$ and $I_3(t)$ for initial conditions near the $m_r = \{1, -1\}$ resonance labelled (a) A, (b) B, (c) and C in Figure 6. For all three trajectories the coupling $\varepsilon_{13} = -0.1$, and the initial conditions $K_1 = 0.5h/2\pi$, $K_2 = 0.0h/2\pi$, $\phi_2 = 0\ \pi$ are held constant, while the initial angle ϕ_1 is chosen to be 1.34π, 1.66π and 2π. Equations 30 and 34 are used to determine the corresponding initial values of I, θ where for the $m_r = \{1, -1\}$ resonance, they are essentially a translation by the constant resonance actions $I_{ir} = J_{ir}$.

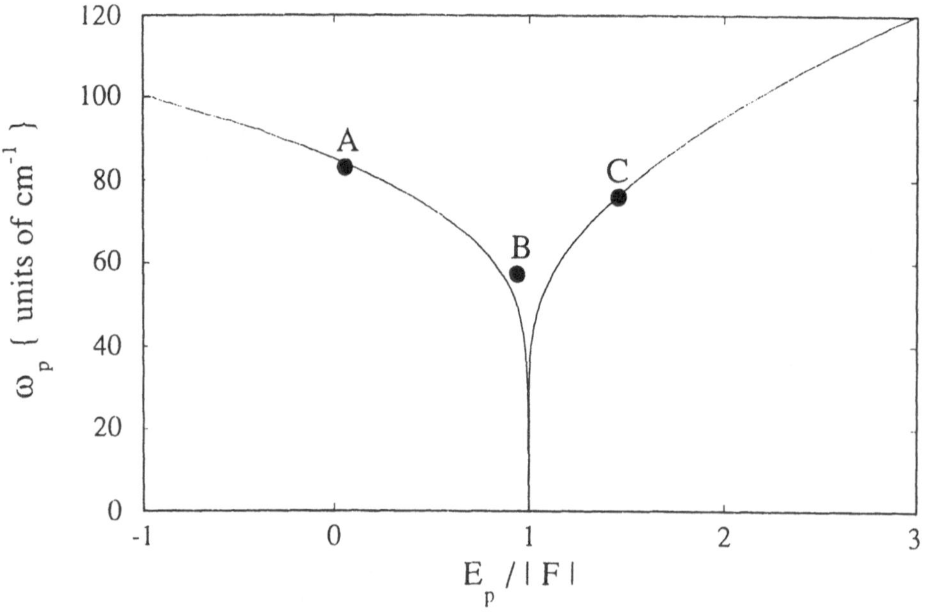

Figure 8. A plot of the pendular frequency as a function of energy. The solid line is given by Equation 38, and compared with the dots labelled A, B, and C numerically determined from the trajectories 7a, b, and c respectively.

by Figures 6 and 7, the actual amplitude of a trajectory, while bounded by the maximum resonance width, can be less than the maximum resonance width and even zero depending on the choice of initial angles. One semi-classical approach to quantum mechanics is to average a quantity of interest over a set of initial angle variables while holding the initial actions constant. The semi-classical resonance width computed by such an average over angles will therefore be less than the maximum resonance width given by Equation 37. This diminished width will exhibit a reduced amount of resonance overlap and therefore correspond to more stable, less chaotic behaviour. It is interesting to speculate that this diminished semi-classical average width corresponds to the lower degree of stochasticity found in quantum systems.

The pendular frequency ω_p obtained from Equation 36 is a function of the pendular energy E_p, and is given by,

$$\omega_p(E_p) = \{2G|F|\}^{1/2}\left\{\frac{\pi}{2k_p}\right\}[F(\gamma_{\max},k_p^2)]^{-1}$$

$$\gamma_{\max} = \left\{\begin{array}{ll} \frac{1}{2}\arccos(-E_p/|F|) & \leftarrow \text{Libration} \\ \pi/2 & \leftarrow \text{Rotation} \end{array}\right\}$$

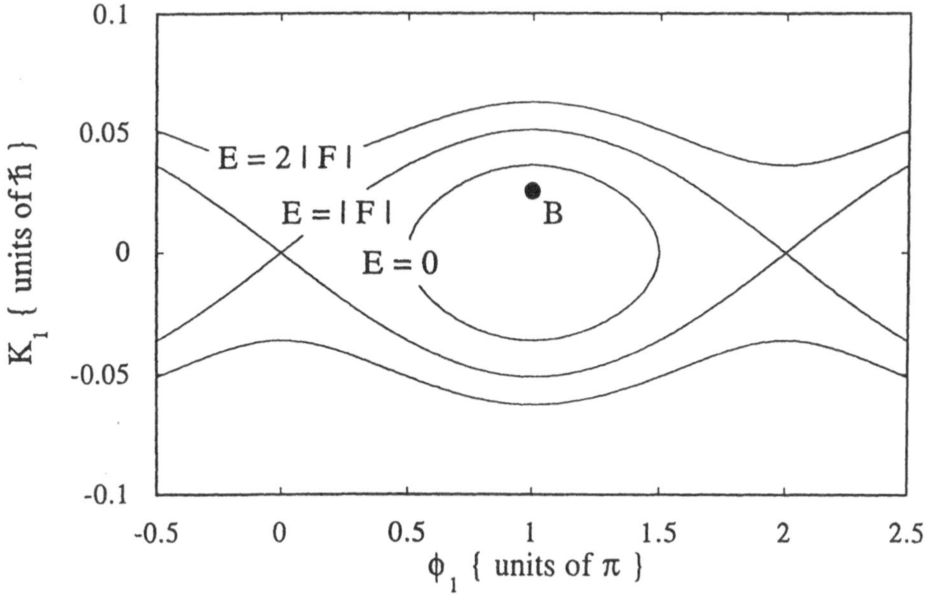

Figure 9. The pendular phase space portrait of the Hamiltonian in Equation 36 around the $m_r = \{2, -3\}$ resonance for C–H and C–C oscillators parametrized by the constants in columns 1 and 2 of Table I, and the kinetic coupling parameter $\varepsilon_{12} = -0.1$. The point labelled B corresponds to the initial conditions $K_1 = (1/4)W_{mr}$ and $\phi_1 = \pi$ used to make the trajectory in Figure 11b. Initial conditions for Figures 11a and c are in the same relative position, however different values of the coupling ε_{12} are used.

$$k_p = \left\{ \frac{2}{E_p/|F| + 1} \right\}^{1/2} \tag{38}$$

where $F(\phi_{\max}, k_p^2)$ is an elliptic integral of the first kind. A plot of ω_p given by Equation 38 is shown in Figure 8 by the solid line, and the frequencies of the numerically generated trajectories in Figure 7 are given by the open circles. Figures 7 and 8 demonstrate that for the $\mathbf{m}_r = \{1, -1\}$ resonance, the pendular approximation alone describes the dynamical behaviour of the system very well. This is because for the $\mathbf{m}_r = \{1, -1\}$ resonance, there are no sub-resonant terms.

A closer look at the trajectories of I_1 and I_3 in Figure 7 reveals that a smaller amplitude, higher frequency oscillation is superimposed on the resonant pendular behaviour. These oscillations correspond to terms in Equation 17 that are labelled with super-resonant Fermi vectors. Because the super-resonant terms are exponentially smaller that the resonant term, the amplitude of this super-resonant oscillation is small compared to the resonant pendular amplitude. Since the $\mathbf{m}_r = \{1, -1\}$ resonance has no sub-resonant terms $\mathbf{I}, \theta \approx \mathbf{J}, \Psi$, and a discussion of the effects of sub-resonant coupling requires a higher order resonance.

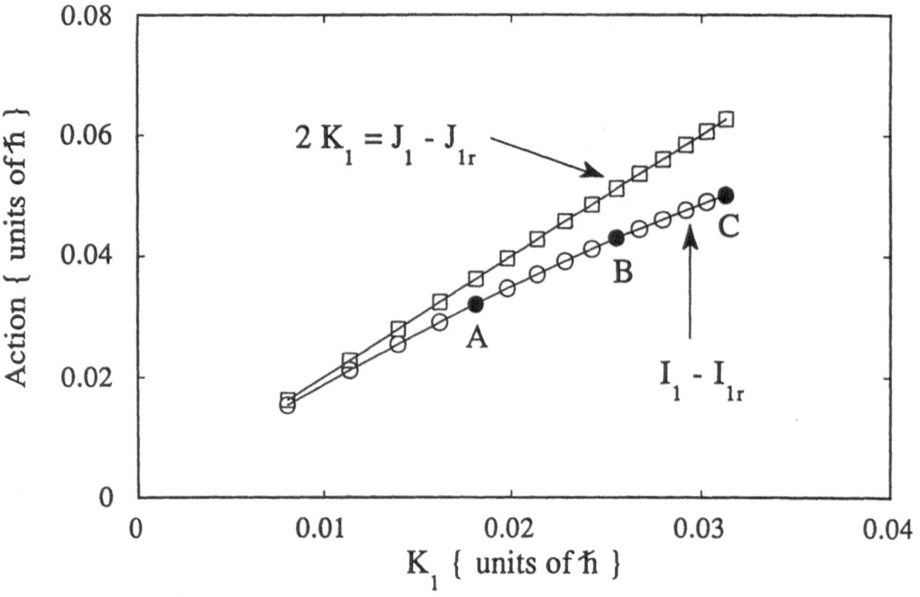

Figure 10. A plot of J_1–J_{1r} and I_1–I_{1r} as a function of K, ϕ initial conditions, $K_1 = (1/4)W_{m_r}$, $\phi_1 = \pi$, $K_2 = 0$, and $\phi_2 = 0$. The coupling $|\varepsilon_{12}|$ is varied from 0.01 to 0.15 in steps of 0.01 and used in Equation 37 to determine K_1. Similar results are obtained when plotting I_2 as a function of K_1. The points labelled A, B, and C correspond to values of the coupling $\varepsilon_{12} = -0.05$, $\varepsilon_{12} = -0.10$, and $\varepsilon_{12} = -0.15$ respectively and label the values of I_1 used to make the trajectories in Figure 11a, b, and c.

4. Sub-Resonant Coupling

Near the $\mathbf{m}_r = \{2, -3\}$ resonance, between two coupled Morse oscillators parametrized by the C–C and C–H acetylenic constants listed in Table I, there are several sub-resonant coupling terms in H_1 given by Equation 17. A pendular phase space portrait of the Hamiltonian in Equation 36 near the $\mathbf{m}_r = \{2, -3\}$ resonance is shown in Figure 9 for a coupling of $\varepsilon_{12} = -0.1$. Using Equations 30, 34 and the values $K_2 = 0$, $\phi_2 = 0$, the initial conditions $K_1 = (1/4)W_{\mathbf{m}_r}$, and $\phi_1 = \pi$ labelled B in Figure 9 are transformed to $\mathbf{J}$, Ψ and then $\mathbf{I}$, θ initial conditions where the corresponding value of I_1 is labelled B in Figure 10. Point B in Figure 10 is one of a series of points obtained by varying the coupling $|\varepsilon_{12}|$ in Equation 37 to determine $K_1 = (1/4)W_{\mathbf{m}_r}$. Figure 10 illustrates how for the $\mathbf{m}_r = \{2, -3\}$ resonance the quantity $J_1 - J_{1r}$ is a factor of two larger than K_1; and how, for the particular choice of initial angles $\phi_1 = \pi$, and $\phi_2 = 0$, the sub-resonant terms in Equation 30 suppress the value of I_1 relative to J_1. Using the initial conditions $K_1 = (1/4)W_{\mathbf{m}_r}$, and $\phi_1 = \pi$ labelled B in Figure 9, the full Hamiltonian $H(\mathbf{I}, \theta) = H_0(\mathbf{I}) + H_1(\mathbf{I}, \theta)$ given by Equations 5 and 17 is used to numerically integrate the trajectory of $I_1(t)$ shown in Figure 11b. The trajectory exhibits a slow oscillation identified as resonant pendular motion with a frequency

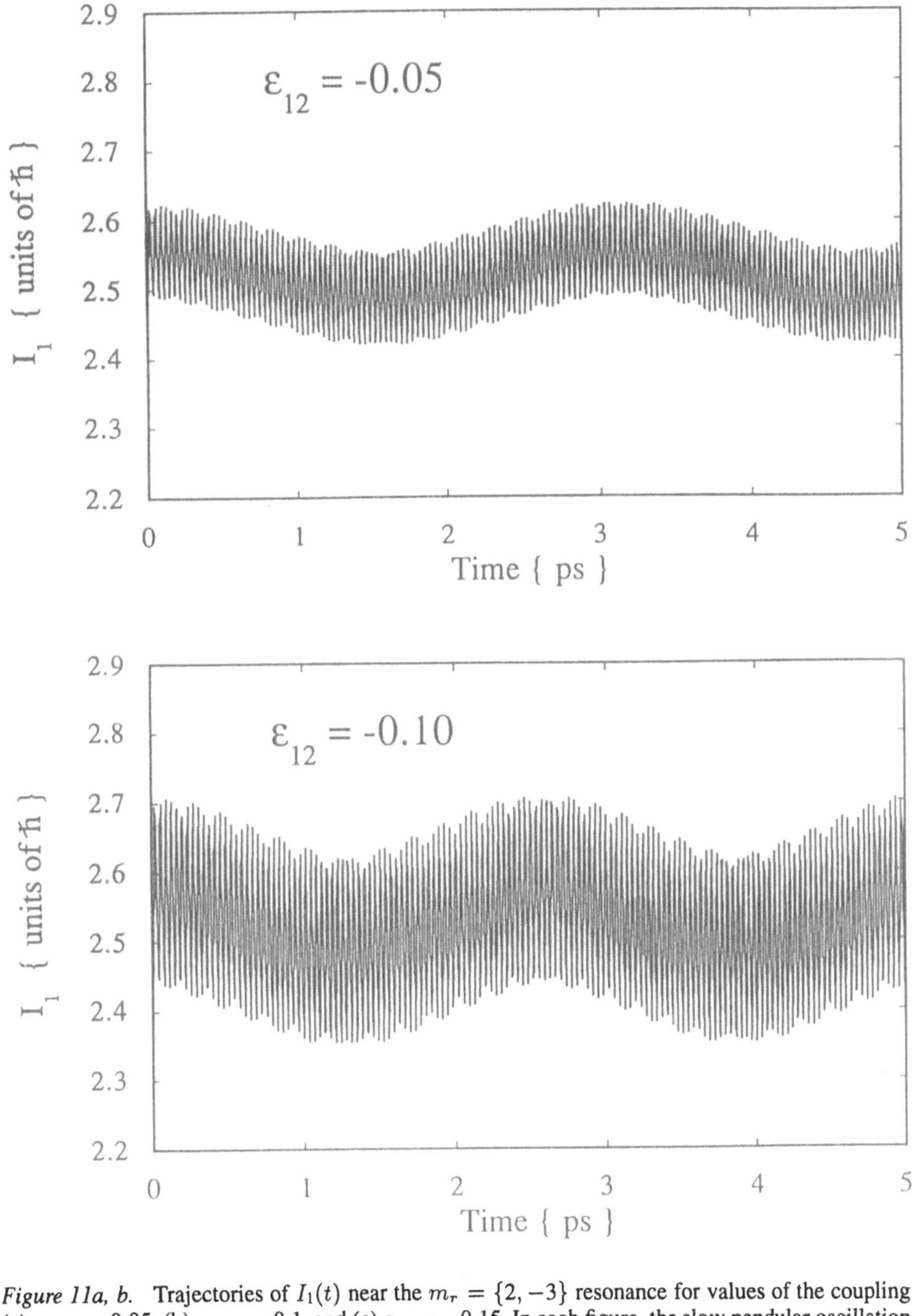

Figure 11a, b. Trajectories of $I_1(t)$ near the $m_r = \{2, -3\}$ resonance for values of the coupling (a) $\varepsilon_{12} = -0.05$, (b) $\varepsilon_{12} = -0.1$, and (c) $\varepsilon_{12} = -0.15$. In each figure, the slow pendular oscillation is modulated by large amplitude high frequency sub-resonant behaviour.

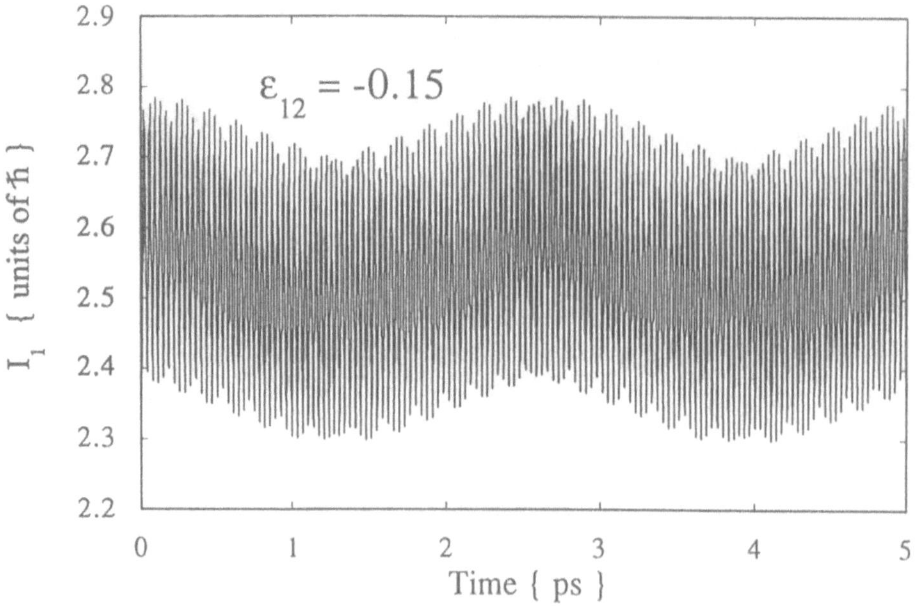

Figure 11c.

and amplitude given by Equations 36 and 37. The high frequency oscillations are the result of sub-resonant terms in Equation 17. Super-resonant terms in Equation 17 also contribute as in Figure 7, however since they are exponentially smaller than the resonant and sub-resonant terms, it is reasonable to neglect their effect. An approximate analytic expression for the trajectories of $I_i(t)$, that includes both the super-resonant and sub-resonant behaviour observed respectively in Figures 7 and 11, can be obtained by combining the canonical transformations in Equations 30 and 34,

$$\begin{aligned} I_1(t) &= J_1(t) - \varepsilon_{12} \sum_{\mathbf{n} \neq p\mathbf{m}_r} n_1 \frac{A_{\mathbf{n}}}{\mathbf{n} \bullet \omega_r} \cos\{\mathbf{n} \bullet \Psi(t)\} \\ &= I_{1r} + n_1 K_1(t) - \varepsilon_{12} \sum_{\mathbf{n} \neq p\mathbf{m}_r} n_1 \frac{A_{\mathbf{n}}}{\mathbf{n} \bullet \omega_r} \\ &\times \cos\{(\mathbf{n} \bullet \omega_r)t + n_1 \Psi_{1o} + n_2 \Psi_{2o}\} \end{aligned} \tag{39}$$

where for two coupled oscillators, the sum over i, j may be dropped. In Equation 39, t is the time, I_{1r} is the first component of the resonance action vector $\mathbf{I}_r$, and the coefficients $A_{\mathbf{n}}$ and Morse oscillator frequencies ω_r, are evaluated at resonance. The time dependence of I_1 enters in Equation 39 in two places; through the resonant pendular behaviour of $K_1(t)$, and through the summation over non-resonant Fermi vectors. The summation over non-resonant Fermi vectors is a function of the angles $\Psi_i(t)$ where to zero order in the coupling $\Psi_i(t) = \omega_{ri}t + \Psi_{io}$,

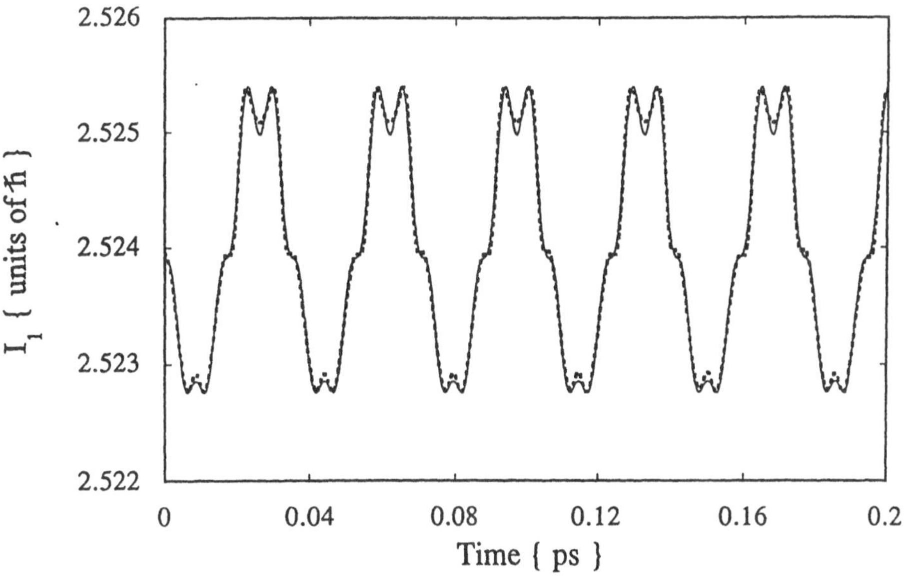

Figure 12. Trajectories of $I_1(t)$ near the $m_r = \{2, -3\}$ resonance for a value of the coupling $\varepsilon_{12} = -0.001$. The trajectory drawn with the dashed line corresponds to the exact, numerically integrated trajectory, and the solid line is obtained from Equation 39.

and Ψ_{io} are the initial angles at time $t = 0$. It can therefore be appreciated that for higher order resonances for which sub-resonant terms exist, there will be non-pendular, sub-resonant excursions in phase space. Since these excursions correspond to the summation over sub-resonant terms in Equation 39, and these sub-resonant terms are exponentially larger than the resonant term, the magnitude of the sub-resonant excursions can significantly exceed the pendular amplitude. Shown in Figure 12 is a comparison between an exact numerically integrated trajectory for $\varepsilon_{12} = -0.001$, and the analytic result given by Equation 39. To compute the analytic trajectory, $K_1(t)$ was set to zero since the pendular frequency is too slow to be observed on the time scale used in Figure 12. The summation in Equation 39 was limited to the four largest sub-resonant terms,

$$n_1 \frac{A_{\mathbf{n}}}{\mathbf{n} \bullet \omega_r} = \begin{array}{c|cccccc} n_1 \backslash n_2 & -3 & -2 & -1 & 1 & 2 & 3 \\ \hline 1 & \cdot & -0.166 & 1.117 & -0.223 & \cdot & \cdot \\ 2 & \mathrm{m}_r & \cdot & 0.113 & \cdot & \cdot & \cdot \\ 3 & \cdot & \cdot & \cdot & \cdot & \cdot & \cdot \end{array} . \tag{40}$$

Figure 12 demonstrates the excellent agreement between the numerical and analytic trajectories and confirms the conclusion that the sub-resonant terms are responsible for the observed non-pendular excursions. A similar comparison is shown in Figures 13a and b for larger coupling and longer times. The approximate trajectories shown in Figure 13 are obtained using the top line in Equation 39 where the time dependent behaviour of $\mathbf{J}$, Ψ is determined by numerically

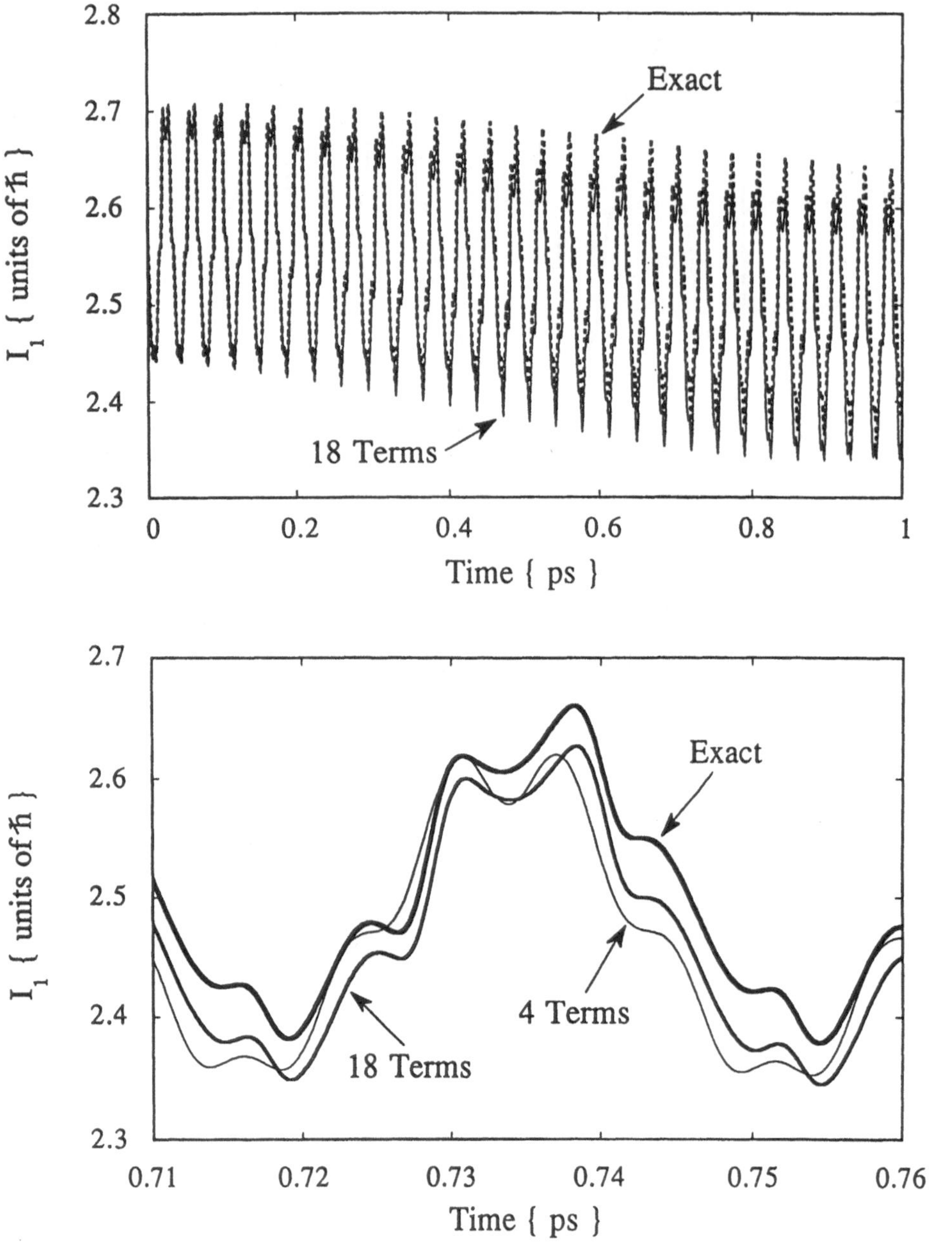

Figure 13. Trajectories of $I_1(t)$ near the $m_r = \{2, -3\}$ resonance for a value of the coupling $\varepsilon_{12} = -0.1$. Both (a) and (b) are magnifications of the trajectory in Figure 11b. In (a) the exact trajectory is drawn with the dashed line, and the approximate trajectory is given by the solid line obtained from Equation 39 where the 18 largest sub-resonant terms are included. In (b) a closer view of the trajectory is shown and compares the exact result with Equation 39 truncated at four and 18 terms.

integrating Equation 35 and transforming the time dependent behaviour of $\mathbf{K}$, ϕ using Equation 34. The frequency in Equation 39 of the $\mathbf{n}$th component of the sub-resonant excursions is independent of the coupling where,

$$f_{\mathbf{n}} = n_1\omega_{1r} + n_2\omega_{2r}. \tag{41}$$

From Figure 11, 12 and 13, the frequency corresponding to the largest sub-resonant term is numerically determined to be 942.9 cm^{-1} and is in excellent agreement with computed result of 943.3 cm^{-1} obtained from Equation 41. Also visible in Figure 12 is at least one other higher frequency component of the sub-resonant behaviour. The frequency of these sub-resonant oscillations is on the order of two orders of magnitude greater than the resonant pendular frequencies of a few wave numbers displayed in Figure 11. A similar treatment of the super-resonant behaviour observed in Figures 7a–c is possible.

The ratio, as a function of the coupling, of the resonant amplitude to sub-resonant amplitude obtained from Equation 39, is a useful measure of the relative importance of the sub-resonant excursions. To compute this ratio we select as a representative pendular trajectory the one given by initial conditions $K_1 = (1/4)W_{\mathbf{m}_r}$, and $\phi_1 = \pi$. It then follows from Equations 18, 34, 36, and 37 that the resonant pendular amplitude is given by,

$$\begin{aligned} S_{\text{res}} &= J_1^+ - J_1^- = 2rK_1 = \frac{r}{2}W_{\mathbf{m}_r} \\ &= \{|\varepsilon_{12}|\}^{1/2}\left\{\frac{2r^2 A_{\mathbf{m}_r}}{G}\right\}^{1/2} \\ &= \{|\varepsilon_{12}|\}^{1/2}\left\{\frac{4r^2\beta_{12}\lambda_1\lambda_2\Gamma_1^{r/2}\Gamma_2^{|s|/2}}{r^2\Omega_1/\kappa_1 + s^2\Omega_2/\kappa_2}\right\}^{1/2} \\ &= \{|\varepsilon 12|\}^{1/2} C_{\text{res}} \\ J_1^{\pm} &= J_{1r} \pm rK_1 \end{aligned} \tag{42}$$

where, for the $\mathbf{m}_r = \{2, -3\}$ resonance between the acetylenic C–H and C–C oscillators, the constant $C_{\text{res}} = 0.324h/2\pi$. The sub-resonant amplitude may be closely approximated by retaining only the largest sub-resonant term in Equation 39, $A_{\mathbf{n}}^{\max}$,

$$\begin{aligned} N_{\text{sub-res}} &= |\varepsilon_{12}| 2n_1 \left|\frac{A_{\mathbf{n}}}{\mathbf{n} \bullet \omega_r}\right| \\ &= |\varepsilon_{12}| \frac{4n_1\beta_{12}\lambda_1\lambda_2\Gamma_1^{n_1/2}\Gamma_2^{|n_2|/2}}{|n_1\omega_{1r} + n_2\omega_{2r}|} \\ &= |\varepsilon_{12}| C_{\text{sub-res}} \end{aligned} \tag{43}$$

where $C_{\text{sub-res}} = 2.234h/2\pi$. A better approximation of $N_{\text{sub-res}}$ is obtained by numerically determining the peak-to-peak amplitude of the analytic trajectory

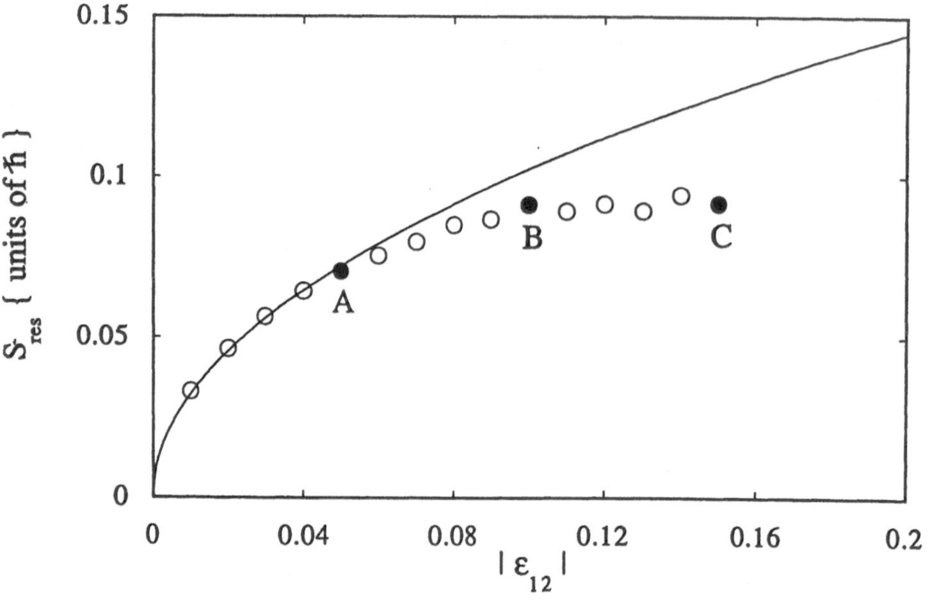

Figure 14. The resonant pendular amplitude is plotted as a function of coupling strength $|\varepsilon_{12}|$ near the $m_r = \{2, -3\}$ resonance. The analytic prediction given by Equation 42 is drawn with a solid line, and numerical results are given by the circles. Values of K_1 and I_1 used to make the trajectories are shown in Figure 10, and the points A, B, and C correspond to the resonant pendular amplitudes of the trajectories in Figures 11a, b, and c.

given by the summation in Equation 39. This amplitude, which includes the contributions of smaller $A_{\mathbf{n}}$ phased by the initial angles, is divided by the absolute value of the coupling to obtain a more refined estimate of $C_{\text{sub-res}}$. Applying this technique to Figure 12, the somewhat larger value of $C_{\text{sub-res}} = 2.642h/2\pi$ is obtained. The ratio $R = S_{\text{res}}/N_{\text{sub-res}}$ can then be expressed as a function of the coupling using Equations 42 and 43,

$$R = \frac{S_{\text{res}}}{N_{\text{sub-res}}} = \{|\varepsilon_{12}|\}^{-1/2} \frac{C_{\text{res}}}{C_{\text{sub-res}}} . \qquad (44)$$

The important result here is that the ratio of resonant to sub-resonant amplitudes is inversely proportional to the square root of the coupling ε_{12}. A series of numerically integrated trajectories around the $\mathbf{m}_r = \{2, -3\}$ resonance is shown in Figures 11a, b, and c for increasing values of the coupling $|\varepsilon_{12}|$. Initial conditions $K_1 = (1/4)W_{\mathbf{m}_r}$, and $\phi_1 = \pi$ are used for each trajectory. It is observed in Figures 11a, b and c that as the coupling is increased, the amplitude of both the resonant and sub-resonant behaviour increases. Predictions for the resonant and sub-resonant amplitudes as a function of the coupling, given by Equations 42 and 43, are shown respectively in Figures 14 and 15 by the solid line. Numerically determined resonant and sub-resonant amplitudes from the trajectories in Figure 11 are also shown in Figures 14 and 15 by the circles labelled A, B, and C.

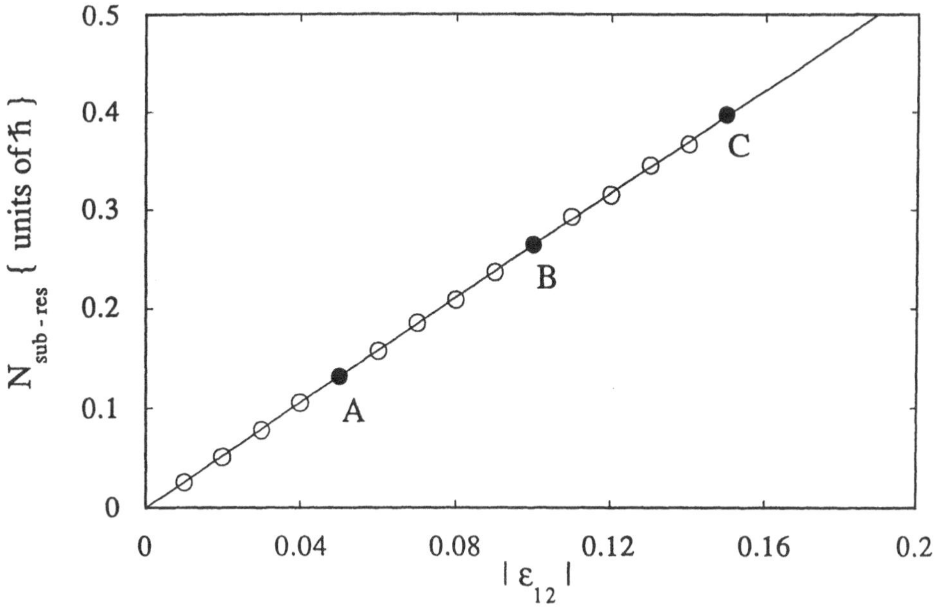

Figure 15. The sub-resonant amplitude is plotted as a function of coupling strength $|\varepsilon_{12}|$ near the $m_r = \{2, -3\}$ resonance. The analytic prediction given by Equation 43 is drawn with a solid line. Sub-resonant amplitudes obtained from numerically integrated trajectories are given by the circles. Values of K_1 and I_1 used to make the trajectories are shown in Figure 10, and the points A, B, and C correspond to the sub-resonant amplitudes of the trajectories in Figures 11a, b, and c.

The unlabelled circles correspond to additional trajectories that are not shown. In Figure 14, the numerically determined resonant amplitudes demonstrate that for small ε_{12}, Equation 42 is in excellent agreement with numerical results, and for larger couplings, Equation 42 provides an upper bound for the pendular amplitude. The prediction for R given by Equation 44 is compared with numerically determined values in Figure 16 and demonstrates once again that Equation 39 describes very well the effects of sub-resonant coupling. From Equation 44 it is possible to estimate the coupling strength required to obtain a trajectory that exhibits predominantly resonant pendular behaviour. For the unrealistically small coupling strength $\varepsilon_{12} = -0.0001$ the ratio R is estimated to be 12 and the corresponding trajectory is shown in Figure 17. Thus while it is possible for extremely small unphysical values of the coupling to ignore sub-resonant effects, for physically relevant values of the coupling, the sub-resonant excursions in phase space will be *significantly* larger than the resonant pendular amplitude and cannot be ignored.

The transformation between **K**, ϕ and **I**, θ can be used to examine the phase space portrait of the **I**, θ variables. Shown in Figure 18a is a pendular phase space portrait of the Hamiltonian in Equation 36 parametrized by Morse oscillator constants in column 1 and 3 of Figure 1, and a coupling of $\varepsilon_{12} = -0.1$. Using Equation 34, the pendular phase space portrait of K_1, ϕ_1 is transformed to a

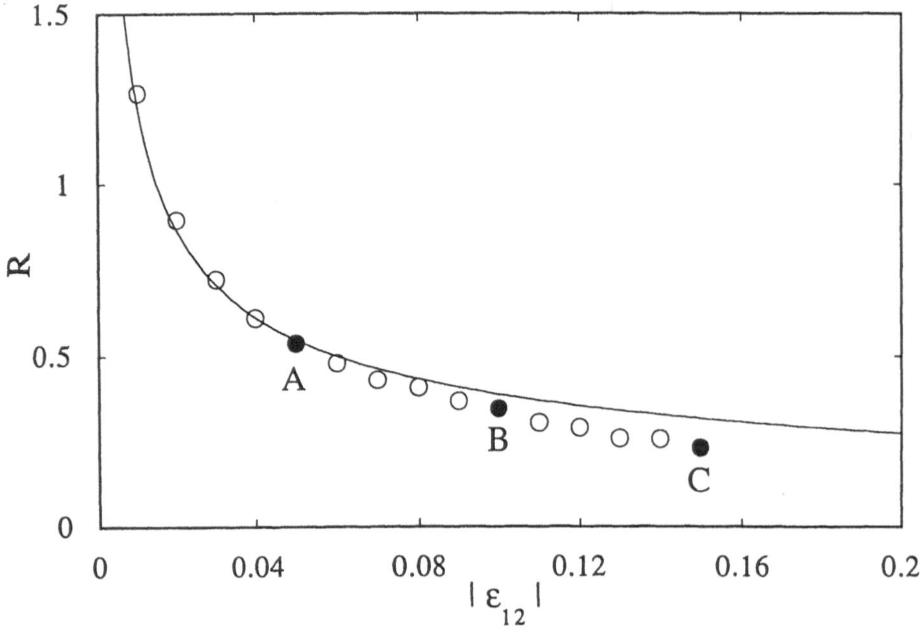

Figure 16. The ratio of resonant to sub-resonant amplitude is plotted as a function of coupling strength $|\varepsilon_{12}|$ near the $m_r = \{2, -3\}$ resonance. The analytic prediction given by Equation 44 is drawn with a solid line. The ratios obtained from numerically integrated trajectories are given by the circles. Values of K_1 and I_1 used to make the trajectories are shown in Figure 10, and the points A, B, and C correspond to the ratios of resonant to sub-resonant amplitudes of the trajectories in Figures 11a, b, and c.

similar portrait of J_1, Ψ_1 and is shown in Figure 18b. To obtain Figure 18b it is necessary to plot in a rotating frame moving with the angular velocity $-2\omega_{2r}/3$. As indicated by Equation 34, Ψ_1 is scaled by a factor of $1/r = 1/2$ while J_1 is scaled by a factor of $|s| = 3$. As a result of the sub-resonant terms in Equation 30, the projection of the trajectory onto the I_1, θ_1 plane displays a large amplitude high frequency oscillation superimposed on the smooth pendular behaviour of J_1, Ψ_1. A plot of only those points that correspond to $\phi_2 = 0$ modulo 4π is shown in Figure 18c and forms a surface of section illustrating how the sub-resonant terms in Equation 30 distort the pendular J_1, Ψ_1 and K_1, ϕ_1 phase space portraits around the $\mathbf{m}_r = (2, -3)$ resonance. Shown in Figure 19 is a comparison between a single librational trajectory of J_1, Ψ_1, the transformed surface of section in the I_1, θ_1 plane, and the exact surface of section in the I_1, θ_1 plane obtained by numerically integrating $H_{\text{tot}}(\mathbf{I}, \theta) = H_0(\mathbf{I}) + H_1(\mathbf{I}, \theta)$ given by Equations 5 and 17.

As the coupling strength increases the degree of distortion also increases and causes the separatrix in Figures 18c and 19 to cross the θ_1 axis at different places. Thus, by holding initial conditions for $\mathbf{I}$, θ constant and increasing the coupling strength, it is possible to induce a transition from librational behaviour inside the

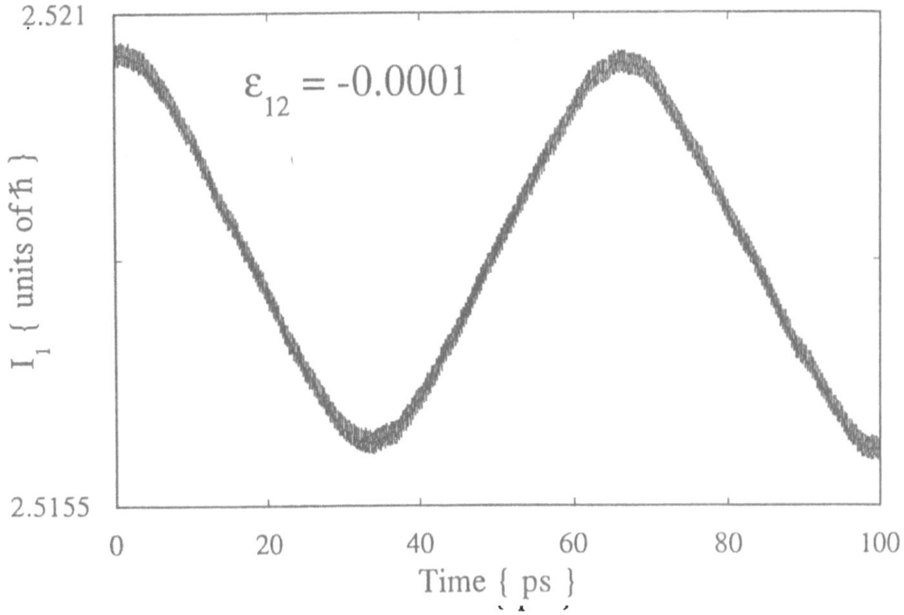

Figure 17. A numerically integrated trajectory of $I_1(t)$ near the $m_r = \{2, -3\}$ resonance for the unrealistically small value of $\varepsilon_{12} = -0.0001$ that corresponds to a value of $R = 12$. Note the very long time of ~ 0.1 ns required to observe the pendular behaviour.

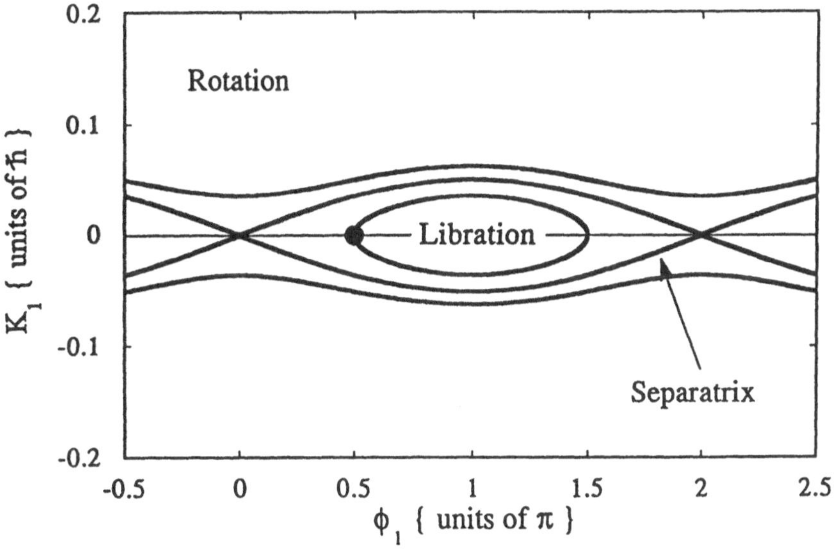

Figure 18a. Phase space portrait of K_1, ϕ_1 near the $m_r = \{2, -3\}$ resonance for a coupling of $\varepsilon_{12} = -0.1$. The separatrix separates librational behaviour from rotational behaviour. The dot represents initial conditions of $K_1 = 0$, and $\phi_1 = \pi/2\{K_2 = 0$ and $\phi_2 = 0\}$, and after transformation by Equations 30 and 34 corresponds to the dots in Figures 18b and c.

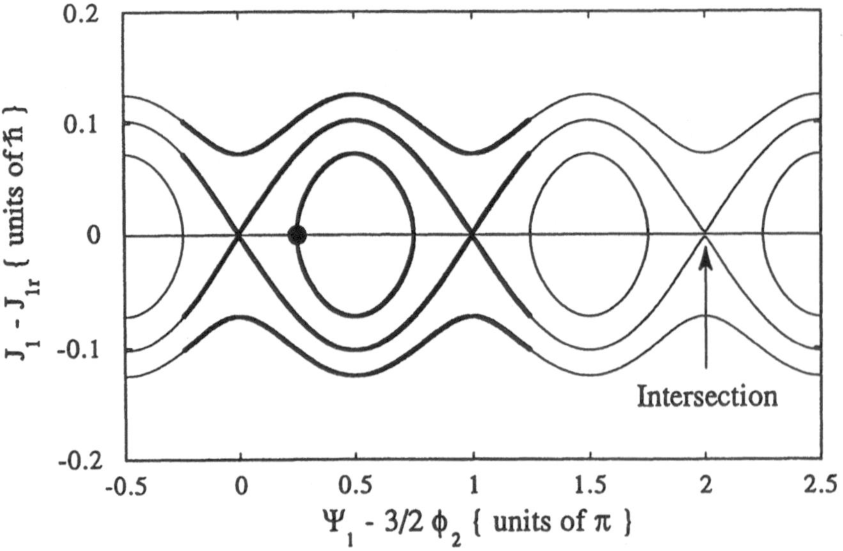

Figure 18b. Phase space portrait of J_1, Ψ_1 obtained by transforming the trajectories in Figure 18a with Equation 34. The thick lines correspond to the trajectories in Figure 18a for $-\pi/2 \leq \phi_1 \leq 5\pi/2$. The plot is made in a rotating frame moving with an angular velocity $-2\omega_{2r}/3$, and is offset by the resonance action $J_{1r} = I_{1r}$.

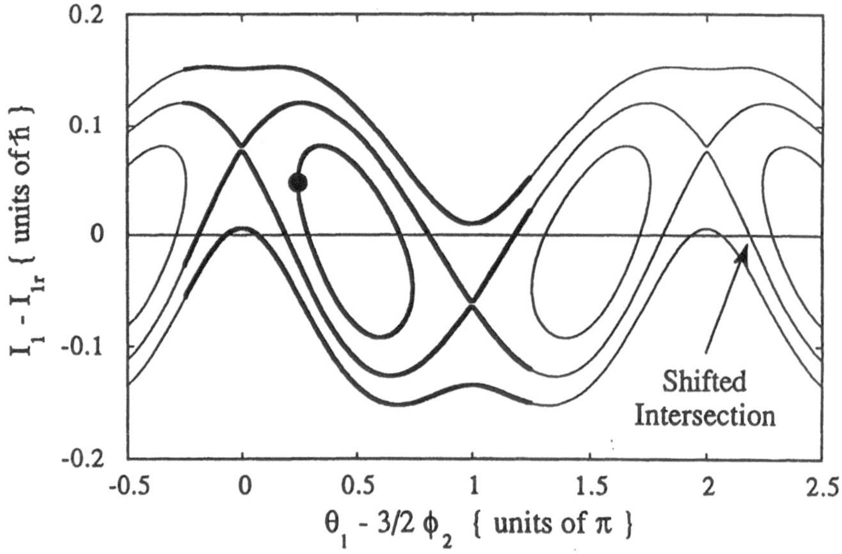

Figure 18c. A surface of section for the variables I_1, θ_1. The section is drawn by transforming trajectories in Figure 18b with Equation 30 where only those points for which $\phi_2 = 0$ modulo 4π are shown. The plot is made in a rotating frame moving with an angular velocity of $-2\omega_{2r}/3$, and is offset by the resonance action I_{1r}. The position of the dot illustrates how, for the choice of initial angles in Figure 18a, the sub-resonant terms increase the value of I_1 relative to J_1. The arrow points to the intersection of the separatrix and the θ_1 axis and when compared to the corresponding intersection shown in Figure 18b, illustrates how the sub-resonant terms shift this intersection.

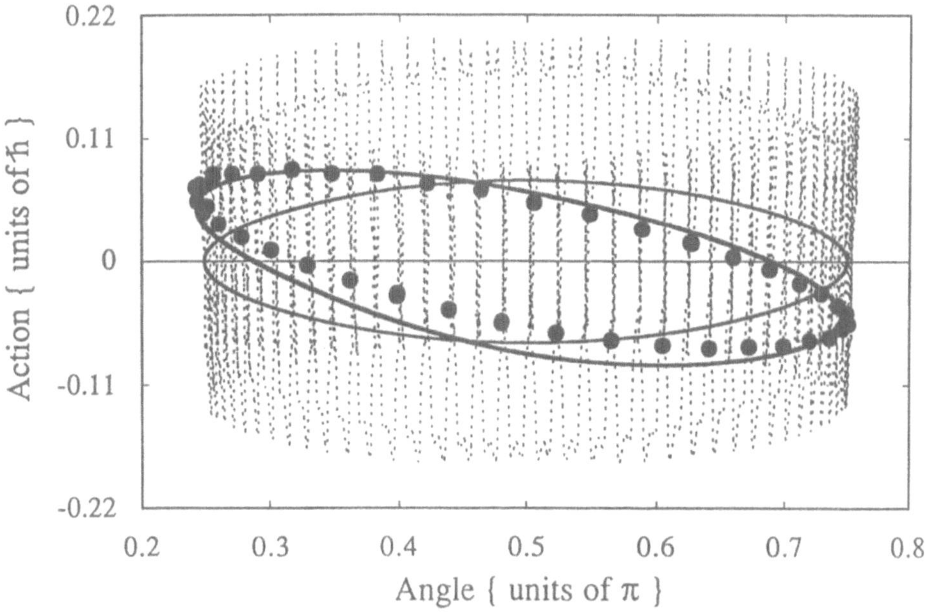

Figure 19. A surface of section for the variables I_1, θ_1. The dashed line is obtained by using Equation 30 to transform the trajectory in Figure 18b labelled with the dot, and projecting the result onto the I_1, θ_1 plane for all values of the angle ϕ_2 in the rotating frame moving with an angular velocity of $-2\omega_r/3$. The inclined trajectory drawn with the thicker solid line corresponds to the trajectory in Figure 18c labelled with the dot. For reference, the trajectory from Figure 18b labelled with the dot is drawn with the thin solid line. The dotted trajectory is the result of numerical integration and compares very well with the prediction of the surface of section given by the thicker solid line. In order to avoid secular phase shifts that accumulate in one period of oscillation $\sim$ 2 ps, the numerical plot is made in a rotating frame offset by $-2\theta_2/3$.

separatrix to rotational behaviour outside the separatrix. This effect can be seen in Figures 20a and b where trajectories around the $\mathbf{m}_r = \{2, -3\}$ resonance are shown for values of the coupling $\varepsilon_{12} = -0.10$ and $\varepsilon_{12} = -0.12$ respectively. The same initial conditions are used for each trajectory and correspond to the dot in Figure 18c projected onto the $I_1 = I_{1r}$ axis. As the coupling increases, the phase space portrait in the I_1, θ_1 plane distorts and moves the separatrix closer to the initial conditions. When the distortion is large enough for the separatrix to cross the initial conditions, the librational trajectory in Figure 20a becomes the rotational trajectory in Figure 20b.

In addition to distorting phase space trajectories, sub-resonant terms will also alter the resonance widths displayed on CE surfaces. As discussed in Section 2, the CE surface and Arnold web are obtained in the zero coupling limit and are therefore independent of the couplings ε_{ij}. In this limit the zero order Hamiltonian $H_0(\mathbf{I})$ is equivalent to $H_0(\mathbf{J})$ as determined by Equation 25. Thus to zero order in the coupling, the CE surface, Arnold web, and resonance actions $\mathbf{I}_r = \mathbf{J}_r$ are

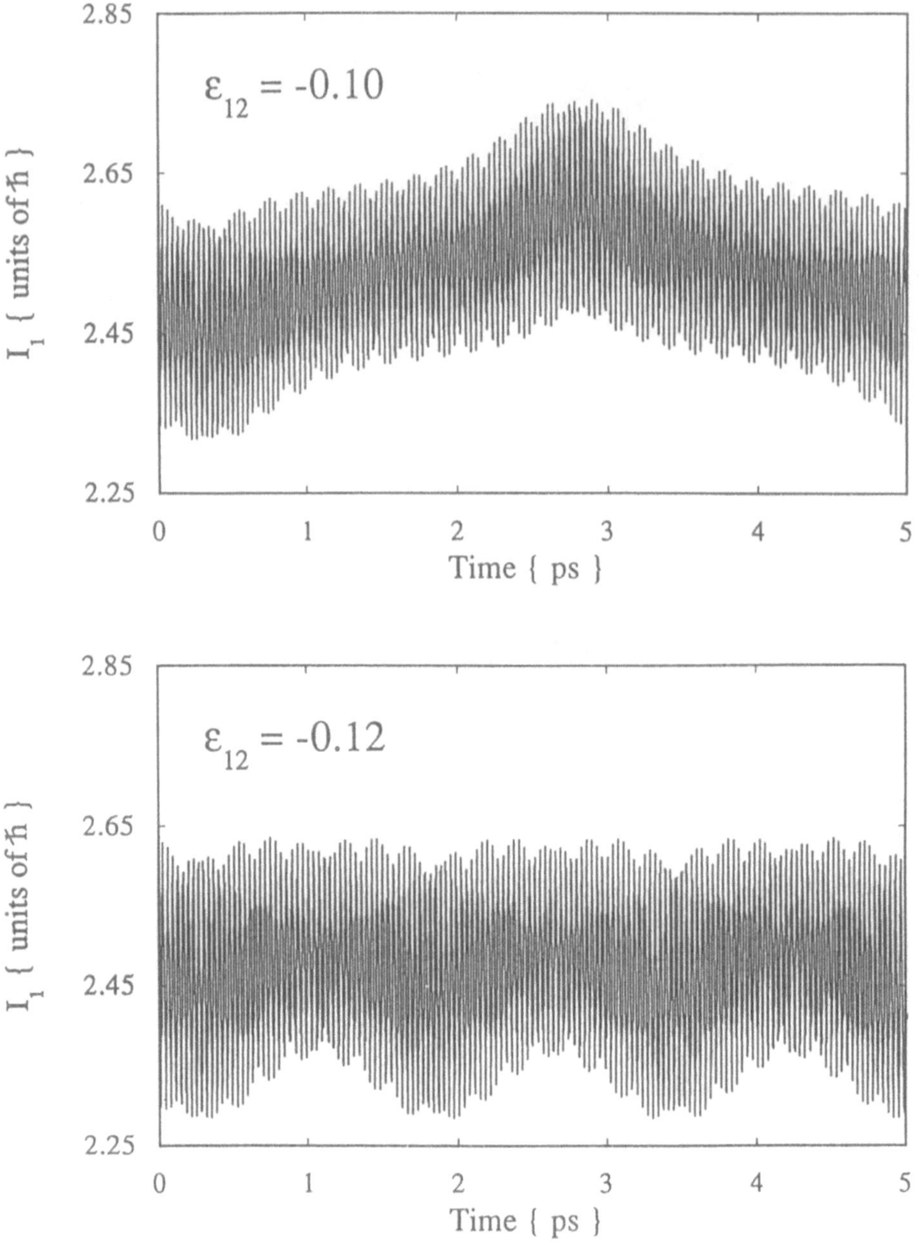

Figure 20. Trajectories of $I_1(t)$ near the $m_r = \{2, -3\}$ resonance for a coupling of (a) $\varepsilon_{12} = -0.10$ and (b) $\varepsilon_{12} = -0.12$. The same initial conditions are used in both figures. The trajectory in (a) exhibits librational motion. As the coupling increases the separatrix shifts its position as illustrated in Figures 18a and b, and the same initial conditions used in (a) result in the trajectory in (b) which, because the separatrix has shifted, now exhibits rotational behaviour.

identical for either set of $\mathbf{I}$, θ and $\mathbf{J}$, Ψ variables. Shown in Figure 21 is a plot of the zero order CE surface for C_2H_2. In the figure the axes are labelled with both $\mathbf{I}$ and $\mathbf{J}$ variables because the zero order Hamiltonian has the same functional dependence on $\mathbf{I}$ and $\mathbf{J}$. Thus when considering the edges of the projected surface and labelled resonance plane intersections drawn using thick lines, they have the same form in either $\mathbf{I}$ or $\mathbf{J}$ action space. The resonance widths however are functions of the coupling as determined by Equation 37 and will have a *different* appearance when plotted in the $\mathbf{I}$ or $\mathbf{J}$ action space. The difference is again the result of sub-resonant terms in the canonical transformation between $\mathbf{I}$, θ and $\mathbf{J}$, Ψ given by Equations 30 and 31. The thin lines in Figure 21 are resonance widths as they appear in $\mathbf{J}$ action space. The realistic C_2H_2 couplings of $\varepsilon_{12} = \varepsilon_{23} =$ -1.0 used to plot the $\mathbf{m}_r = \{2, -3, 0\}$, and $\mathbf{m}_r = \{0, 2, -3\}$ resonance widths are too large for a perturbative treatment of the Hamiltonian in Equation 17, and are chosen only to make the widths clearly visible on the scale of the figure. A smaller value of $\varepsilon_{13} = -0.1$ is used to plot the $\mathbf{m}_r = \{1, 0, -1\}$ resonance width and qualitatively mimics a small potential coupling between the H–C and C–H oscillators [26]. To plot the resonance widths as shown in Figure 21, the width as a function of resonance actions given by Equation 37 is transformed using Equation 34 to compute the widths in terms of $\mathbf{J}$ as a function of $\mathbf{J}_r$,

$$\begin{aligned} J_i^{\pm} &= \pm r W_{\mathbf{m}_r} + J_{ir} \\ J_j^{\pm} &= K_2 \pm s W_{\mathbf{m}_r} + J_{jr}. \end{aligned} \tag{45}$$

The resonance widths computed from Equation 45 and illustrated in Figure 21 are time independent. This is in direct contrast to the representation of the widths on the CE surface defined by the zero order variables $\mathbf{I}$, θ where the resonance widths will exhibit the same large amplitude high frequency sub-resonant motion exhibited by the trajectories in Figures 11a, b, and c. The specific time dependent behaviour of the resonance widths in $\mathbf{I}$ action space is determined by substituting Equation 45 into the top line of Equation 39,

$$\begin{aligned} I_i^{\pm}(t) &= J_i^{\pm} - \varepsilon_{ij} \sum_{\mathbf{n} \neq p\mathbf{m}_r} n_i \frac{A_{\mathbf{n}}}{\mathbf{n} \bullet \omega_r} \cos\{\mathbf{n} \bullet \Psi(t)\} \\ &= J_i^{\pm} - \varepsilon_{ij} \sum_{\mathbf{n} \neq p\mathbf{m}_r} n_i \frac{A_{\mathbf{n}}}{\mathbf{n} \bullet \omega_r} \cos\{(\mathbf{n} \bullet \omega_r)t + n_i \Psi_{io} + n_j \Psi_{jo}\} \end{aligned} \tag{46}$$

where $I_j^{\pm}(t)$ is obtained in similar fashion. The difference between the time dependence of the trajectories $I_i(t)$ given by Equation 39, and the time dependence of the widths $I_i^{\pm}(t)$ is that in Equation 46, the pendular oscillation of $J_i(t)$ is replaced by the time independent value $J_i^{\pm}(t)$. However, Equation 46 retains the same sub-resonant time dependence on the angles $Y_i(t)$ exhibited in Equation 39. As mentioned before, the lowest order resonances $\mathbf{m}_r = \{\ldots, 1, \ldots, -1, \ldots\}$ have no sub-resonant terms and therefore their widths will appear the same when plotted on both $\mathbf{I}$ and $\mathbf{J}$ CE surfaces. However, for all higher-order resonances,

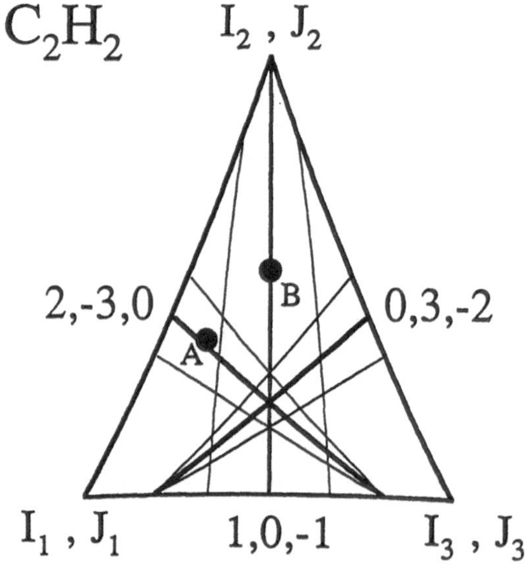

Figure 21. A projection of the $E = 19260\ \mathrm{cm}^{-1}$ C_2H_2 CE surface onto the plane defined by the intersection of the CE surface and the I_1, I_2, and I_3 axes. The surface corresponds to a position $\sim 1/3$ of the way between the origin and the E_1 surface in Figure 2. The points labelled A and B correspond to initial values of I used in later figures. Point B lies roughly in the region of action space accessed by SEP experiments [17].

resonance widths plotted in the action space spanned by the zero order actions **I** will oscillate in time as determined by the sub-resonant terms in Equation 46. It should be stressed that in the small coupling limit, the observed sub-resonant modulation of the resonance widths in the action space spanned by the zero order actions **I** occurs only because of the particular choice of **I**, θ as canonical variables. By transforming from **I**, θ to **J**, Ψ using Equation 30, the sub-resonant modulation of the widths is eliminated. However, the resulting time independent resonance widths shown in Figure 21 correspond to the exchange of action between resonant **J** modes and not between the **I** modes. Never the less, in **I** action space the resonance widths defined by Equation 45 serve as a useful reference but care should be taken not be confuse them with the true time dependent widths given by Equation 46.

For three coupled oscillators the coupling Hamiltonian H_1 is given by,

$$\begin{aligned} H_1 &= \varepsilon_{12} \sum_{\mathbf{n}_{1,2}} A_{\mathbf{n}_{1,2}} \cos(\mathbf{n}_{1,2} \bullet \theta) \\ &\quad + \varepsilon_{23} \sum_{\mathbf{n}_{2,3}} A_{\mathbf{n}_{2,3}} \cos(\mathbf{n}_{2,3} \bullet \theta) \\ &\quad + \varepsilon_{13} \sum_{\mathbf{n}_{1,3}} A_{\mathbf{n}_{1,3}} \cos(\mathbf{n}_{1,3} \bullet \theta), \end{aligned} \tag{47}$$

where the summation over i, j in Equation 17 has been explicitly carried out. Each of the three Fourier series in Equation 47 is identified as the coupling Hamiltonian H_1 between the ith and jth zero order modes. Thus, by selecting two of the three couplings ε_{ij} to be zero, Equation 47 can be reduced to the coupling Hamiltonian between two oscillators. The CE surfaces in Figures 2 and 3a correspond to the three local stretch modes of C_2H_2. The action-angle variables of the H–C oscillator, the C–C oscillator, and the C–H oscillator are respectively $I_1, \theta_1, I_2, \theta_2$, and I_3, θ_3. The two-dimensional problems considered above for the $\mathbf{m}_r = \{1, -1\}$ resonance between two C–H oscillators, and the $\mathbf{m}_r = \{2, -3\}$ resonance between a C–H and a C–C oscillator, can be considered as a special cases of the larger three-dimensional problem of the stretch modes of C_2H_2 where certain coupling terms have been set to zero. For example, by selecting $\varepsilon_{12} = \varepsilon_{23} = 0$, Equation 47 reduces to two kinetically coupled C–H oscillators, and for $\varepsilon_{23} = \varepsilon_{13} = 0$, Equation 47 reduces to kinetically coupled C–H and C–C oscillators. The connection between the full three-dimensional problem and the two-dimensional special cases can also be seen by comparing Figures 1a and b with Figure 2. For example: (i) The projection of the cube of physically bound states in Figure 2 onto the I_1, I_2 and I_1, I_3 planes corresponds respectively to the highlighted square in Figure 1a and rectangle in Figure 1b; (ii) The intersections of the CE surfaces and the I_1, I_3 plane in Figure 2, evaluated at $I_2 = 0$ for $E < \kappa_3\Omega_3/4$ and at $I_2 = \kappa_2/2$ for $E > \kappa_3\Omega_3/4$, correspond to the CE surfaces shown in Figure 1a evaluated respectively at E and $E - \kappa_3\Omega_3/4$; (iii) The intersection of the $\mathbf{m}_r = \{1, 0, -1\}$ resonance surface with the I_1, I_3 plane corresponds to the 1:1 resonance in Figure 1a; and (iv) The same correspondence exists between the intersections of the CE and resonant surfaces with the I_1, I_2 plane in Figure 2 (and by symmetry the I_3, I_2 plane), and Figure 1b.

When considering sub-resonant coupling in the three-dimensional case, up to three sets of sub-resonant terms, one from each of the three Fourier series, can appear in Equation 47. For an isolated resonance $\mathbf{m}_r$ between the ith and jth modes, there will be sub-resonant terms that belong to the same Fourier series as the resonant term $A_{\mathbf{m}_r}$, and sub-resonant terms that belong to the other two Fourier series. In general we call the sub-resonant terms that belong to the same Fourier series as the resonant term, collaborative sub-resonances, and sub-resonances that belong to the other Fourier series, cross sub-resonances. The effect of collaborative and cross sub-resonances is best illustrated by considering the total time derivative of the action variables $\mathbf{I}$,

$$
\begin{aligned}
\dot{\mathbf{I}} \;&=\; -\frac{\partial H}{\partial \theta} = \varepsilon_{12} \sum_{\mathbf{n}_{1,2}} A_{\mathbf{n}_{1,2}} \sin(\mathbf{n}_{1,2} \bullet \theta) \\
&+ \varepsilon_{23} \sum_{\mathbf{n}_{2,3}} \mathbf{n}_{2,3} A_{\mathbf{n}_{2,3}} \sin(\mathbf{n}_{2,3} \bullet \theta)
\end{aligned}
$$

$$+ \varepsilon_{13} \sum_{\mathbf{n}_{1,3}} \mathbf{n}_{1,3} A_{\mathbf{n}_{1,3}} \sin(\mathbf{n}_{1,3} \bullet \theta). \tag{48}$$

Equation 48 indicates that a trajectory in action space will move in the direction determined by a sum over Fermi vectors $\mathbf{n}_{ij}$. By definition, each Fermi vector $\mathbf{n}_{ij}$ lies in one of the three mutually orthogonal I_i, I_j planes. Therefore each Fourier series in Equation 48 is able to influence the motion of the trajectory only in a single plane determined by the subscripts i, j. Furthermore, since the coupling is assumed to be small, the trajectory must remain approximately in the CE surface. Thus, for example, an isolated resonance in the summation over $\mathbf{n}_{1,2}$, will cause a trajectory in action space to exhibit pendular motion in the CE surface back and forth across the resonance while remaining parallel to the I_1, I_2 plane. In addition, the collaborative sub-resonances that belong to the same Fourier series, will induce large amplitude, high frequency, sub-resonant excursions that must also remain parallel to the I_1, I_2 plane. On the other hand, cross sub-resonances in the summation over the $\mathbf{n}_{2,3}$ and $\mathbf{n}_{1,3}$ Fermi vectors will induce sub-resonant excursions that are not restricted to the I_1, I_2 plane but rather are restricted to the I_2, I_3 and I_1, I_3 planes. As a result, the cross sub-resonant excursions will cause the trajectory to move with a component perpendicular to the I_1, I_2 plane defined by the isolated resonance.

As a more specific example we will consider an isolated $\mathbf{m}_r = \{2, -3, 0\}$ resonance corresponding to the point A in Figure 21. By selectively turning off the coupling, it is possible to examine the effect of collaborative and cross sub-resonant terms separately. For example, by setting ε_{23} and ε_{13} to zero, only the collaborative sub-resonant excursions are exhibited by the trajectory shown in Figure 22a which has been plotted in the full three-dimensional $\mathbf{I}$ action space. Trajectories exhibiting only cross sub-resonant excursions are shown in Figures 22b and c for $\varepsilon_{12} = \varepsilon_{23} = 0$ and $\varepsilon_{23} = \varepsilon_{13} = 0$ respectively. The apparent motion of the trajectories in Figures 22a, b, and c out of the I_1, I_2, I_2, I_3, and I_1, I_3 planes respectively is the result of projecting the trajectories onto the plane defined by the CE surface intersections with the coordinate axes. The trajectory obtained when all three couplings are turned on is shown in Figure 22d and forms a generalized Lissajous figure thereby illustrating that both collaborative and cross sub-resonant excursions significantly modify the resonant canonical pendular behaviour. During the $\sim$ 0.1 ps period used to plot Figures 22a, b, c, and d, all the observed oscillations are the result of high frequency sub-resonant terms since on this time scale the small value of the coupling $\varepsilon_{12} = -0.1$ causes resonant pendular motion that is too slow to be observed.

The relative magnitude of the $\mathbf{m}_r = \{2, -3, 0\}$ collaborative sub-resonant excursions illustrated in Figure 22a is larger, by roughly a factor of two, than the $\mathbf{m}_r = \{0, 3, -2\}$ cross sub-resonances shown in Figure 22b. This can be qualitatively understood by examining Equation 43 where, for example, the magnitude of the $\mathbf{n} = \{1, -1, 0\}$ sub-resonant term increases as $\mathbf{n} \bullet \omega$ is evaluated

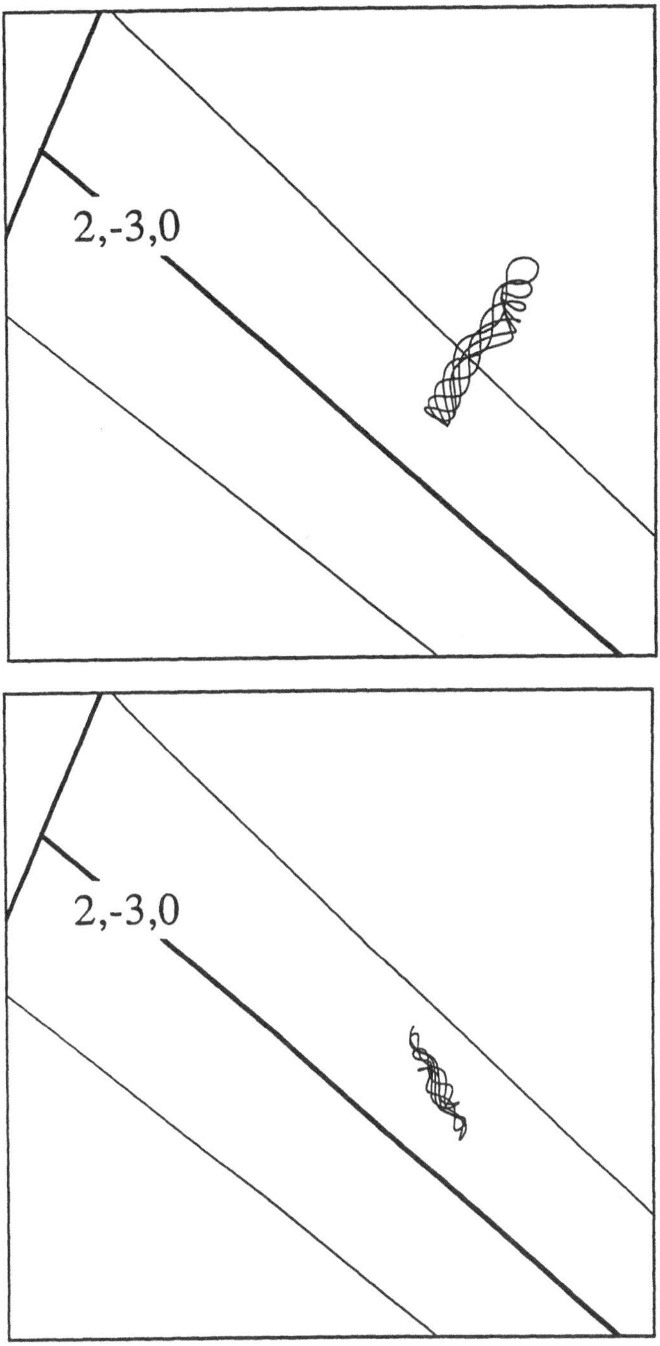

Figure 22a, b. Projections of numerically integrated trajectories on a C_2H_2 CE surface in the action space spanned by I_1, I_2, and I_3. The coupling parameters and integration times are (a) $\varepsilon_{12} = -0.1$, $\varepsilon_{23} = 0$, $\varepsilon_{13} = 0$, $t = 0.1$ ps, (b) $\varepsilon_{12} = 0$, $\varepsilon_{23} = -0.1$, $\varepsilon_{13} = 0$, $t = 0.1$ ps (c) $\varepsilon_{12} = 0$, $\varepsilon_{23} = 0$, $\varepsilon_{13} = -0.1$, $t = 0.1$ ps, and (d) $\varepsilon_{12} = -0.1$, $\varepsilon_{23} = -0.1$, $\varepsilon_{13} = -0.1$, $t = 0.2$ ps. Initial actions labelled A in Figure 21 are used for each trajectory .The figures are magnified by a factor of 10 relative to Figure 21 where the thick lines correspond to the edge of the projected CE surface and resonance plane intersections. The thin lines correspond to artificial resonance widths drawn by neglecting sub-resonant terms in Equation 46.

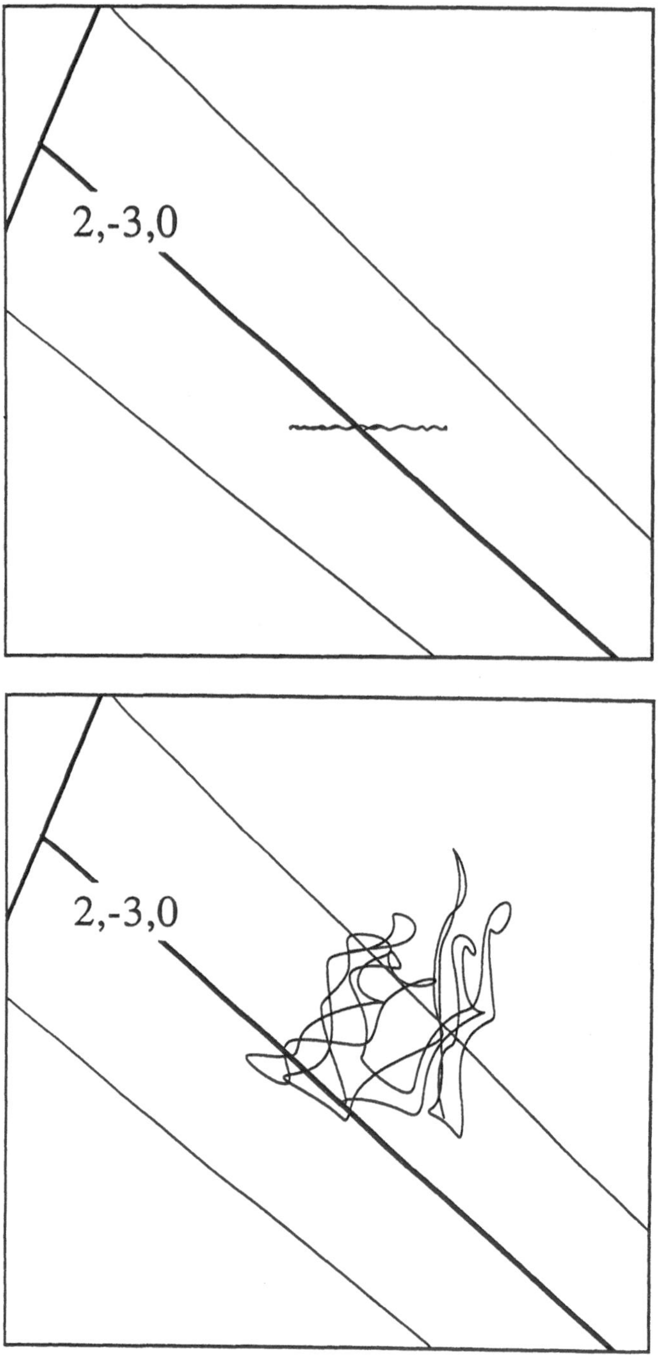

Figure 22c, d.

closer to the $\mathbf{m}_r = \{1, -1, 0\}$ resonance. For a symmetric choice of initial conditions, the sub-resonant amplitude corresponding to the $\mathbf{m}_r = \{2, -3, 0\}$ is expected to be the same as the sub-resonant amplitude corresponding to the $\mathbf{m}_r = \{0, 3, -2\}$ resonance. However, because asymmetric initial actions labelled A in Figure 21 are used to obtain Figures 22a and b, the trajectories are located in action space closer to the $\mathbf{m}_r = \{1, -1, 0\}$ resonance than to the $\mathbf{m}_r = \{0, 1, -1\}$ resonance. Therefore the $A_{1,-1,0}$ sub-resonant term corresponding to the $\mathbf{m}_r = \{2, -3, 0\}$ resonance is larger than the $A_{0,1,-1}$ sub-resonant term corresponding to the $\mathbf{m}_r = \{0, 3, -2\}$ resonance. A similar argument applies for the remaining sub-resonant terms, and therefore the $\mathbf{m}_r = \{2, -3, 0\}$ sub-resonant excursions are expected to be larger than the $\mathbf{m}_r = \{0, 3, -2\}$ sub-resonant excursions. It should be noted that an asymmetric choice of initial angles is also used to make Figures 24a and b, however, it is observed that this asymmetry plays a minor role in the relative amplitudes.

For $n > 2$ coupled oscillators, trajectories are no longer bounded by KAM tori. As a result, it is possible for the trajectory to drift along the $n - 2$ dimensional resonant surfaces of the Arnold web. This behaviour, called Arnold diffusion, provides a mechanism for exploring phase space where the rate of diffusion vanishes only in the limit of zero coupling. This gives rise to the chemically interesting prospect that IVR may in some way be a manifestation of Arnold diffusion where the trajectory wanders along one resonance, encounters an intersection, and then drifts along a new resonance. While under the influence of one resonance, IVR is characterized by the exchange of action between a particular set of modes, and under the influence of the new resonance, IVR is characterized by the exchange of action between a different set of modes. Numerical studies of linear C_2H_2 have shown that while in principle, Arnold diffusion may take place, it will do so at a time scale that is greatly exceeds typical time scales of observed IVR.

5. Large Coupling

Many molecules are linear or nearly linear and will by virtue of their large bond angles exhibit large coupling dynamics. An important example is C_2H_2 which has a linear configuration in its ground electronic state. The dynamics of C_2H_2 and other linear molecules will be dominated by the lowest order resonant and sub-resonant terms in the Fourier series expansion of the coupling Hamiltonian given by Equation 17. For large values of the coupling found in linear molecules, the first-order perturbative analysis outlined in the preceding sections will break down. The particular value of the coupling at which the perturbative analysis is no longer valid depends on the particular resonant Fermi vector $\mathbf{m}_r$ that is being considered. For example, near a lowest order resonance $\mathbf{m}_r = \{1, \ldots, -1, \ldots\}$, all non-resonant Fermi vectors are super-resonant and are therefore exponentially smaller than $\mathbf{m}_r$. Since both resonant and super-resonant terms in a particular sum over n_{ij}

are multiplied by the same value of ε_{ij}, the ratio $A_{\mathbf{n}_{ij}}/A_{\mathbf{m}_r} < 1$ is independent of the coupling. When considering the lowest order $\mathbf{m}_r = \{\ldots, 1, \ldots, -1, \ldots\}$ resonances, it is therefore a valid approximation for arbitrarily large values of ε_{ij}, to truncate the Fourier series expansion in Equation 17 by neglecting super-resonant terms. The approximate Hamiltonian in Equation 28 is thereby directly obtained without the use of the type two generating function given in Equations 21 through 24. It should be noted that, as illustrated in Figure 4, the ratio $A_{\mathbf{n}_{ij}}/A_{\mathbf{m}_r}$ increases as the zero order actions in the ith and jth bonds increases, and the quality of the approximation made by neglecting the super-resonant terms is thereby diminished as the zero order energy of the coupled oscillators increases.

The transformation given by Equations 32, 33 and 34 is canonical for arbitrary large values of the coupling. Thus, for the lowest order $\mathbf{m}_r = \{\ldots, 1, \ldots, -1, \ldots\}$ resonances, the Hamiltonian in Equation 28 can be reexpressed in terms of the $\mathbf{K}$, ϕ variables. The pendular approximation will therefore break down in the large coupling limit only when the values of K_1, and K_2 defined by Equation 34 are no longer small, and A_{mr} can no longer be evaluated at resonance. Since these conditions are required to obtain the approximate pendular Hamiltonian given by Equations 35 and 36, in the limit of large coupling the pendular Hamiltonian will no longer describe the dynamics of the system. It should be noted that by neglecting the super-resonant terms near the lowest order $\mathbf{m}_r = \{\ldots, 1, \ldots, -1, \ldots\}$ resonances the Hamiltonian in Equation 28 is directly obtained without expanding H_0 around $\mathbf{J}$ in Equation 25. Therefore the coupling ε_{ij} proportional to the difference $\mathbf{I}$–$\mathbf{J}$ given by Equation 24, *is not* specifically required to be small and relatively large values of ε_{ij} will not invalidate the first-order perturbation treatment. As a result, near an isolated lowest order resonance, qualitative pendular behaviour is observed even for large values of the coupling $\varepsilon_{ij} \sim -1$ [27].

For all higher order resonances, there are exponentially larger sub-resonant terms which for physically realistic values of the coupling may not be neglected. In contrast to the lowest order resonances, near a higher order resonance, the Taylor's series expansion in Equation 25 must be employed to obtain the transformed Hamiltonian in Equation 28. Therefore the coupling ε_{ij} proportional to the difference $\mathbf{I}$–$\mathbf{J}$ given by Equation 24, *is* specifically required to be small and large values of ε_{ij} *will* invalidate the first-order perturbation treatment. For large values of the coupling ε_{ij}, the Taylor's series expansion in Equation 25 cannot be legitimately truncated to first order in $\mathbf{I}$–$\mathbf{J}$, and the identification $B_{\mathbf{n}_{ij}} = -A_{\mathbf{n}_{ij}}$ in Equation 27 is no longer valid. Thus, near a higher order resonance in the large coupling limit, it is not possible to determine the values of $B_{\mathbf{n}_{ij}}$ that define the transformation to the approximate Hamiltonian in Equation 28, and the pendular Hamiltonians in Equations 35 and 36 will not describe the dynamics of the system. Therefore near a higher order resonance, the validity of first-order perturbation theory is more sensitive to the magnitude to the coupling than near the lowest order resonances. In the large coupling limit, it is reasonable to assume that near higher order resonances, action-angle variables do not exist and that the large sub-resonant excursions in phase space are generic for any choice of modes. Thus, in the large coupling limit, the simplest modes to work with are those described by the zero order variables **I**, θ.

The relative influence of sub-resonant, resonant, and super-resonant Fourier coefficients on the dynamics may be examined by directly integrating the expanded Hamiltonian in Equation 17 while various $\zeta_{\mathbf{n}_{ij}}$ are artificially set to zero and others are set to one. As a specific example, acetylenic constants in Table I will be used in Equation 17 to examine classical trajectories of C_2H_2 where the C–H, C–C coupling $\varepsilon_{12} = \varepsilon_{23}$ is set to the physically realistic value of -1, and the smaller coupling between the two C–H bonds $\varepsilon_{13} = -0.1$ is chosen to realistically mimic a small potential coupling [26].

Shown in Figure 23a is an exact 1 ps trajectory plotted on the **I**, θ CE surface obtained by numerically integrating $H_{\text{tot}} = H_0 + H_1$ where H_0 is given by Equation 5 and H_1 is given in closed form by the top line in Equation 17. By using the closed form of H_1 to compute the trajectory, the contributions of all sub-resonant, resonant, and super-resonant terms in the Fourier series expansion of H_1, are included. The primary contribution to the dynamics shown in Figure 23a comes from the lowest order resonance $\mathbf{m}_r = \{1, 0, -1\}$, between the two C–H bonds, and sub-resonances corresponding to the higher order $\mathbf{m}_r = \{2, -3, 0\}$, and $\mathbf{m}_r = \{0, 3, -2\}$ resonances. This is illustrated by the trajectory shown in Figure 23b obtained by numerically integrating H_{tot} where H_1 is given by a Fourier series expansion in Equation 17 that is truncated to exclude the resonant and super-resonant terms labelled by Fermi vectors with components $|n_i| \geq 3$. The effect of the $\mathbf{m}_r = \{1, 0, -1\}$ resonance on the dynamics is moderated by the relatively small value of $\varepsilon_{13} = -0.1$. A trajectory obtained using the Fourier series expansion of H_1, where only the higher order resonant and super-resonant terms labelled by Fermi vectors with components $|n_i| \geq 3$ are included, is shown in Figure 23c. This trajectory remains confined to the immediate neighbourhood of the initial actions. The close correspondence between Figures 23a and b demonstrate the dominant role of the lowest order resonant and sub-resonant terms in the dynamics, while Figure 23c illustrates that in the large coupling limit, higher order resonant and super-resonant terms play a relatively minor role.

A clearer comparison of Figures 23a and b is shown in Figures 24a and b where I_1 is chosen as reasonable representative of the dynamics and plotted as a function of time. Both trajectories display the same qualitative behaviour, and nearly the same quantitative behaviour. The trajectory in Figure 24a is obtained by integrating the exact Hamiltonian and therefore includes the effect of all possible resonances, while the trajectory in Figure 24b is obtained by integrating the same sub-resonant Hamiltonian used to make Figure 23b. The close agreement between these trajectories implies that the infinite number of higher order resonances included in the exact Hamiltonian make a negligible contribution to the dynamics

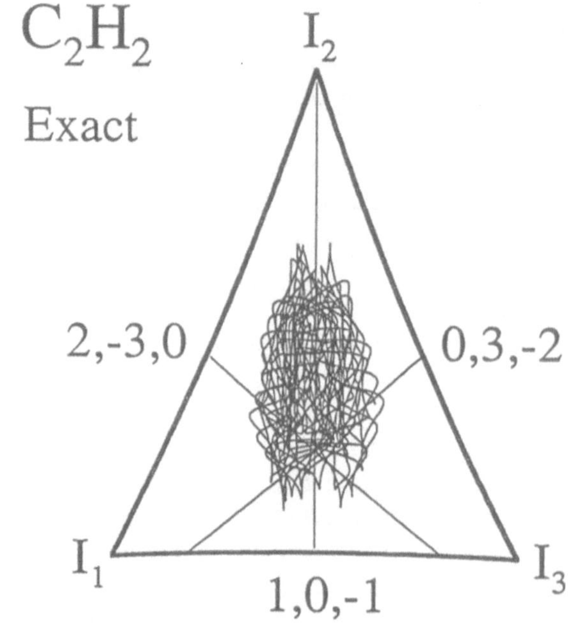

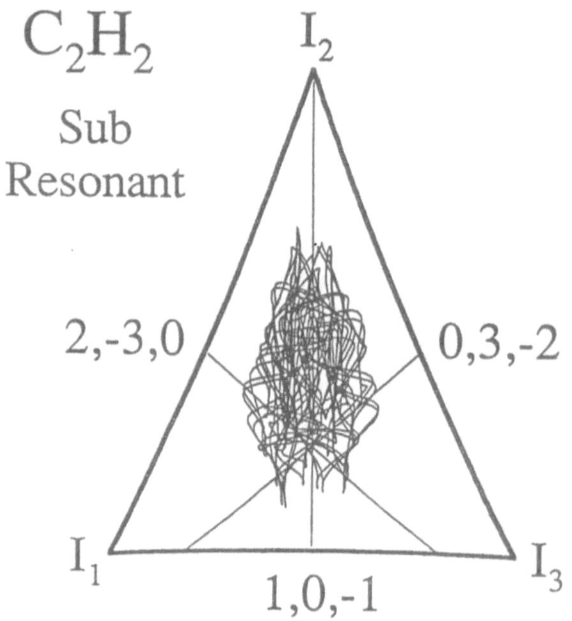

Figure 23a, b. Projections of numerically integrated trajectories on a C_2H_2 CE surface in the action space spanned by I_1, I_2, and I_3 using an (a) exact, (b) sub-resonant, and (c) super-resonant Hamiltonian. Initial actions for all three trajectories lie in the region labelled B in Figure 21. Each trajectory is integrated for 1 ps and realistic values of the coupling $\varepsilon_{12} = -1$, $\varepsilon_{23} = -1$, and $\varepsilon_{13} = -0.1$ are used. Since the trajectories are projections onto the plane defined by the vertices of the CE surface, large excursions of the trajectory off the CE surface are not clearly observed.

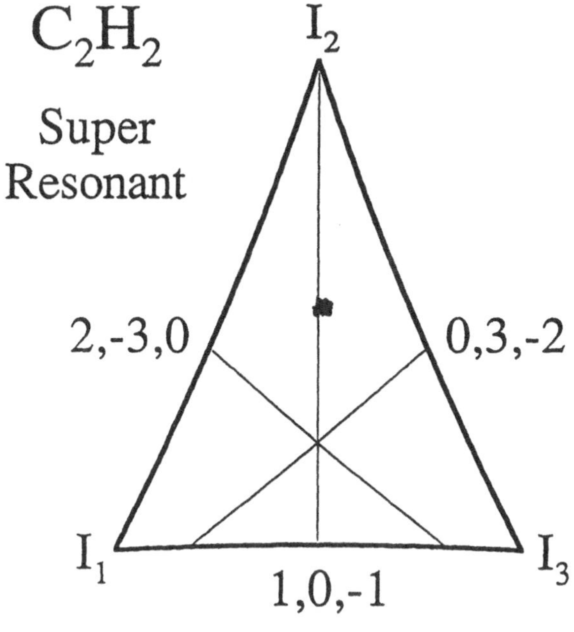

Figure 23c.

on a time scale of 1 ps. It therefore possible to draw the conclusion that even though the number of resonances is everywhere dense in the CE surface, their decreasing widths render them ineffective in influencing the behaviour of the coupled oscillators.

It is useful to examine the individual resonance terms in the absence of all other terms in the Fourier series. Shown in Figure 25a is a trajectory near an isolated $\mathbf{m}_r = \{1, 0, -1\}$ resonance obtained by numerically integrating $H_{\text{tot}} = H_0 + H_1$ where H_0 is given by Equation 5 and,

$$H_1 = 2\varepsilon_{31}\beta_{31}\lambda_3\lambda_1 \frac{-1}{|-1|} \Gamma_3^{1/2}\Gamma^{|-1|/2} \cos(\theta_3 - \theta_1). \tag{49}$$

Equation 49 is obtained by truncating the summation over i, j in Equation 17 to include only $i = 1$, $j = 3$, and truncating the sum over $\mathbf{n}_{13}$ to include only the resonant Fermi vector $\mathbf{m}_r = \{1, 0, -1\}$. The inclusion of collaborative super-resonant terms is observed to make a negligible change to the trajectory. Using initial actions labelled B in Figure 21, the trajectory in Figure 25a illustrates that the small kinetic coupling between the C–H bonds causes action to be exchanged in accord with the canonical pendular model. The $\mathbf{m}_r = \{1, 0, -1\}$ resonance

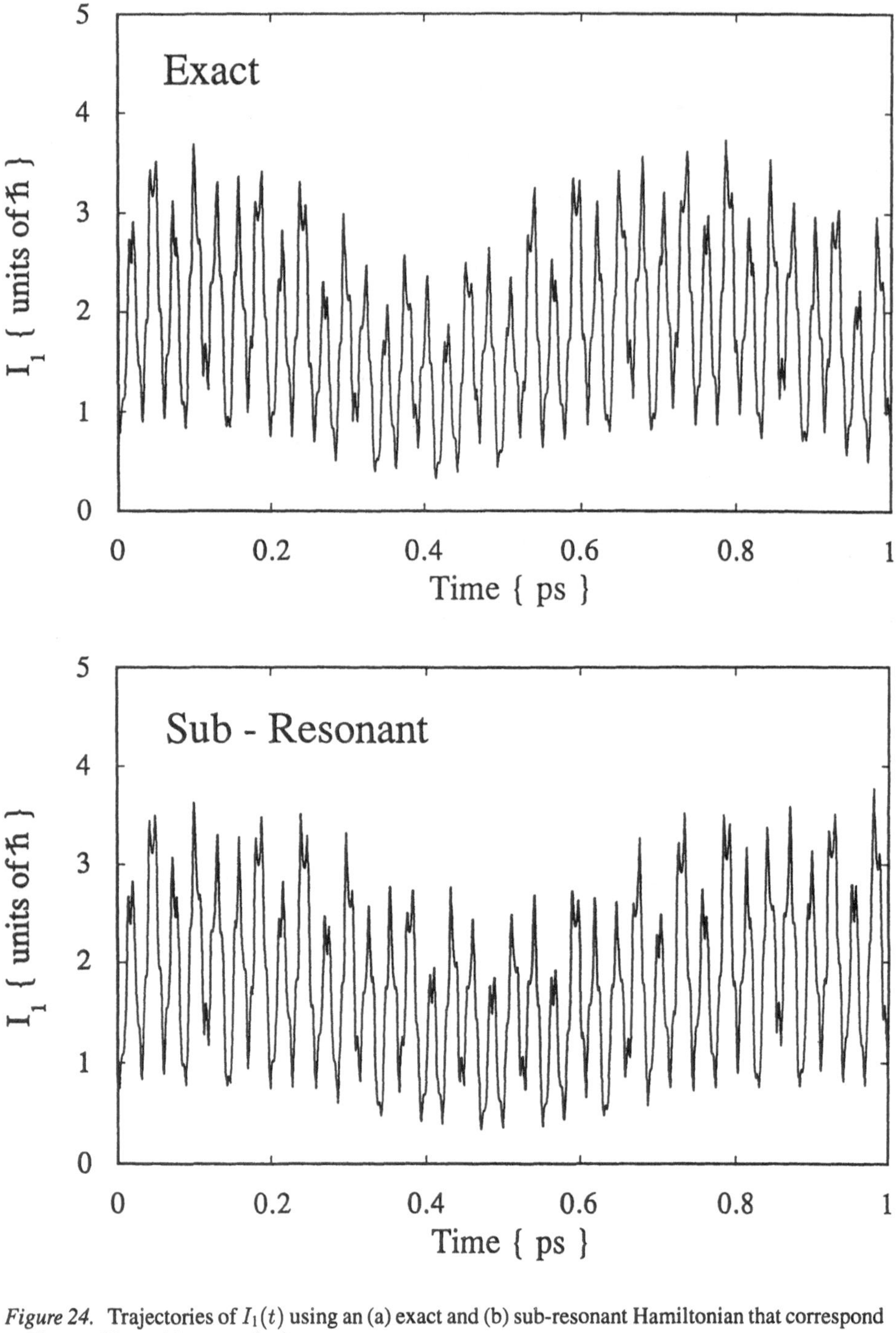

Figure 24. Trajectories of $I_1(t)$ using an (a) exact and (b) sub-resonant Hamiltonian that correspond to Figures 23a and b respectively.

width drawn in Figure 25a is valid in **I** action space because for the Hamiltonian in Equation 49, there are no sub-resonant terms in Equation 46. The trajectory in Figure 25a was made using a set of angles analogous to point A in Figure 6 that do not yield the maximum amount of action exchange. Larger values of ε_{13} will increase the amplitude and frequency of action exchange as determined by Equations 37 and 38, and eventually for couplings larger than 1 the pendular approximation will no longer be applicable. A similar examination of the $\mathbf{m}_r = \{2, -3, 0\}$ resonance is possible by numerically integrating $H_{\text{tot}} = H_0 + H_1$ where H_0 is given by Equation 5 and,

$$H_1 = 2\varepsilon_{12}\beta_{12}\lambda_1\lambda_2 \frac{-3}{|-3|} \Gamma_1^{2/2}\Gamma_2^{|-3|/2} \cos(2\theta_1 - 3\theta_2). \tag{50}$$

Using the same initial conditions used to make Figure 25a, the resulting trajectory is shown in Figure 25b and demonstrates that far from resonance the amplitude of the pendular motion obtained from the truncated Hamiltonian is very small as expected. However, if the collaborative sub-resonant terms are included in H_1 given by Equation 50 then the trajectory is dramatically altered as shown in Figure 25c where the collaborative sub-resonances induce large amplitude high frequency motion parallel to the I_1, I_2 plane. These large sub-resonant excursion are expected to be generic for any set of angle-action variables in the large coupling limit. Near the $\mathbf{m}_r = \{1, 0, -1\}$ resonance, these large amplitude excursions will correspond to cross sub-resonant motion, and will drive the trajectory both along and across the $\mathbf{m}_r = \{1, 0, -1\}$ resonance with amplitudes large enough to invalidate the pendular approximation about the $\mathbf{m}_r = \{1, 0, -1\}$ resonance. By appealing to the C_v symmetry of C_2H_2, it is clear that the trajectories obtained by including only the $\mathbf{m}_r = \{0, 3, -2\}$ resonant term in H_1 will be identical to Figures 25b and c.

Using the trajectories in Figure 25, is possible to make several qualitative statements about the dynamics of C_2H_2 for realistically large couplings illustrated in Figure 23a. First, the large vertical motion represents the exchange of action between the two C–H bonds with the C–C bond and is caused by the sub-resonant terms of the higher order $\mathbf{m}_r = \{2, -3, 0\}$ and $\mathbf{m}_r = \{0, 3, -2\}$ resonances; second, there is a lower amplitude, lower frequency horizontal oscillation that corresponds to the exchange of action between the two C–H bonds and results from the small kinetic coupling term ε_{13}; and third, it is possible to qualitatively influence the sub-resonant behaviour by selecting appropriate initial conditions. For example, as shown in Figures 22 and 25, the collaborative i, j sub-resonant behaviour will induce phase space excursions parallel to the I_i, I_j plane while cross sub-resonances will cause motion parallel to a plane orthogonal to the I_i, I_j plane. For C_2H_2, it is possible to select initial angles $\theta_1 = \theta_3$ such that the $\mathbf{m}_r = \{2, -3, 0\}$ and $\mathbf{m}_r = \{0, 3, -2\}$ resonances are in phase. This causes very large amplitude, high frequency exchange of action between the two C–H and the C–C bonds as shown in Figure 26a that is driven primarily along the

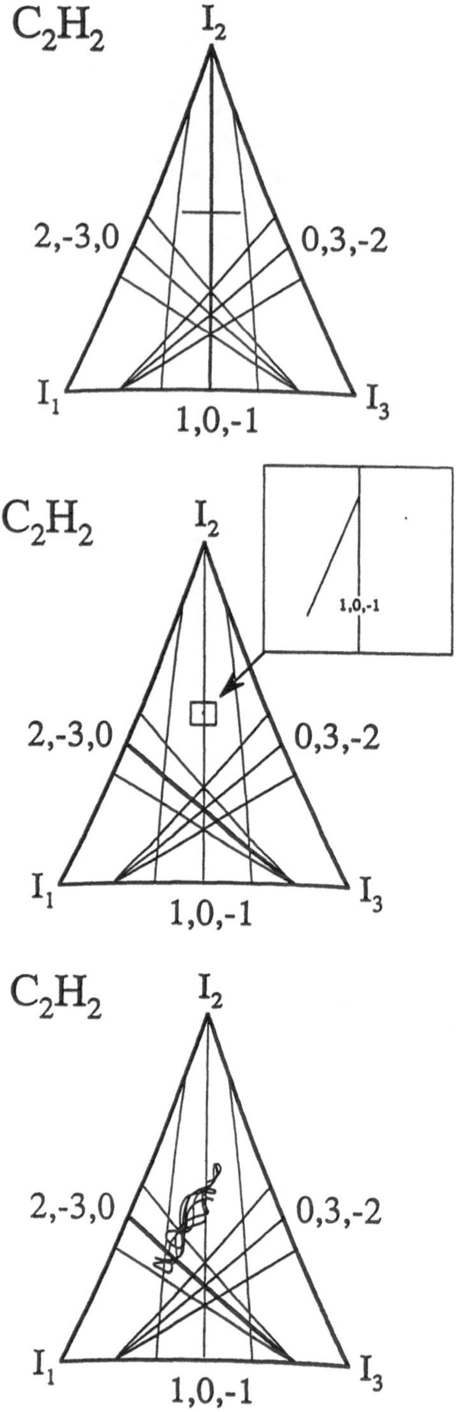

Figure 25. Projections of numerically integrated trajectories on a C_2H_2 CE surface in the action space spanned by I_1, I_2, and I_3 using H_1 given by (a) Equation 49, (b) Equation 50, and (c) Equation 50 plus collaborative sub-resonant terms. The trajectories in (a), (b), and (c) are drawn using SEP initial conditions labelled B in Figure 21 and an integration times of 0.45 ps, 0.10 ps, and 0.10 ps respectively.

$\mathbf{m}_r = \{1, 0, -1\}$ resonance. By using different initial conditions $\theta_1 = \theta_3 + \pi$, it is possible to shift the phase of the cross sub-resonances thereby inducing the qualitatively different behaviour shown in Figure 26b where the vertical extent of the trajectory is reduced while motion perpendicular to the $\mathbf{m}_r = \{1, 0, -1\}$ resonance is enhanced.

The selective turning off of the sub- and super-resonant terms artificially eliminates all but the resonant terms and thereby reproduces the results of the perturbative canonical transformation from the $\mathbf{I}$, θ to the $\mathbf{J}$, Ψ variables, which when valid also eliminates all but the resonant terms. Thus for example, using initial actions given by A in Figure 21, the numerical integration of Equation 50 results in the trajectory shown in Figure 27a that clearly oscillates back and forth across the $\mathbf{m}_r = \{2, -3, 0\}$ resonance where the $\mathbf{m}_r = \{2, -3, 0\}$ resonance width drawn in Figure 27a is valid in $\mathbf{I}$ action space because for the Hamiltonian in Equation 50, there are no sub-resonant terms in Equation 46. It should be stressed however that this trajectory is obtained by ignoring all other resonant terms, and both collaborative and cross sub-resonant terms, and is therefore purely artificial. The same initial conditions are used to obtain the trajectory in Figure 27b for H_1 that in addition to the resonant contribution, also includes the collaborative sub-resonant terms. The resulting sub-resonant excursions are qualitatively similar to those in Figure 25b and illustrate that in the large coupling limit, their magnitudes are relatively insensitive to the choice of initial actions. Although Equation 43 is valid only in the small coupling limit, it qualitatively suggests that the amplitude of the sub-resonant excursions will decrease only if the sub-resonant terms are evaluated sufficiently far from their corresponding resonances. Since all resonance planes intersect at the resonance centre, the distance between resonances increases with decreasing energy. This is illustrated in Figure 1b where the distance between the 1:1 and 2:3 resonances increases as the origin is approached from the resonance centre. Thus, for example, by using initial actions that correspond to a total energy of $\sim$ 1 quantum of excitation in the C–C bond, it is possible to further separate the $\mathbf{m}_r = \{1, -1, 0\}$ resonance and the $\mathbf{m}_r = \{2, -3, 0\}$ resonance thereby reducing the contribution of $A_{1,-1,0}$ to the sub-resonant excursions. The same argument applies equally well to the relative positions of the $\mathbf{m}_r = \{0, 3, -2\}$ resonance and the $\mathbf{m}_r = \{0, 1, -1\}$ resonance and therefore the magnitude of $A_{0,1,-1}$ will also be diminished. This effect is illustrated by the trajectory in Figure 28 obtained by integrating $H_t = H_0 + H_1$ where the total energy is reduced to one quantum of the C–C stretch and H_1 is given by the closed form in the top line of Equation 17. At this energy the dominant effects are still the slow resonant exchange of action between the C–H bonds and the rapid sub-resonant exchange of action between the C–C and C–H bonds, however, relative to the amplitude of resonant oscillation between the C bonds, the sub-resonant amplitude has been reduced. It should be noted that at this energy, the $\mathbf{m}_r = \{2, -3, 0\}$ resonance does not intersect the physically bound CE surface.

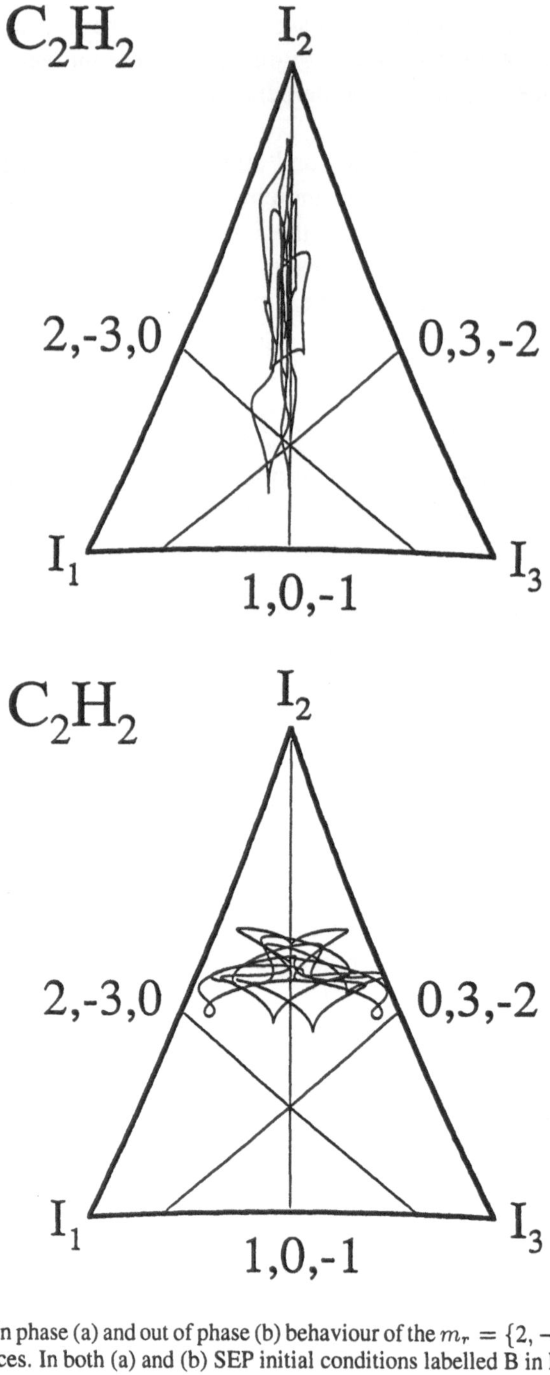

Figure 26. In phase (a) and out of phase (b) behaviour of the $m_r = \{2, -3, 0\}$ and $m_r = \{0, 3, -2\}$ sub-resonances. In both (a) and (b) SEP initial conditions labelled B in Figure 21 are used where in (a) $\theta_3 = \theta_1 + 0.1$, and in (b) $\theta_3 = \theta_1 + 0.1 + \pi$. Both trajectories are integrated for 0.15 ps. The factor of 0.1 is included to break the symmetry between the to C–H oscillators.

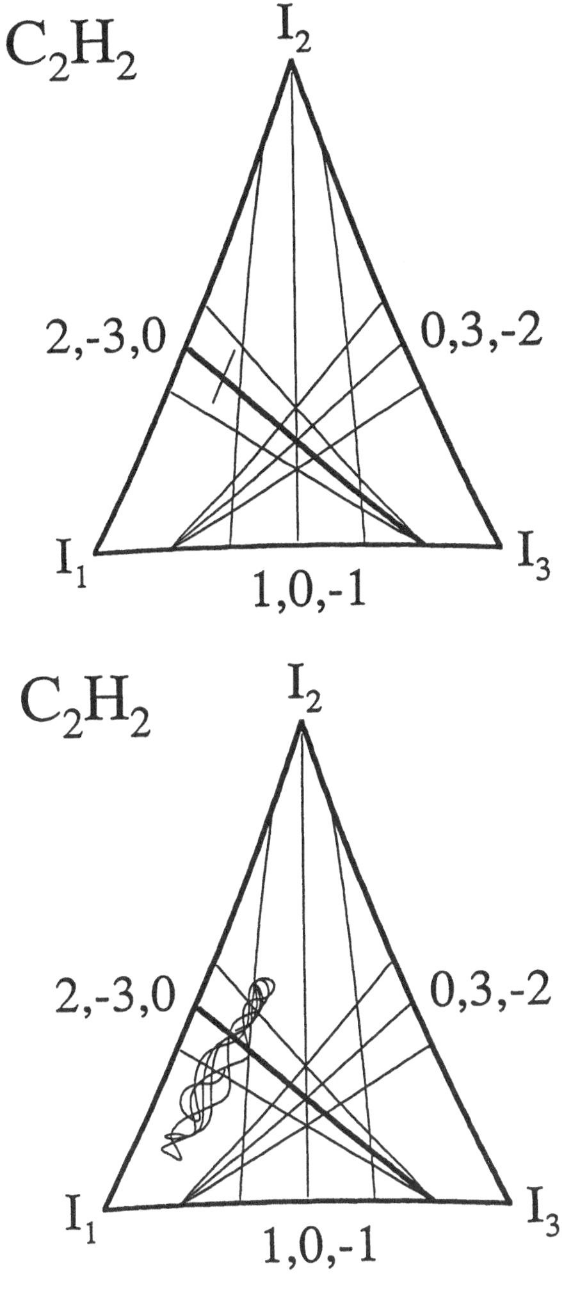

Figure 27. Projections of numerically integrated trajectories on a C_2H_2 CE surface in the action space spanned by I_1, I_2, and I_3 using H_1 given by (a) Equation 50, and (b) Equation 50 plus collaborative sub-resonant terms. The trajectories are drawn using initial actions labelled A in Figure 21. Resonance widths shown in (a) are valid since Equation 50 has no sub-resonant terms. The trajectory in (a) was integrated for 0.45 ps with a set of initial angles analogous to point B in Figure 9 that do not yield the maximum amount of action exchange. The trajectory in (b) was integrated for 0.1 ps.

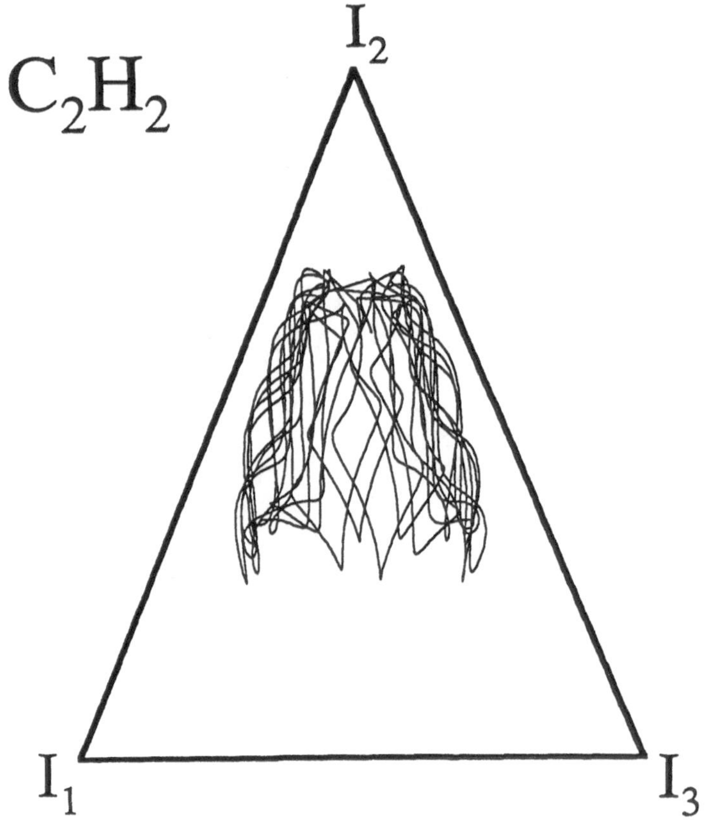

Figure 28. A projection of a numerically integrated trajectory on the E = 1600 cm^{-1} C_2H_2 CE surface in the action space spanned by I_1, I_2, and I_3. The surface corresponds to $\sim$ 1 quanta in the C–C stretch mode and sits nearly at the origin in Figure 2.

6. Conclusion

A system of two or more uncoupled oscillators is described by a zero order Hamiltonian H_0 that is a function only of the actions of the individual oscillators. These zero order local mode actions span an action space within which H_0 can be used to construct a constant energy (CE) surface. In the presence of non-linearity, resonance conditions that are also functions of the zero order actions define resonance surfaces that intersect the CE surface thereby forming the Arnold web. For zero order Hamiltonians, bilinear in the action variables, the CE surfaces are a series of concentric ellipsoids, while the corresponding resonance surfaces are linear in the actions and intersect at the resonance centre which coincides with the centre of the concentric CE surfaces. Trajectories of the zero order actions are conveniently displayed on the CE surface and, in the presence of a coupling perturbation H_1, will display the dynamics of the exchange of action between the zero order modes. These dynamics have been analysed in the small coupling limit with the use of perturbative techniques, and in the large coupling limit where the perturbative analysis is not applicable. For the case of small coupling it is demonstrated that trajectories of the zero order actions, typically chosen to correspond to the local modes of the system, are subject to sub-resonant excursions. For realistic values of the coupling these sub-resonant excursions are typically larger than the resonance width obtained in the pendular approximation. It is possible however, for small values of the coupling, to eliminate the sub-resonant excursions by transforming to new nearly separable modes. It is these new modes, and not the original zero order modes, that exhibit the canonical pendular motion in action space and provide a superior basis within which to understand the dynamics of the coupled oscillators. The quantum analog of the transformation to these new modes is a unitary transformation to a new basis characterized by new quantum numbers. Our classical considerations therefore suggest that new quantum states and numbers are possible [28, 29] for weakly coupled Morse oscillators.

For larger couplings as found in linear molecules such as C_2H_2, the perturbative approach breaks down, and it is reasonable to expect that the sub-resonant excursions will be generic for any choice of zero order basis. In this case, the dynamics of the zero order actions are dominated by sub-resonant excursions that exhibit high frequency large amplitude exchange of action between the zero order modes.

7. Acknowledgement

This work was supported by the Air Force Office of Scientific Research, AFOSR, and the U.S.-Israel Binational Science Foundation, BSF, Jerusalem, Israel. The Fritz Haber Research Centre is supported by the Minerva Gesellschaft für die Forschung, mbH, Munich, Germany.

References

[1] N. Grant Cooper (Ed.): *From Cardinals to Chaos*, Cambridge University Press, Cambridge (1989).
[2] B.V. Chirikov: *Phys. Rep.* **52** (1979) 263.
[3] G.M. Zaslavsky: *Phys. Rep.* **80** (1981) 157; R.Z. Sagdeev, D.A. Usikov and G.M. Zaslavski: *Nonlinear Physics From the Pendulum to Turbulence and Chaos*, Harwood Academic Publishers, New York (1988).
[4] A.J. Lichtenberg and M.A. Lieberman: *Regular and Stochastic Motion*, Springer-Verlag, New York (1983).
[5] M. Tabor: *Chaos and Integrability in Nonlinear Dynamics*, Wiley, New York (1989); S.N. Rasband: *Chaotic Dynamics of Nonlinear Systems*, Wiley, New York (1989).
[6] J. Ford: *Adv. Chem. Phys.* **24** (1973) 155.
[7] M.V. Berry and M. Tabor: *Proc. R. Soc. Lond.* **A349** (1976) 101.
[8] R.A. Marcus: *Discussions Faraday Soc.* **55** (1973) 34; D.M. Wardlaw and R.A. Marcus: *Adv. Chem. Phys.* **70** (1987) 231; D.W. Noid, M.L. Koszykowski and R.A. Marcus: *Annu. Rev. Phys.* **32** (1981) 267.

[9] S.A. Rice: *Adv. Chem. Phys.* **47** (1981) 117; P. Brumer: *Adv. Chem. Phys.* **47** (1981) 201.
[10] J. Jortner, R.D. Levine and B. Pullman (Eds.): *Mode Selective Chemistry*, Kluwer, Dordrecht (1991); E.W. Schlag and M. Quack (Eds.): "Intermolecular Processes", *Ber. Bunsenges. Phys. Chem.* **92** (1988) 3; J. Manz, C.S. Parmenter, R.M. Hochstrasser and G.L. Hofacker (Eds.): "Mode Selectivity in Unimolecular Reactions", *Chem. Phys.* **139** (1989) 1.
[11] S.K. Gray, S.A Rice and D.W. Noid: *J. Chem. Phys.* **84** (1986) 3745; R.T. Skodje, M.J. Davis: *J. Chem. Phys.* **88** (1988) 2429; M.J. Davis: *J. Chem. Phys.* **85** (1985) 1016; M.J. Davis: *J. Chem. Phys.* **86** (1987) 3978.
[12] T. Uzer: *Phys. Rep.* **199** (1991) 75.
[13] D.W. Oxtoby and S.A. Rice: *J. Chem. Phys.* **65** (1976) 1676.
[14] C. Jaffe and P. Brumer: *J. Chem. Phys.* **73** (1980) 5646.
[15] E.L. Sibert, W.P. Reinhardt and J.T. Hynes: *J. Chem. Phys.* **77** (1982) 3583.
[16] C.C. Martens and G.S. Ezra: *J. Chem. Phys.* **87** (1987) 284; G.S. Ezra, C.C. Martens and L.E. Fried: *J. Phys. Chem.* **91** (1987) 3721; C.C. Martens, M.J. Davis and G.S. Ezra: *Chem. Phys. Lett.* **142** (1987) 519.
[17] Y.M. Engel and R.D. Levine: *Chem. Phys. Lett.* **164** (1989) 270.
[18] The use of action-angle variable also makes the correspondence with quantum mechanics particularly simple: The vibrational quantum numbers are the action variables measured in units of $h/2\pi$. The zero order Hamiltonian $H_0(I)$ can thus be inferred from a Dunham-type fit of the energy as an analytic function of vibrational quantum numbers to the molecular spectrum. Of course, such a fit will be increasingly less perfect as the energy is increased, due to the breakdown of the zero order separable approximation.
[19] D.E. Weeks and R.D. Levine: *Phys. Letts. A*, **167** (1992) 32.
[20] G.E. Ewing: *J.Phys. Chem.* **91** (1987) 4662.
[21] J.A. Beswik and J. Jortner: *Adv. Chem. Phys.* **47** (1981) 363.
[22] Although typically not done, it is possible to employ standard higher order perturbative techniques to reintroduce the sub-resonant and super-resonant terms ignored by the simple process of averaging (see for example Ref. 4, p. 107). These terms then give rise to secondary islands, the largest of which correspond to the sub-resonant terms.
[23] Loosely speaking, higher order resonances can 'borrow width' from lower order ones thereby increasing their effective coupling strength. The effective width of a higher order resonance is therefore increased by lower order resonances which 'mix into' the higher order one.
[24] T.A. Holme and R.D. Levine: *Chem. Phys.* **131** (1989) 169.
[25] I. Benjamin, O.S. van Roosmalen and R.D Levine: *J. Chem. Phys.* **81** (1984) 3352; O.S. van Roosmalen, I. Benjamin and R.D. Levine: *J. Chem. Phys.* **81** (1984) 5986.
[26] G. Strey and I.M. Mills: *J. Mol. Spec.* **59** (1976) 103.
[27] R.P. Muller, J.S. Hutchinson and T.A. Holme: *J. Chem. Phys.* **90** (1989) 4582.
[28] R. Fleming and J.S. Hutchinson: *Comp. Phys. Comm.* **51** (1988) 13.
[29] M. Iwai and R.D. Levine: *Phys. Rev.* **A42** (1990) 3991.
[30] Y.S. Li, R.M. Whitnell, K.R. Wilson and R.D. Levine, *J. Phys. Chem.* **91** (1993) 3647.

SUBJECT INDEX

TOPICS IN MOLECULAR ORGANIZATION AND ENGINEERING

1. J. Maruani (ed.): *Molecules in Physics, Chemistry, and Biology.* Vol. 1: General Introduction to Molecular Sciences. 1988 ISBN 90-277-2596-9
2. J. Maruani (ed.): *Molecules in Physics, Chemistry, and Biology.* Vol. 2: Physical Aspects of Molecular Systems. 1988 ISBN 90-277-2597-1
3. J. Maruani (ed.): *Molecules in Physics, Chemistry, and Biology.* Vol. 3: Electronic Structure and Chemical Reactivity. 1989 ISBN 90-277-2598-5
4. J. Maruani (ed.): *Molecules in Physics, Chemistry, and Biology.* Vol. 4: Molecular Phenomena in Biological Sciences. 1989 ISBN 90-277-2599-3
5. E. Schoffeniels and D. Margineanu: *Molecular Basis and Thermodynamics of Bioelectrogenesis.* 1990 ISBN 0-7923-0975-8
6. A. Lund and M. Shiotani (eds.): *Radical Ionic Systems.* Properties in Condensed Phases. 1991 ISBN 0-7923-0988-X
7. P.I. Lazarev (ed.): *Molecular Electronics.* Materials and Methods. 1991 ISBN 0-7923-1196-5
8. E. Rizzarelli and T. Theophanides (eds.): *Chemistry and Properties of Biomolecular Systems.* 1991 ISBN 0-7923-1393-3
9. L.A. Montero and Y.G. Smeyers (eds.): *Trends in Applied Theoretical Chemistry.* 1992 ISBN 0-7923-1745-9
10. M.T. Pope and A. Müller (eds.): *Polyoxometalates: From Platonic Solids to Anti-Retroviral Activity.* 1994 ISBN 0-7923-2421-8
11. N. Russo, J. Anastassopoulou and G. Barone (eds.): *Properties and Chemistry of Biomolecular Systems.* 1994 ISBN 0-7923-2666-0
12. Y.G. Smeyers (ed.): *Structure and Dynamics of Non-Rigid Molecular Systems.* 1994 ISBN 0-7923-2774-8

KLUWER ACADEMIC PUBLISHERS – DORDRECHT / BOSTON / LONDON